Science: Philosophy, History and Education

Scope of the Series

This book series serves as a venue for the exchange of the complementary perspectives of science educators and HPS scholars. History and philosophy of science (HPS) contributes a lot to science education and there is currently an increased interest for exploring this relationship further. Science educators have started delving into the details of HPS scholarship, often in collaboration with HPS scholars. In addition, and perhaps most importantly, HPS scholars have come to realize that they have a lot to contribute to science education, predominantly in two domains: a) understanding concepts and b) understanding the nature of science. In order to teach about central science concepts such as "force", "adaptation", "electron" etc, the contribution of HPS scholars is fundamental in answering questions such as: a) When was the concept created or coined? What was its initial meaning and how different is it today? Accordingly, in order to teach about the nature of science the contribution of HPS scholar is crucial in clarifying the characteristics of scientific knowledge and in presenting exemplar cases from the history of science that provide an authentic image of how science has been done. The series aims to publish authoritative and comprehensive books and to establish that HPS-informed science education should be the norm and not some special case. This series complements the journal Science & Education http://www.springer.com/journal/11191. Book Proposals should be sent to the Publishing Editor: Claudia Acuna. Email: Claudia.Acuna@springer.com.

More information about this series at http://www.springer.com/series/13387

Hagop A. Yacoubian • Lena Hansson
Editors

Nature of Science for Social Justice

Springer

Editors
Hagop A. Yacoubian
Department of Education
Lebanese American University
Beirut, Lebanon

Lena Hansson
Department of Mathematics and Science Education
Kristianstad University
Kristianstad, Sweden

ISSN 2520-8594 ISSN 2520-8608 (electronic)
Science: Philosophy, History and Education
ISBN 978-3-030-47259-7 ISBN 978-3-030-47260-3 (eBook)
https://doi.org/10.1007/978-3-030-47260-3

This Springer imprint is published by the registered company Springer Nature Switzerland AG.
The registered company address is: Gewerbestrasse 11, 6330 Cham, Switzerland

Foreword

This book was conceived a few years ago during the ESERA conference in Dublin. While reading the conference programme, I noticed a session organized by Lena and Hagop on nature of science and social justice. I knew both of them personally, I was well aware of the discussions and the debates about teaching and learning nature of science, but I had never thought about the connection between nature of science and social justice, nor had I read anything on the topic. Honestly believing that as long as I live, I learn, I went to that session to see what it was about. I became so interested that Lena, Hagop and I had already agreed to have the present book published in this series by the end of the lunch break of the same day. Needless to say, I am very glad that you are now holding it into your hands, and I hope that you will find it as useful as I did.

An important issue in science education is the tendency of teachers to stick with content knowledge, forgetting that if their students are to become citizens who are literate about science, they will need more than that. Content knowledge is of course important, but students also need to appreciate how science is done, what are the features of scientific knowledge and scientific practices, how the respective contexts matter, and a lot more. However, it is also important to reflect more broadly about the relation between science and society. Science and society do not simply influence each other, nor do they simply interact, rather they are co-constructed. As a result, students need to be able to understand and reflect upon such ideas.

This is of course far from easy to achieve, as it requires special skills by our science teachers, who are not trained as philosophers, historians or sociologists. Worse than that, those science teachers who come to teacher education programmes after finishing their undergraduate studies in science often focus on content knowledge because they have rarely had the opportunity during their studies to have any metascientific reflections of any kind. This is why it is left to teacher educators to cultivate new mentalities and pave the way for teachers' forays in the humanities and social sciences. Do we always succeed? No, of course not. But we can always provide teachers with stimuli and ideas that will make them able to identify ideas and topics that are important for education.

Social justice and nature of science topics are already present in education. However, the contributions to this volume nicely bring these topics together and provide a lot of material for reflection. As the editors note, they also raise important questions: Why should school science aiming at social justice address nature of science? How can school science address nature of science for social justice? What nature of science-related content, skills and attitudes are required when aiming at social justice? I hope that the present volume will serve as the means to promoting reflection on such questions and a motivation for future discussion and reflection.

Kostas Kampourakis
Series Editor

Preface

The initiation of our dialogue goes back to the summer of 2016, when we first met in Flensburg, Germany, at the European regional conference of the *International History, Philosophy and Science Teaching Group*. At the time, we both felt passionate about the need of an agenda that situates the discourse on nature of science (NOS) in school science along a social justice (SJ) pathway.

We continued our dialogue over the next few months and enlarged our circle by having other colleagues join the discussions. The result was the summer 2017 symposium that we organized in Dublin, Ireland, at the biennial conference of the *European Science Education Research Association*. The symposium entitled "Nature of Science for Social Justice" aimed to create a dialogue among colleagues regarding how NOS in school science could contribute to SJ.

The idea to develop this volume started in Dublin. We were fortunate to have Kostas Kampourakis, our dear colleague and the Springer series editor of *Science: Philosophy, History and Education*, in our audience. Right after the symposium, the three of us started exploring the possibility of an edited volume in the series. A few months later, we had already signed a contract with Springer and had expanded our circle of colleagues to continue the dialogue. We are grateful that a number of scholars accepted our invitation to bring their contribution to this volume.

The present volume aims at creating space for scholars to engage in a dialogue on NOS for SJ, with the purpose of advancing the existing discussion and creating new avenues for research. We do not aim to claim that the volume is comprehensive in addressing all problems pertinent to NOS for SJ. In fact, the volume is an attempt to bring two agendas of research, namely NOS and SJ, closer to each other with the purpose of shedding light on how and why they can and should mutually contribute to each other. The volume does not aim at reaching consensus and final answers, and that explains the lack of a conclusion chapter. Instead, we aim for a dialogue around a theme that we believe is important.

We have adopted an inclusive policy. The present volume, comprised of a set of theoretical and empirical chapters, draws upon different frameworks of NOS and SJ. Accordingly, the authors use different conceptualizations and approaches as they explore issues that in different ways relate to the following questions: *Why* should a school science aiming at SJ address NOS? *What* NOS-related content, skills and attitudes form the basis when aiming at SJ? *How* can school science address NOS for SJ?

We hope that the different starting points and conceptualizations on NOS and SJ serve the aim of empowering educators and researchers to reflect on their own conceptualizations. Our hope is that the dialogue initiated in this volume can facilitate further dialogue, formulate new questions and explore fresh avenues of research that aims at bringing NOS and SJ agendas closer to each other.

As editors, we contributed equally to the development of this volume. Initially, in late 2017, we set the vision of the volume and invited colleagues to contribute with chapters. In our work, we took a micro and a macro approach. The micro approach involved working with the authors to ensure the alignment of their visions with the vision of the volume. We also organized an internal peer review process, which helped provide constructive feedback to the authors. The macro approach involved making sure that the overall volume is focused, coherent and organized.

We are grateful to all the authors for their patience and persistence with us throughout this project. Not only did they develop their chapters but also contributed in reviewing other chapters. All the chapters of this volume have undergone a rigorous peer review, whereby a panel of 2–3 reviewers reviewed each chapter in addition to us. The number of times each chapter was reviewed ranged from one to three.

We are also thankful for Kostas Kampourakis, for his continuous support and advice to us throughout the project, as well as the external reviewers of this volume, whose identities are anonymous to us, yet we believe their invaluable comments helped improve the quality of the final product.

A word of gratitude also goes to our home universities, namely the Lebanese American University and Kristianstad University for providing us with all the needed support for completing this work.

Finally, we are both grateful to our families; this work would not have been possible without their support.

Hagop A. Yacoubian
Lena Hansson

Contents

About the Editors

Hagop A. Yacoubian is an Associate Professor of Education at the Lebanese American University. He holds a PhD in Secondary Education from the University of Alberta in Canada. His research generally involves exploring how future citizens can be empowered so that they develop a critical mindset and engage in democratic decision-making processes. One particular focus involves the preparation of scientifically literate future citizens and how the teaching of the nature of science, within the context of school science education, can facilitate such a preparation. His publications have appeared in various international academic journals. He is the author of the book *Critical Thinking at the Armenian Schools in Lebanon* (Haigazian University Press, 2016) and the co-editor of a special issue of the *Canadian Journal of Science, Mathematics and Technology Education* on "Rethinking Education for Citizenship" (2015). He is a reviewer for a number of international journals in science education and is an active member of several research organizations. He has consulted schools and projects as well as trained teachers and school administrators in the Middle East, Europe and North America. Previously he has also worked as secondary school biology teacher and administrator as well as teacher educator in Lebanon and Canada.

Lena Hansson is a Professor of Science Education at Kristianstad University, Sweden. She is trained as a secondary physics and mathematics teacher and has a PhD in Science Education. Her main research interest is in the intersection between nature of science perspectives and cultural perspectives on science teaching. One of her interests regards how different images of science, communicated in science class, include and exclude different groups of students. Such images of science, associated with science by students, could be related to worldview and/or ideology, or general views about nature of science. Lena has an interest in research which is close to school practice, but also in more theoretical as well as policy issues. She has published in various international academic journals, such as *Science Education, International Journal of Science Education* and *Science & Education*, and in several international research anthologies. She serves as a reviewer for different international journals and is an active member of the organizations *ESERA* and *IHPST*. She serves as a director at the *IHPST* council (2017–2021). In addition to research, Lena is also engaged in research-based developmental projects and pre- and in-service teacher training, aiming at bridging the gap between science education research and school practice.

About the Contributors

Agustín Adúriz-Bravo was born in Buenos Aires, Argentina. He holds a degree in physics teaching from the Universidad de Buenos Aires and a PhD in didactics of science (science education) from the Universitat Autònoma de Barcelona (Spain). He is currently Senior Researcher of the CONICET (National Research Council in Argentina) and Associate Professor at the Institute CeFIEC of the Universidad de Buenos Aires. He is head of the Research Group GEHyD, which focuses on the various contributions of the philosophy and history of science to science education. His research interests include: nature of science, science teacher education in the philosophy of science, epistemics of scientific argumentation, construction and validation of historical narratives, and scientific models and modelling.

Douglas Allchin is a historian and philosopher of science and science educator at the University of Minnesota. His primary research focuses on disagreement and error in science, and on applying history to nature-of-science education. He is author of *Teaching the Nature of Science: Perspectives and Resources* (2013) and *Sacred Bovines: The Ironies of Misplaced Assumptions in Biology* (2017).

Lucy Avraamidou is a Rosalind Franklin Fellow and Associate Professor of Science Education at the University of Groningen in the Netherlands. She holds a PhD in science education from the Pennsylvania State University. Prior to her current appointment she worked at King's College London and the University of Nicosia. Her research is associated with theoretical and empirical explorations of what it means to widen and diversify STEM participation in school and out-of-school settings. At the heart of the account of her research is an exploration of minoritized individuals' identity trajectories and negotiations with the use of narrative and life-history methods.

Jesse Bazzul is a Lecturer in the Department of Secondary Education at Maynooth University, Ireland. He believes imaginative work in education is needed more than ever to find new collective ways of living together. A salient aspect of Jesse's work is finding transdisciplinary ways of engaging the sciences, but more importantly expanding the ethical and political scope usually attributed to science and science education. One thing that seems certain to Jesse is that the current ecological crisis has radically changed the reasons for doing science education in that questions of collective existence are much more integral to science practices and knowledges. Jesse recently published a co-edited volume with Dr. Christina Siry entitled: *Critical voices in science education research: narratives of hope and struggle*.

J. Lawrence Bencze (Larry Bencze) (PhD, MSc, BSc, BEd) is an Associate Professor (Emeritus) in Science Education at the University of Toronto (1998–present). Prior to this role, he worked for 15 years as a science teacher and as a science education consultant in Ontario, Canada. His research programme emphasizes critical analyses – drawing on history, philosophy, sociology, etc. – of science and technology, explicit teaching about problematic power relations and student-led research-informed and negotiated socio-political actions to address personal, social and environmental harms associated with fields of science and technology. Recent publications include two edited books about proactive citizenship. He also is co-editor of an open-source activist journal (goo.gl/ir7YRj). ORCID#: 0000-0002-0700-9580; Website: www.lbencze.ca; Research Site: www.stepwiser.ca

Lynette C. Carter (Lyn Carter) lectured science and technology education to undergraduate primary and secondary teacher education students in the School of Education at Australian Catholic University's Melbourne Campus. She also worked in postgraduate education particularly in the areas of research methodologies and contemporary issues in curriculum. Lyn's research and scholarship explores new articulations of science education valuing cultural diversity, ecological sustainability and social justice in a globalized world. More specifically, Lyn's interests are in science educational policy and curriculum studies, with a particular emphasis on the effects and consequences of globalization and neoliberalism as it reshapes science education for the twenty-first century.

Zoubeida R. Dagher is a professor of science education at the School of Education and a Faculty Fellow at the Center for Science, Ethics, and Public Policy, the University of Delaware, and is an elected Fellow to the American Association for the Advancement of Science. Dr. Dagher teaches undergraduate science curriculum courses and doctoral courses in the educational leadership and learning sciences programs. Her research focuses on the nature of scientific methods and practices and representations of scientific epistemology in science curriculum and instruction. Her academic degrees include a BS in science education from the Lebanese American University, MA in education from the American University of Beirut and a PhD in science education from the University of Iowa.

Sibel Erduran is a Professor of Science Education at the University of Oxford and a Fellow of St Cross College. She also holds Visiting Professorships at the University of Oslo, Norway, and Zhejiang Normal University, China. She is an Editor for

International Journal of Science Education, Section Editor for *Science Education* and has served on the Executive Boards of ESERA, NARST and IHPST. Her research interests include nature of science in science education. She is currently the Principal Investigator of Project Calibrate and OARS Project. Her recent books include *Argumentation in Chemistry Education: Research, Policy and Practice* (Royal Society of Chemistry) and *Transforming Teacher Education through the Epistemic Core of Chemistry: Empirical Evidence and Practical Strategies* (Springer).

Andreia Guerra is a Senior Researcher and full Professor at Graduate Program Science, Technology and Education from CEFET/RJ, Rio de Janeiro, Brazil. She holds a PhD in History of Science for Science Education. Granted as CNPq (Brazilian Research Council) associate researcher, her main research themes include: Science Education; History, Philosophy and Sociology for Science Education; and Cultural Studies. She is the President of IHPST Group (2019–2021), co-editor of a Brazilian journal on physics teaching (CBEF), author of several papers about science education published in Brazilian and international journals, and author of several books concerning history of science for the general public. She is also member of science education research associations (IOSTE, ABRAPEC (BR), IHPST, ESERA).

Iamni Torres Jager is a Science Teacher at SEEDUC/RJ, Rio de Janeiro, Brazil. She holds a Master in Science, Technology, and Education (CEFET/RJ). Her main research themes include: History and Philosophy of Science for Science Education; Gender and Science; Science Education for prison inmates. She is a member of Science Education Research Associations (IHPST and SBEnBio).

Ebru Kaya is an Associate Professor of Science Education at Bogazici University, Turkey. She was a visiting scholar at the University of Bristol, UK, and at National Taiwan Normal University, Taiwan. Her research interests include nature of science in science education. She received the Young Scientist Awards from the Science Academy and TUBITAK in Turkey and the Outstanding Paper Award from EASE in Japan. She currently serves as the Nature of Science Strand Co-coordinator for ESERA and previously served on the NARST Outstanding Paper Award Committee. She is the co-author of the book *"Transforming Teacher Education through the Epistemic Core of Chemistry: Empirical Evidence and Practical Strategies"* published by Springer.

Stein Dankert Kolstø is professor of science education at the University of Bergen, Norway. He received his PhD in science education from the University of Oslo in 2000. Since then he has taught courses for pre- and in-service science teachers. Before taking up PhD, he was a teacher of science and mathematics in secondary schools for several years. His research interest is related to scientific literacy, in specific school science as preparation for democratic participation and science teaching involving inquiry-based learning. Kolstø has cooperated with science teachers in many R&D projects. In recent years he has taken a special interest in science teaching using dialogue activities and to develop students' ability to construct and critically evaluate science based arguments involved in socioscientific issues.

Lotta Leden is Senior Lecturer in Science Education Research at Kristianstad University, Sweden. She has worked as a science and mathematics teacher in grades 6–9 (ages 12–16) for many years. She has also worked as a special needs teacher with the assignment to teach students in need of supplemental support as well as to supervise the regular teacher. Presently she is involved in teacher education aimed at preschool and secondary level. Furthermore, she educates pre-service special needs teachers. Her research focuses on the teaching of nature of science (NOS) in Swedish pre-school and compulsory school (K-12). The research is often conducted in close collaboration with the practitioners.

Isabel Martins is Professor of Science Education at NUTES-Institute of Science and Health Education, Federal University of Rio de Janeiro. She taught Physics in secondary schools in Brazil before acquiring her PhD from University College London Institute of Education. She has led and collaborated in several international research projects and served in editorial boards of leading Science Education journals. Her research interests concern the recontextualizations of science-related discourses in curriculum policy documents, in textbooks, in popular science texts as well as in classroom contexts, and their role in the promotion of scientific literacy. She published extensively in refereed journals and co-edited academic books and served as keynote speaker in international conferences.

Cristiano B. Moura is a Researcher and Science Teacher at CEFET/RJ, Rio de Janeiro, Brazil. He holds a PhD in Science, Technology and Education (CEFET/RJ). His main research themes include: History and Philosophy of Science for Science Education, Curriculum Theory, and Post-Colonial theories. He was elected to serve as Teacher Member of IHPST Council (2019–2021). He is a reviewer (ad-hoc or permanent) for several Brazilian, Latin-American and international journals of science education, section editor (HPSS for Science Teaching) of a Brazilian journal on chemistry education, and member of science education research associations (IHPST, IOSTE, ABRAPEC (BR) and some others).

Meshach B. Ogunniyi is an Emeritus Professor of Science Education at the University of the Western Cape, South Africa. He played leadership roles in three universities, the last being as Director/UNESCO Chair, University of the Western Cape, South Africa. He has published over 200 refereed articles and supervised over 80 theses. He is well known for his Contiguity Argumentation Theory. He was the Founding Editor of the *African Journal of Mathematics* and *Science and Technology Education* and member of several editorial boards. He has received several awards, for example, the Best Teacher Award, UWC (2008), Life-time Research Award of the Southern African Association for Research in Mathematics, Science and Technology Education (2014), and the Distinguished Contribution to Research Award of NARST (2015).

Alejandro P. Pujalte was born in Buenos Aires, Argentina. He holds a degree in Agronomics from the Universidad de Buenos Aires and a degree in Education from the Universidad Nacional de Quilmes. He has a Diploma in Constructivism and Education from FLACSO (Facultad Latinoamericana de Ciencias Sociales) and a PhD in social and human sciences from the Universidad Nacional de Quilmes. He is currently Professor at the Universidad Nacional de Lomas de Zamora. He develops research in science teacher education at the Institute CeFIEC of the Universidad de Buenos Aires, as a member of the Research Group GEHyD. His research interests includes nature of science, specifically in the field of images of science and scientists.

Rita Vilanova is Associate Professor of Science Education at NUTES-Institute of Science and Health Education, Federal University of Rio de Janeiro (UFRJ), where she coordinates the Science and Health Education Graduate Studies Programme. She got her PhD at Pontificia Universidade Católica in Rio de Janeiro and has led and participated in a number of funded research and outreach projects. She serves in the board of UFRJ's Teacher Education Programme and has published in refereed journals. Her research interests are teacher education, curriculum, science and health education, university–school relations and community of practices.

Chapter 1
Nature of Science for Social Justice: Why, What and How?

Lena Hansson and Hagop A. Yacoubian

1.1 Introduction

"Nature of Science" (NOS) and "Social Justice" (SJ) are vivid areas in contemporary science education research. There are different conceptualizations of NOS and SJ, giving rise to divergent research agendas. NOS and SJ research areas have mostly been separate tracks, with only a few contributions across each other. The aim of this volume is to bring NOS and SJ research closer together, explore the possibilities that might arise, and start a dialogue on the characteristics of *NOS for SJ*. In this chapter, we prioritize SJ as an overall aim of science education and shed light on how NOS teaching can contribute to that aim.

Both NOS and SJ research, in different ways, challenge traditional school science (e.g. Zacharia and Barton 2004) and add perspectives and new questions to science education research. NOS scholarship challenges teaching traditions where science is taught merely as facts (Leden et al. 2017), and where myths about science continue to be propagated (McComas 1998, 2020). It questions the image of science communicated as "sort[ing] things crisply into black and white, true and false, without any 'shades of grey,' partial conclusions or residual uncertainties" (Allchin

The original version of this chapter was revised: This chapter was previously published as non-open access. It has now been changed to open access under a CC BY 4.0 license. The correction to this chapter is available at https://doi.org/10.1007/978-3-030-47260-3_14

L. Hansson (✉)
Department of Mathematics and Science Education, Kristianstad University, Kristianstad, Sweden
e-mail: lena.hansson@hkr.se

H. A. Yacoubian
Department of Education, Lebanese American University, Beirut, Lebanon
e-mail: hagop.yacoubian@lau.edu.lb

H. A. Yacoubian, L. Hansson (eds.), *Nature of Science for Social Justice*, Science: Philosophy, History and Education,
https://doi.org/10.1007/978-3-030-47260-3_1

2003, p. 333). Thus, discussing issues about NOS in science classrooms challenges traditional science teaching not only due to incorporating new content, but also because of disputing those binary notions of true/false or black/white that are part of traditional school science, as different perspectives exist to approaching many NOS issues (see also Leden et al. 2017). SJ literature, for its part, challenges traditional school science through visions that problematize many aspects of what characterizes traditional school science. This includes how science and school science are viewed, but also includes the roles of teachers and students (see, for example, the characteristics of "Critical School Science" in Zacharia and Barton 2004). That being said, NOS and SJ research studies have mostly been on parallel research tracks.

One main focus of NOS research involves the teaching of what science is; how knowledge is developed within science; and in what ways societal, cultural and human elements are involved in these knowledge processes (e.g. Allchin 2017; Erduran and Dagher 2014; Hodson 2014; Kampourakis 2016; Lederman 2007; Matthews 2012; McComas 2020). Thus, NOS research explores how different historical, philosophical and sociological perspectives on science might be included in the teaching of science at different educational levels. However, there are different frameworks and viewpoints within NOS research that prioritize various perspectives in the science classroom. In fact, NOS research is influenced in different ways by current societal discourse, as well as by ongoing debates within the science study research fields (e.g. Rudolph 2000).

This ongoing debate is a key reason why different historical, philosophical and sociological perspectives are emphasized at various degrees by different NOS scholars. The debates have resulted in various conceptualizations and boundaries of NOS within the recommendations for NOS teaching. One example of these differences is whether only epistemic and cognitive issues should be considered part of NOS, or whether ideological and ethical issues should also be included. Such differences in NOS research mirror ongoing discussions among philosophers of science on whether science is free from contextual values (a concept used by Longino 1990), such as norms and interests being "unrelated to the cognitive aims of science" (Curd et al. 2013, p. 184). Alternatively, if such values are part of science then they may or may not be compatible with scientific objectivity. See Curd et al. (2013) for a discussion on rationality, values and objectivity in science.

Related to these ongoing debates are different views that have been raised explicitly or implicitly by science education scholars in relation to NOS research. These views range from NOS research is not doing enough to break the reproduction of positivism and/or taken-for-granted socio-cultural heritages in the teaching of science, to the view that NOS research is going too far in picturing science as contextually embedded and thereby ending up in relativism. Such discussions within science education research "tend to settle on the conflict between the apparent relativism of constructivist accounts of science on one hand and the rationalist/realist views of science on the other" (Rudolph 2000, p. 414). As editors of this book, we consider the ongoing scholarly debates on the various conceptualizations of NOS important and unavoidable due to the ongoing contentious debates within fields such as philosophy and sociology of science. The authors in this volume use different conceptualizations of NOS in their chapters, ranging from a discourse focusing on epistemic conceptions to ones that embody contextual values. We consider the exploration of different frameworks and perspectives on NOS as well as their consequences for teaching and learning as part of the dialogue we are aiming for.

SJ has been discussed in general as well as in specific relation to education. There is quite a large body of contemporary scholarship on education for SJ in general education literature. Scholars engaged in this line of research often address issues about access, power and equity in relation to education. In science education, SJ as an aim was underplayed for some time (Reiss 2003). There is a large body of science education literature on related aims such as "scientific literacy", but the latter does not necessarily incorporate a vision of SJ. Reiss (2003) argues that "the scientific literacy movement /…/ offers too narrow a vision of what science education might achieve" (Reiss 2003, p. 153). Similarly, Barton (2003) describes how the scientific literacy movements have been criticized for using a "deficit model" when discussing the needs of marginalized students. In recent years, there has been an increasing focus on SJ in science education research. Atwater et al. (2014), Barton (2003), Barton and Upadhyay (2010), and Reiss (2003) are just a few examples of scholars in this line of research.

As is the case with NOS, SJ has been conceptualized in different ways and from different perspectives. We acknowledge that there is no consensus on a definition of SJ. Instead, there are different meanings and political ideologies attached to the concept (Zajda et al. 2006). In the scholarly literature, the concept is under discussion and different conceptualizations exist and have been developed according to different traditions (see Jost and Kay (2010), Pérez-Garzón (2018) and Zajda et al. (2006) for the history of the concept). Nevertheless, "[m]ost conceptions of social justice refer to an egalitarian society that is based on the principles of equality and solidarity, that understands and values human rights, and that recognizes the dignity of every human being" (Zajda et al. 2006, p. 1). Along the same lines, Bell (2016) writes:

> Social justice is both a goal and a process. The *goal* of social justice is full and equitable participation of people from all social identity groups in a society that is mutually shaped to meet their needs. The *process* of attaining the goal of social justice should also be democratic and participatory, respectful of human diversity and group differences, and inclusive and affirming of human agency and capacity for working collaboratively with others to create change…. (Bell 2016, p. 3).

As editors, our point of departure is aligned with that of Bell (2016) and Zajda et al. (2006). Thus, for the purpose of this volume, we use a rather broad conceptualization of SJ that encompasses SJ as a goal and as a process. SJ as a goal targets, for instance, social, political, economic, cultural and gender equity in science teaching and in broader society. SJ as a process involves the contributions that NOS teaching can bring to that goal. We view SJ as a vision crucial for acting against inequality, injustice, environmental destruction, marginalization and hegemony prevalent in today's world, and view education conducive for SJ as being integral for achieving such a vision. Having said this we acknowledge that (science) education for SJ is a political choice. As is the case with NOS, the authors in this volume use different SJ frameworks as their starting point. They also emphasize different aspects of SJ in their chapters. We consider those different conceptualizations and emphases crucial for the dialogue and exploration that this volume aims to achieve.

We argue that if (science) education conducive to SJ is prioritized then NOS teaching and learning cannot stay away from achieving a similar vision.

In this volume, we aim to explore what NOS for SJ could mean, how NOS teaching can contribute to SJ aims, and what NOS content and teaching methods are appropriate when addressing NOS for SJ. Thus, we attempt to bring NOS and SJ agendas closer to each other with the purpose of shedding light on how they can contribute to one another. In particular, we argue for the need to explore: *Why* should school science aimed at SJ address NOS? *What* NOS-related content, skills and attitudes form the basis when aiming at SJ? *How* can school science address NOS for SJ? Related to all of these three questions is the issue of inclusion. Highlighting this perspective in relation to the three questions, we acknowledge that even though the science curriculum and NOS in school science are (in principle) for everyone, they can have the tendency to marginalize and exclude certain groups of students. Therefore, inclusion should be in the foreground when addressing the three questions.

In this introduction chapter, we provide our perspective as editors and initiate a discussion on what NOS for SJ can mean, and what should characterize NOS teaching. It is worth noting that we do not intend to make grandiose claims about what NOS for SJ entails, as it is too early to derive such generalizations. Instead, together with the authors of this volume, we take a humble step and aim to highlight a theme that we believe is important. We also initiate a research-based dialogue on the topic, hoping that the science education community can take it from here, continue the discussion, further refine the construct of NOS for SJ, and develop a solid repertoire of what it entails. At the end of the chapter we provide an overview of the volume and identify some of the main arguments that the authors make as they embark upon this dialogue.

1.2 Nature of Science (NOS)

Many science education researchers and a number of science education policy and curriculum documents around the world highlight the importance of NOS teaching and learning (e.g., Hodson 2014). There are different conceptualizations of NOS (e.g., Allchin 2017; Erduran and Dagher 2014; Hodson 2008; Kampourakis 2016; Lederman 2007; Matthews 2012; McComas 2017). Some of these conceptualizations include general tenets, which are thought to be valid across different science disciplines and on which there is some consensus among scholars in philosophy, sociology and the history of science as well as in science education (e.g., Lederman 2007; McComas 2017). Others argue for the need of contextualization and broader perspectives on NOS (Allchin 2011), or describe NOS through different features that are unique or can be shared over different science disciplines (Erduran and Dagher 2014; Matthews 2012).

These various conceptualizations mirror differences with respect to what NOS content is viewed appropriate, possible and/or important to teach in compulsory schooling. They also mirror disagreements on NOS among philosophers, sociologists and historians of science, as well as among scientists. As discussed above, conceptualizations of NOS are dependent upon the emphasis given to different historical, philosophical and sociological perspectives, as well as different visions for school science. Thus, NOS could be viewed as a boundary object (Star and Griesimer 1989) in the sense that even though there is agreement among NOS scholars on the importance of including NOS in the teaching of science, there is disagreement regarding the reasons as to *why* this is important, *what* NOS perspectives should be taught, and *how* this should be done. From this standpoint, different reasons are provided in the literature for the inclusion of NOS perspectives in school science (Lederman 2007; Matthews 1994). The reasons include teaching NOS for its own sake to give students the opportunity to learn more nuanced images of science, but also as a means to reach other goals. These goals are related to different visions of scientific literacy (Roberts 2007; Sjöström and Eilks 2018). Frequently mentioned goals include NOS contributing to conceptual understanding, fostering increased interest in science, and teaching science for citizenship.

Part of the NOS literature is concerned with the broadening and problematizing of traditional, stereotypical images of science and scientists frequently communicated in the teaching of science, in textbooks and in the media (e.g., Allchin 2013; McComas 1998). Stereotypical images include scientists being pictured as white men, working alone in laboratory, wearing a lab coat and glasses, and sometimes pictured as the mad genius or at other times more like a superhero (e.g., Chambers 1983; Sjøberg 2000). In addition, stereotypical images of science are communicated both explicitly—for example, as part of a positivistic way to view scientific knowledge and knowledge production—and implicitly. For example, teaching science as a large number of facts may implicitly communicate that scientific knowledge is not open to change. Neither does the "fact tradition" (Zacharia and Barton 2004) of science teaching communicate the processes that has led to these "facts". Accordingly, humans involved in the making of these facts often remain hidden. Another example is that usually neither limits of science nor other ways of knowing are discussed in science classrooms. This results in students frequently associating science with scientism (Hansson and Lindahl 2010). These kinds of stereotypical images of science are communicated in science classrooms even though such views are not part of formal curricula. Instead, they constitute a hidden curriculum in many science classrooms around the world (Hansson 2018). There is a need for understanding what NOS content and teaching approaches can break the reproduction of this hidden curriculum.

Other reasons over why NOS should be part of compulsory science education include developing students' conceptual learning, increasing their interest in science, and fostering the preparation of active citizens. Thus, NOS is viewed as contributing very different visions and overall aims for science teaching, ranging from cultivating future scientists and engineers—what Aikenhead (2006) calls a pipeline science teaching—to fostering activism and SJ. However, with some exceptions (such as Kolstø 2001), NOS literature fails to elaborate on the consequences of these different visions and aims. In line with this, Hodson (2014) argues that "[t]here are numerous goals for science education (and education in general) that can, will and *should* impact on decisions about the NOS content of lessons" (p. 945). Thus, there is a need to scrutinize NOS content, but also teaching approaches with different specific aims for the teaching of science in mind. In this volume, we focus on the goal of SJ and explore what NOS content should be taught, as well as how and why.

1.3 Social Justice (SJ)

There is a large body of research related to issues of SJ in general education literature (such as Bell 2016; Zajda et al. 2006). Most often, scholars advocating an education for SJ acknowledge injustices in society and school. They also acknowledge failures to address the marginalization of different student groups based on gender, race, and sexuality, as well as cultural, social and economic background. These scholars often address issues about access, power and equity. Through their recommendations, they hope to achieve some change. Not only in academic literature, but also curricula and other policy documents from different parts of the world focus on SJ in their visions. For example, UN states that "the pursuit of social justice for all is at the core of our global mission to promote development and human dignity" (https://www.un.org/en/events/socialjusticeday/). Working towards an education for SJ means taking a critical stance in relation to different aims of education and the means for achieving those aims. In line with Bell (2016), we highlight the importance of equal participation for all groups in society, equitable distribution of resources, and the need for social actors who have a sense of their own agency and are socially responsible. In addition, in line with Carlisle et al. (2006), we acknowledge the importance of critical perspectives and social action.

Aligned with such a conceptualization of SJ, Keddie (2012) emphasizes a commitment to "the emancipatory possibilities of education" (p. 12) and describes how educators have to work "against the grain of the discourses that impede marginalised students' schooling success" (p. 4). Another example from general education literature is Hackman (2005), who argues that "[t]o be most effective, social justice education requires an examination of systems of power and oppression combined with a prolonged emphasis on social change and student agency in and outside of the classroom" (p. 104). The notion of oppression builds upon Freire's (1970) thoughts; he argues for an education aimed at transforming society by empowering students to take action against oppression. The movement of critical pedagogy, based on Freire's work, is grounded on a vision of SJ and equity. Founded on critical theory, critical pedagogy is dedicated to the alleviation of human suffering and acknowledges that education is political (Kincheloe 2008).

In recent years, more scholars have explored and argued for goals of school science directly related to SJ. There are a number of scholarly books dealing with science education and SJ. Examples include *Activist Science and Technology Education* by Bencze and Alsop (2014) and *Multicultural Science Education: Preparing Teachers for Equity and Social Justice* by Atwater et al. (2014). Special issues of academic journals targeting SJ in science education have also been developed. Among those are "Teaching and Learning Science for Social Justice" in *Equity and Excellence in Education* by Barton and Upadhyay (2010) and "Rethinking Education for Citizenship" in the *Canadian Journal of Science, Mathematics and Technology Education* by Yacoubian and Bazzul (2015).

As in general education literature, SJ in science education literature is conceptualized in different ways and scholars have emphasized different aspects of it. For example, Butler et al. (2014), in the introduction to the volume *Multicultural Science Education: Preparing Teachers for Equity and Social Justice*, state that "[c]hallenging the status quo in how science has been traditionally taught is the first step in changing the outcomes of 'who does science'" (p. 3). This can serve as an example of how science education can contribute in the processes towards SJ by opening up science for more student groups than has traditionally been the case. Along the same lines, Brickhouse and Kittleson (2006) argue for a re-visioning of both science and science education:

> We envision a science that is responsible for supporting the interests, goals, and needs of a diverse population. /…/ To achieve this kind of science, educators must provide a critical education in science to everyone so that these science related communities are inclusive of everyone, particularly those who have historically benefited least from science. This critical education includes an education in the substance of science, but also in its epistemologies and social relations (p. 204).

Other examples include studies focused on possibilities and challenges for different groups in relation to science education, and yet others argue for the need to empower students to be prepared to act in society, and conditions for different social and cultural groups with respect to the teaching and learning of science. For example, using a Freirean perspective, Santos (2009) discusses critical pedagogy, its aim of changing society and its implications for science teaching. Hodson (2011) argues for educating students to become citizens able to engage in sociopolitical activism. Along similar lines, Bencze et al. (2018) argue for science teaching aimed at ecojustice and activism, while Barton and Yang (2000) discuss "cultural power" in relation to science education with the starting point in an ethnographic case study of a homeless, inner-city family.

A number of scholars have explored curricular ideas and contexts in which SJ as an aim can be included in science teaching. Examples include Reiss (2003), who suggests including issues about food, nuclear power and individual differences in teaching units; and Dimick (2012), who uses SJ in an environmental science class. Aikenhead and Michell (2011) elaborate on what a science curriculum recognizing indigenous knowledge could look like, while Bencze (2017) describes an instructional framework—STEPWISE—to be used when teaching for action in relation to controversial, societal and environmental issues. In addition, certain science educators call for having SJ as the guiding framework of science teacher education (e.g., Finkel, 2018). Examples include Atwater et al. (2014), who discuss and provide examples on how to prepare teachers to teach for marginalized student groups. Similarly, Varelas et al. (2018) report on an empirical research involving programmes that equip science teachers with justice-oriented pedagogy. In addition, science educators refer to different teaching approaches that can facilitate SJ. Among those approaches are culturally responsive science teaching (Hernandez et al. 2013), justice centred science pedagogy (Morales-Doyle 2018), transformative science education (Codrington 2014), and science education as a means for fostering student ownership (O'Neill 2010).

It is worth noting that SJ as an overarching aim might include slightly different foci in relation to the teaching of science. These include scrutinizing issues of access and equity within the teaching of science, problematizing perspectives on science itself in the teaching of science, and empowering students to use science as a tool when taking action in relation to societal issues. The latter could have different political/ideological starting points as long as it includes a vision of SJ (see above and Zajda et al. (2006) for a description of how SJ has been connected to different political views).

In summary, in recent years SJ has been discussed more frequently as an aim of education, in general, and of science education, in particular. There are multiple visions of SJ among science education researchers, which also give rise to divergent research agendas. There is a growing number of studies exemplifying content and teaching approaches in science education aiming for SJ. However, the precise consequences for content and teaching approaches are in need of further scrutiny. In the section that follows, we look more closely into one particular area of science education—NOS teaching—and explore its role and characteristics in school science aimed towards facilitating SJ.

1.4 NOS for SJ

This volume aims to explore the relevance of NOS in school science for SJ and outline the consequences that SJ could have on NOS research and practice. We aim to generate a dialogue on the characteristics of NOS teaching for SJ through bringing together a group of scholars, each having different starting points. The dialogue focuses on the three key questions presented earlier: *Why* should school science aimed at SJ address NOS? *What* NOS-related content, skills and attitudes form the basis when aimed towards SJ? *How* can school science address NOS for SJ? The contributors to this dialogue use different approaches and focus on partly different themes, but in one way or another they all contribute with answers and issues related to these questions.

Consequently, NOS and SJ research both contribute to the development of research-based answers to the three key questions. Considering that the questions *Why? What? How?* are important to frame discussions related to the characteristics of a curriculum (e.g., Dillon 2009; Duit 2015), a research-based dialogue focusing on them could have the potential of contributing towards a curriculum of *NOS for SJ*. Figure 1.1 shows a proposed model of how we conceptualize this dialogue. The arrows in the figure are bidirectional to emphasize that NOS research and SJ research both contribute to one another and to the generation of possible answers,

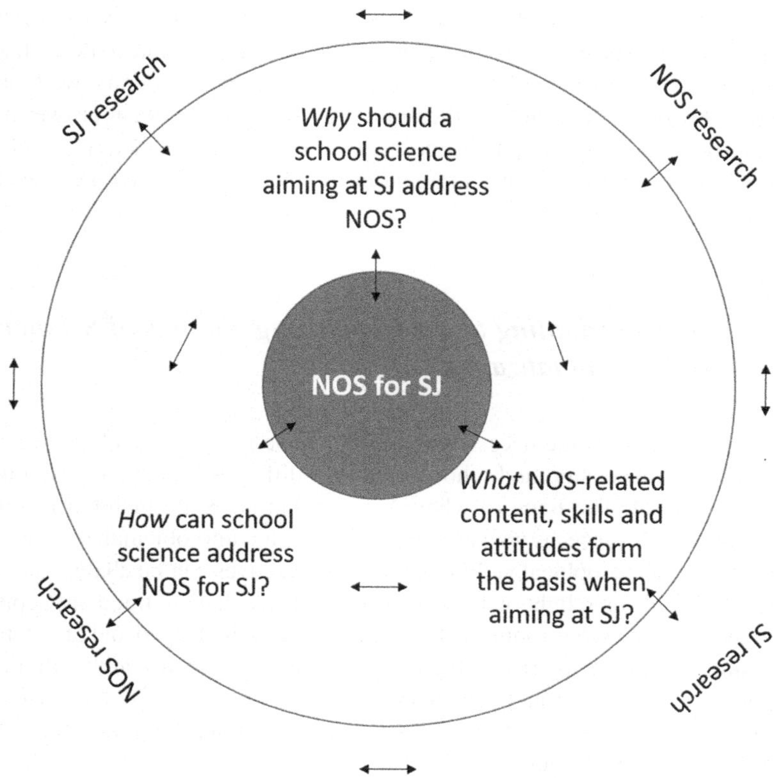

Fig. 1.1 Proposed model for conceptualizing the dialogue in this volume

suggestions and new questions related to the three overall questions, thus contributing to conceptualizing *NOS for SJ*. These answers, suggestions and new questions all contribute to NOS and SJ research areas. The arrows in between the three questions indicate that the questions are interdependent rather than being independent.

The rest of this introduction chapter aims to initiate a dialogue and illustrate the need to further explore the three questions. Based on relevant literature, we expand on three overarching aims of NOS that we believe are central to developing a rationale for *NOS for SJ*. These aims could be considered some possible answers to the *why* question in Fig. 1.1. They are (1) NOS contributing to problematizing notions of science and conceptualizations of NOS, (2) NOS contributing to challenging who science and science education are for (the issue of access), and (3) NOS contributing to empowering students in society. We do not claim that these three aims are the only ones. In addition, we acknowledge that separating them here is purely an analytical choice. In fact, they are intertwined and when considered together have the potential to challenge taken-for-granted answers to the three questions, as well as suggest possibilities of new questions and answers. This is partly because the three aims complement each other but also because they can be viewed as educational means for another educational end (education for social justice). There is certainly a need to continue exploring these aims separately. However, there is also a need to study their synergistic effect, as the latter has the potential to contribute with a rich research base to the agenda of NOS for SJ. In the following sections, we illustrate how present lines of NOS and SJ literature can contribute to begin answering the three questions and raising further questions related to the characteristics of *NOS for SJ*. We also argue for the need to continue exploring the three questions along different lines of research.

1.4.1 NOS Contributing to Problematizing Notions of Science and Conceptualizations of NOS

SJ literature highlights the necessity to problematize the power and special status that science has in society (and school), and how different people are positioned in relation to science. The issue of whose science and NOS are pictured in science teaching are of relevance, as well as how traditional and unproblematized precedencies to specific philosophical and ideological perspectives can be challenged. NOS research can shed some light on how notions of science that are frequently communicated in the science classroom can be problematized, challenged and broadened in the teaching of science. SJ research could contribute to NOS research with respect to providing directions through which to problematize notions of science within the teaching of science. This could contribute to re-evaluating the goals, content and approaches for NOS teaching.

For example, research has shown that even though there are different perspectives on NOS, science classrooms most often and implicitly communicate a narrow span of worldviews and NOS perspectives (Proper et al. 1988; Hansson 2018). For example, positivistic viewpoints are often reproduced in science classrooms around the world. Zacharia and Barton (2004) state that traditional school science is coupled with positivism where "scientific knowledge is an objective representation of how the world works" (p. 203). In line with a SJ agenda, NOS research problematizes and challenges the notion of science as being objective and value free, and instead positions science as a human activity that is embedded in cultural and societal structures, traditions and values. In doing so, many science educators have challenged and suggested alternatives to the positivistic views frequently communicated in the science classroom. Examples include Lederman (2007), who has offered a list of NOS ideas (tenets), Allchin (2013), who has acknowledged the role of culture on scientific claims, and Dagher and Erduran (2016), who have considered the importance of social and cultural values as well as political power structures in shaping scientific knowledge. As discussed earlier, it is worth reminding here that science education researchers also debate consequences of the philosophical discourse on the role of contextual values in science, including issues of reliability and objectivity.

Along the same lines, most often only a narrow span of perspectives is offered in the science classroom with respect to diverse worldviews and ideologies (Hansson 2018). This has been given less attention by NOS researchers, with the exception of those engaged in exploring the relationship between science and religion. One example of such a narrow span of perspectives is the frequent communication of scientism in science teaching (Taber 2013), with the consequence that students come to associate science with scientism (Hansson and Lindahl 2010). It is possible to combine scientism with science, but there are many other worldviews (including many religious ones) that are also possible to combine with science. However the latter is often not communicated to students. NOS researchers have argued for the necessity to include the limits of science as part of NOS teaching (McComas 2020; Hansson 2020).

In addition, scholarly literature has highlighted how indigenous knowledge about nature is marginalized in traditional school science. The relation between indigenous knowledge and science is an important area of research in science studies and science education. Debate on the role of culture in science and science education is a contentious area, which derives from competing positions that scholars have regarding NOS. The different positions range from situating culture and its role outside epistemic grounds (e.g., Matthews 1994; Siegel 1997), to arguing for multicultural science (e.g., Snively and Corsiglia 2001) and thus encouraging cross-cultural perspectives (Lee et al. 2012), and building decolonizing bridges (e.g., Aikenhead and Ogawa 2007) between western science and indigenous knowledge.

From another perspective, research on ideologies in science education shows that neoliberal perspectives are frequently communicated in science education (e.g. Bencze and Carter 2011). Selby (2014) indicated how science teaching tends to communicate "ever upward progress" and "technological fix" (p. 172). Along similar lines, Yacoubian (2018) showed how different science curriculum and policy documents highlight the importance of scientific literacy through a neoliberal lens, ensuring that developing future citizens' abilities of decision making serve specific neoliberal agendas. There is still need for research to scrutinize how to handle relationships between science and ideology in the teaching of science.

NOS research has focused on problematizing and providing suggestions on how this narrow span of notions of science and conceptualizations of NOS might be addressed in the science classroom. However, more research is needed. In order for NOS to increase its relevance from a SJ perspective, we argue that NOS scholars can challenge themselves, being inspired by SJ literature but also from research in science studies to further problematize their conceptualizations of NOS. For example, looking at today's science teaching: *Whose NOS perspectives are being taught? What social and political structures contribute to strengthening rather than challenging stereotypical images and power structures?* And scrutinizing the NOS research itself: *What needs to be added, changed, and removed from present suggestions of NOS perspectives that should be taught in school?* SJ literature sheds light on different perspectives and why some (rather than others) are reproduced in the teaching of science. This is highly relevant for NOS research. In addition, issues such as the kind of NOS curriculum that we are aiming for becomes relevant: *Should positivism be replaced with other small-span views of NOS, or should pluralism be the target with respect to NOS perspectives in the science classroom? What are the opportunities and constrains of different alternatives from a SJ perspective? How could these perspectives of NOS be taught in relation to different student groups?*

1.4.2 NOS Contributing to Challenging Who Science and Science Education Are for (The Issue of Access)

SJ literature has shown that groups, such as students of colour and female students, continue to be marginalized at schools and in science classrooms. There are many and complex reasons for marginalization. We focus here on one aspect: how does providing distorted images of science and scientists in the teaching of science facilitate the reproduction of inclusion and exclusion patterns in relation to both science teaching and science?

Science textbooks portray stereotypical images of scientists—picturing scientists mostly as male and from Western countries (Vesterinen et al. 2013; Yacoubian et al. 2017). Such stereotypical images are problematic in terms of the implicit messages given to learners regarding who a scientist is and for whom science is designed. Research also shows that both students (Sjøberg 2000; Villar and Guppy 2015) and science teachers (McDuffie 2001) frequently have stereotypical images of scientists.

However, it is not only the images of scientists but also those of science communicated in schools which are highly relevant for the issue of access and the current marginalization of student groups. Science studies research fields have shed light on various perspectives and ways to understand science, but as discussed previously this is rarely communicated to students. This has consequences for student groups who are unable to identify with science in the ways it is presented. For example, students who do not identify with positivistic and scientistic notions of science may have a hard time identifying the relevance of science to them (Hansson and Lindahl 2010). Communicating a very narrow span of NOS perspectives in the teaching of science (including worldview and ideology, such as modernism, technological optimism and neoliberalism) limits the possibilities for students to engage in science in ways meaningful for them. Not providing a broader span of views about science has the consequence that science becomes distorted for many students.

NOS teaching can contribute to a more inclusive school science if it shows students diverse views of how science works rather than prioritizing one view over another. For example, Stenmark (2004) states that it is important to pay attention to *what* values and *what* worldviews are present in different phases of scientific research, and *which* ones are absent. Consequently, he argues that it is essential to provide students with examples of people with different ideologies and worldviews who are actively engaged in scientific research. Thus, if we want to increase the possibilities for students to identify more with science and scientists, science teaching should show how science can be shaped by, and understood from, various worldviews, ideologies and philosophical perspectives. Similarly, it is important to offer the possibility for students to see that people with different backgrounds, interests, abilities, and personalities might find science meaningful despite any sociocultural structures that have limited and still do limit many groups' access to science.

Thus, we argue that in order to break patterns of reproduction and expand on who science education is for, it is important to study how traditions, values, interests, and habits communicated by the scientific community might exclude students. This is in line with what is suggested by Andrée and Hansson (2014). NOS research has the potential to contribute perspectives, as well as ways to challenge and problematize traditions, habits, images and values frequently associated with science and scientists. As described previously, such research is present, yet there is still the need to do more work to help understand the details and how broader perspectives of science could find a way into the science classroom. This line of research can contribute towards challenging oppressive traditions that exclude many young people from science. Research that sheds light on the problems of presenting narrow images and traditions is necessary, yet equally important is research exploring possible teaching approaches that challenges these narrow images and highlights possible ways of supporting teachers. In relation to all of this, SJ highlights new perspectives for NOS researchers such as the value of considering grounds for marginalization other than gender and ethnicity (which are more commonly discussed in relation to marginalization). For example, worldview, ideology, and special needs are other grounds in need of further consideration. Interesting questions to further explore include: *How do images of NOS communicated in science classrooms limit access and meaningful engagement in science for different student groups, in different parts of the world? What NOS content and teaching methods could contribute to breaking patterns of marginalization?*

1.4.3 NOS Contributing to Empowering Students in Society

Over the last few decades, science for citizenship has been increasingly discussed in the science education literature as a goal for science teaching. This goal is in addition to traditional ones such as learning science concepts and models for their own sake or preparing students for the next educational level and science careers. For example, a large body of literature deals with science-technology-society (STS) approaches (e.g., Aikenhead 1986), science-technology-society-environment (STSE) approaches (e.g., Pedretti 2005), and 'socio-scientific issues' (SSI) approaches (e.g. Ratcliffe and Grace 2003; Sadler et al. 2007; Zeidler et al. 2005). Even though there are a variety of aims under SSI teaching (Simonneaux 2014), one frequently mentioned aim is to empower students to make decisions on societal issues. The latter is important from a democratic standpoint (Driver et al. 1996; Yacoubian 2018). The aim of empowering students to make use of their knowledge in and about science is related to some conceptualizations of SJ. For example, there are scholars arguing for science education as a means of preparing students to take action on societal issues in their role as citizens (e.g. Bencze 2017; Bencze and Alsop 2014; Hodson 2011). Along the same lines, Bazzul (2015) argues for rethinking how social and political issues are approached in science classrooms, where consensus and non-activist choices prevail. Discussion on the aims of science teaching echoes ongoing discussions in general education literature. For example, Giroux (2019) states "that education is not just about job training and product manufacturing, but also about matters of civic engagement, critical thinking, civic literacy and the capacity for democratic agency, action, and change" (p. 149).

Several scholars have pointed out the high relevance of NOS in science curricula oriented towards societal issues (e.g. Aikenhead 2006; Allchin 2017; Driver et al. 1996; Hodson 2011, 2014; Kolstø 2001; Ryder 2001; Yacoubian 2015). It is argued that knowledge of different aspects of NOS is important to be able to make decisions on socio-scientific issues or to take sociopolitical action. Scientific content relevant to socio-scientific issues is often contemporary and ongoing, rather than being part of established science. NOS knowledge is necessary to be able to understand such things as why scientists do not always agree on interpretations or why they are sometimes accused of being biased. The nature of contemporary science performed in universities, private companies or military organizations, as well as issues about research funding are highly relevant to science curricula oriented towards societal issues. In addition, NOS knowledge helps students understand the limits of science and how science can contribute (and not contribute) to decisions made on socio-scientific issues (Hansson 2020). These are examples of NOS perspectives that might empower students to take part in societal and individual decision making where knowledge of science content and its processes are important.

There are a few examples of scholars suggesting specific NOS issues or teaching approaches for NOS in an SSI/activism-oriented curriculum. One example is Kolstø (2001), who suggests a framework consisting of content-transcending topics, most of which can be viewed as NOS topics (at least in its broad definition). Another example is Yacoubian (2015), who suggests a framework aimed at guiding citizens to think critically with NOS when engaging in decision making on SSI. However, further in-depth research is needed to continue scrutinizing questions such as *What NOS concepts should be given priority in science teaching with the aim of empowering students in relation to societal issues?* and *What could such teaching look like?* Thus, scholars need to move forward from arguing that NOS has an important role, to elaborating on the details of what characterizes NOS teaching that empowers students in relation to societal issues and with SJ as an overall aim.

1.5 Overview of the Current Volume

The aim of this chapter has been to initiate a dialogue on NOS for SJ. We started this dialogue by illustrating how previous research contributes to answering the three questions—why, what and how—and thus supplies to an agenda of NOS for SJ. We have also suggested issues in need of further research. In their respective chapters, the authors use a variety of approaches and make different recommendations concerning what NOS for SJ is and what it entails. They suggest answers and propose new questions that in different ways relate to the three overall questions (why, what and how).

The intent of this volume is not to reach agreement. On the contrary, we believe that a diversity of ideas and approaches enrich the dialogue and create a healthy pathway towards the agenda of NOS for SJ. Thus, the volume could be regarded as a platform upon which the authors engage in a scholarly discussion on what NOS for SJ can mean, and what issues need to be taken into consideration in future research and practice. Our hope is that the volume will contribute some answers, but also highlight new questions. In the paragraphs that follow, we provide a brief overview of the chapters, leaving it for the reader to engage in critical reading and to draw conclusions.

In Chap. 2, ***From Nature of Science to Social Justice: The Political Power of Epistemic Lessons***, Douglas Allchin defends the conventional epistemic approach to NOS by articulating its relevance when applied to scientific claims that inform the distribution of privilege, profit and power. He critiques current approaches to NOS as being "internalist, idealized and decontextualized", and argues for the importance of societal epistemic dimensions that include resolving issues of expertise and credibility, remedying errors due to cultural bias, coping with scientific uncertainties, and counterbalancing conflicts of interest. Such issues, Allchin argues, are important to achieve SJ.

Zoubeida R. Dagher proposes a vision that brings NOS and SJ approaches closer to each other. In Chap. 3, ***Balancing the Epistemic and Social Realms of Science to Promote Nature of Science for Social Justice***, Dagher argues that holistic NOS frameworks that balance social and epistemic domains of science have higher potential and flexibility for engaging learners in reflection on SJ themes compared to frameworks that focus on epistemic statements. Such holistic frameworks, she continues, can link standard-based science and community-based issues.

In ***Capitalism, Nature of Science and Science Education: Interrogating and Mitigating Threats to Social Justice*** (Chap. 4), J. Lawrence Bencze and Lynette C. Carter argue against product-oriented science education, which they claim benefits capitalists and provides an idealized image of science and capitalism. Focusing on the capitalist influences on the fields of science and technology and their adverse effects on SJ and environmental health, the authors provide their vision of a socially-just NOS education through narratives of approaches in school science education and science teacher education. They argue for the preparation of critical and action-oriented citizens who can evaluate systems of power as well as contribute to improving situations.

In his chapter ***Political Entanglement and the Changing Nature of Science*** (Chap. 5), Jesse Bazzul makes a democratic argument and sets forth a "dissensus view of science"—one that recognizes pluralistic NOS as being indispensable for bringing SJ. He argues for politicizing as well as broadening the scope of questions often raised by the proponents of the consensus view of addressing NOS. His proposed questions enable learners to appreciate Western Modern Science in relation to other disciplines and ways of knowing.

Chapter 6, ***Does Research on Nature of Science and Social Justice Intersect? Exploring Theoretical and Practical Convergence for Science Education***, by Sibel Erduran, Ebru Kaya and Lucy Avraamidou discusses how NOS and NOS instruction can contribute to SJ. Drawing from the work of political philosophers such as Rawls and Miller, the authors show some overlapping themes between SJ and NOS. Among those themes are diversity, respect, equity, identity, ethos, opportunity and economic fairness. The authors make recommendations for curriculum and instruction.

In Chap. 7, ***A Discursive Analysis of Relationships Between Nature of Science and Citizenship Education: The Case of Brazilian Science Textbook Evaluation Policies***, Rita Vilanova and Isabel Martins report on the results of a critical discourse analysis of policy documents dealing with textbook evaluation and distribution in Brazil. Being particularly interested in exploring the relationship between NOS and citizenship education, the authors find that even though NOS is treated as a key element in education for citizenship, the focus of NOS is on epistemological dimensions of science rather than on sociopolitical, economic and historical dimensions. The authors find this problematic from a citizenship perspective.

The next chapter (Chap. 8) is written by Cristiano Moura, Iamni Torres Jager and Andreia Guerra. In ***Teaching About Sciences in/ for the Global South: Lessons from a Case Study in a Brazilian Classroom***, the authors argue against a critical use of NOS frameworks derived from the Global North within the context of the Brazilian classroom. Using postcolonial theories and an example from a Brazilian classroom in a women's prison, the authors search for a socially conscious alternative for addressing NOS, which can engage learners with meaningful and contextual learning of NOS. They argue for an approach that is considerate to the asymmetries of power inside knowledge and throughout history, and that is sensitive to the particularities of the Brazilian contexts and that promotes SJ.

The chapter entitled ***Tapping the Potential of Ubuntu for a Science that Promotes Social Justice and Moral Responsibility*** (Chap. 9), by Meshach Ogunniyi, highlights the importance of the African philosophy of Ubuntu within the community of scientific practice. Ogunniyi underscores the potential of Ubuntu to promote human virtues such as humanness, communalism, interdependence, equity, SJ, and moral responsibility. He shows how science can shift from what he characterizes as a "weapon of oppression" to being more relevant to the lives of students, particularly focusing on the South African experience.

The chapter ***Teaching Robust Argumentation Informed by the Nature of Science to Support Social Justice. Experiences from Two Projects in Lower Secondary Schools in Norway*** (Chap. 10), by Stein Dankert Kolstø, elucidates design principles that can be used to empower students to think critically and to produce evidence-based arguments based on their understandings of NOS. The author then discusses the challenges and successes in guiding students through this process by reporting on the implementation and results of two projects.

In Chap. 11, ***Social Images of Science and of Scientists, and the Imperative of Science Education for All***, Agustin Adúriz-Bravo and Alejandro P. Pujalte argue for teaching NOS using a more democratic conception of NOS as well as addressing the social image of science and scientists in a critical way. The authors argue that the traditional pedagogies of science classrooms hinder the possibility of engaging in explicit reflective discussions on NOS. This results in the students forming stereotypes about science and scientists. It also inhibits empowering students and aiming for SJ.

In the next chapter, Chap. 12, entitled ***Images of Scientists in Textbooks Aimed at Students in Need of Supplemental Support – An Analysis of Adjustments***, Lena Hansson and Lotta Leden report on an analysis of Swedish science textbooks that are adjusted for students who require supplemental support. Using content analysis, the authors compare images of scientists in general textbooks to those found in adjusted ones. The results point to a number of different kinds of adjustments, and Hansson and Leden show how the images of scientists, in different ways, are narrowed in the adjusted textbooks. The authors discuss the different adjustments from a SJ perspective.

Hagop A. Yacoubian in Chap. 13, entitled ***Turning Unwanted Stereotypes about Scientists into Nature of Science Learning Experiences that Foster Social Justice***, argues for a proactive approach in dealing with stereotypes about science and scientists. He suggests the use of stereotypes as resources for teaching and learning in ways that empower learners to critically challenge them. A corrective approach that aims to decrease unwanted stereotypes or expose learners to counter-stereotypes facilitates the goal of SJ. Yet, Yacoubian discusses why those efforts need to be supplemented with a proactive approach, and how the latter facilitates SJ not only as a goal but also as a process.

References

Aikenhead, G. (1986). The content of STS education. *STS Research Network Missive, 2*(3), 18–23.

Aikenhead, G. S. (2006). *Science education for everyday life: Evidence-based practice*. New York: Teachers College Press.

Aikenhead, G., & Michell, H. (2011). *Bridging cultures: Indigenous and scientific ways of knowing nature*. Ontario: Pearson Education.

Aikenhead, G. S., & Ogawa, M. (2007). Indigenous knowledge and science revisited. *Cultural Studies of Science Education, 2*, 539–620. https://doi.org/10.1007/s11422-007-9067-8.

Allchin, D. (2003). Scientific myth-conceptions. *Science Education, 87*(3), 329–351.

Allchin, D. (2011). Evaluating knowledge of the nature of (whole) science. *Science Education, 95*(3), 518–542.

Allchin, D. (2013). *Teaching the nature of science. Perspectives and Resources*. St. Paul: SHiPS Education Press.

Allchin, D. (2017). Beyond the consensus view: Whole science. *Canadian Journal of Science, Mathematics and Technology Education, 17*(1), 18–26.

Andrée, M., & Hansson, L. (2014). Recruitment campaigns as a tool for social and cultural reproduction of scientific communities: A case study on how scientists invite young people to science. *International Journal of Science Education, 36*(12), 1985–2008.

Atwater, M. M., Russell, M., & Butler, M. B. (Eds.). (2014). *Multicultural science education: Preparing teachers for equity and social justice*. Dordrecht: Springer.

Barton, A. C. (2003). *Teaching science for social justice*. New York: Teachers College Press.

Barton, A. C., & Upadhyay, B. (Eds.). (2010). Teaching and Learning Science for Social Justice (special issue). *Equity and Excellence in Education, 43*(1).

Barton, A. C., & Yang, K. (2000). The culture of power and science education: Learning from Miguel. *Journal of Research in Science Teaching, 37*(8), 871–889.

Bazzul, J. (2015). Towards a politicized notion of citizenship for science education: Engaging the social through dissensus. *Canadian Journal of Science, Mathematics and Technology Education, 15*(3), 221–233.
Bell, L. A. (2016). Theoretical foundations for social justice education. In M. Adams, L. A. Bell, D. J. Goodman, & K. Y. Joshi (Eds.), *Teaching for diversity and social justice* (3rd ed., pp. 3–26). New York: Routledge.
Bencze, L. (Ed.). (2017). *Science and technology education promoting wellbeing for individuals, societies and environments: STEPWISE* (Vol. 14). Cham: Springer.
Bencze, J. L., & Alsop (Eds.). (2014). *Activist science and technology education*. Dordrecht: Springer.
Bencze, L., & Carter, L. (2011). Globalizing students acting for the common good. *Journal of Research in Science Teaching, 48*(6), 648–669.
Bencze, L., Reiss, M. J., Sharma, A., & Weinstein, M. (2018). STEM education as "Trojan horse": Deconstructed and reinvented for all. In L. Bryan & K. Tobin (Eds.), *13 Questions: Reframing education's conversation: Science* (pp. 69–87). New York: Peter Lang.
Brickhouse, N. W., & Kittleson, J. M. (2006). Visions of curriculum, community, and science. *Educational Theory, 56*(2), 191–204.
Butler, M. B., Atwater, M. M., & Russell, M. L. (2014). Introduction: Culture, equity, and social justice for science teacher educators. In M. M. Atwater, M. Russell, & M. B. Butler (Eds.), *Multicultural science education* (pp. 1–7). Dordrecht: Springer.
Carlisle, L. R., Jackson, B. W., & George, A. (2006). Principles of social justice education: The social justice education in schools. *Project, Equity & Excellence in Education, 39*(1), 55–64.
Chambers, D. W. (1983). Stereotypic images of the scientist: The draw-a-scientist test. *Science Education, 67*(2), 255–265.
Codrington, J. (2014). Sharpening the lens of culturally responsive science teaching: A call for liberatory education for oppressed student groups. *Cultural Studies of Science Education, 9*(4), 1015–1024. https://doi.org/10.1007/s11422-013-9543-2.
Curd, M., Cover, J. A., & Pincock, C. (2013). *Philosophy of science: The central issues* (2nd ed.). New York: W.W. Norton.
Dagher, Z. R., & Erduran, S. (2016). Reconceptualizing the nature of science for science education: Why does it matter? *Science & Education, 25*(1–2), 147–164.
Dillon, J. T. (2009). The questions of curriculum. *Journal of Curriculum Studies, 41*(3), 343–359.
Dimick, A. S. (2012). Student empowerment in an environmental science classroom: Toward a framework for social justice science education. *Science Education, 96*(6), 990–1012. https://doi.org/10.1002/sce.21035.
Driver, R., Leach, J., Millar, R., & Scott, P. (1996). *Young people's images of science*. Philadelphia: Open University Press.
Duit, R. (2015). Didaktik. In R. Gunstone (Ed.), *Encyclopedia of science education* (pp. 325–327). Dordrecht: Springer.
Erduran, S., & Dagher, Z. (2014). *Reconceptualizing the nature of science for science education: Scientific knowledge, practices and other family categories*. Dordrecht: Springer.
Finkel, L. (2018). Infusing social justice into the science classroom: Building a social justice movement in science education. *The Journal of Educational Foundations, 31*(1–2), 38–56.
Freire, P. (1970). *Pedagogy of the oppressed*. New York: Bloomsbury Publishing.
Giroux, H. (2019). Toward a pedagogy of educated hope under casino capitalism. *Pedagogía y Saberes, 50*, 147–151.
Hackman, H. W. (2005). Five essential components for social justice education. *Equity & Excellence in Education, 38*(2), 103–109.
Hansson, L. (2018). Science education, indoctrination, and the hidden curriculum. In M. R. Matthews (Ed.), *History, philosophy and science teaching* (pp. 283–306). Cham: Springer.
Hansson, L. (2020). Teaching the limits of science with card sorting activities. In W. F. McComas (Ed.), *Nature of science in science instruction: Rationales and strategies*. Cham: Springer.

Hansson, L., & Lindahl, B. (2010). I have chosen another way of thinking. *Science & Education, 19*(9), 895–918.
Hernandez, C. M., Morales, A. R., & Shroyer, M. G. (2013). The development of a model of culturally responsive science and mathematics teaching. *Cultural Studies of Science Education, 8*(4), 803–820.
Hodson, D. (2008). *Towards scientific literacy: A teachers' guide to the history, philosophy and sociology of science*. Rotterdam: Sense Publishers.
Hodson, D. (2011). *Looking to the future. Building a curriculum for social activism*. Rotterdam: Sense Publishers.
Hodson, D. (2014). Nature of science in the science curriculum: Origin, development, implications and shifting emphases. In M. R. Matthews (Ed.), *International handbook of research in history, philosophy and science teaching* (pp. 911–970). Dordrecht: Springer.
Jost, J. T., & Kay, A. C. (2010). Social justice: History, theory, and research. In S. T. Fiske, D. T. Gilbert, & G. Lindzey (Eds.), *Handbook of social psychology* (pp. 1122–1165). Hoboken: Wiley.
Kampourakis, K. (2016). The "general aspects" conceptualization as a pragmatic and effective means to introducing students to nature of science. *Journal of Research in Science Teaching, 53*(5), 667–682.
Keddie, A. (2012). *Educating for diversity and social justice*. New York: Routledge.
Kincheloe, J. (2008). *Critical pedagogy primer*. New York: Peter Lang.
Kolstø, S. D. (2001). Scientific literacy for citizenship: Tools for dealing with the science dimension of controversial socioscientific issues. *Science Education, 85*(3), 291–310.
Leden, L., Hansson, L., & Redfors, A. (2017). From black and white to shades of grey. *Science & Education, 26*(5), 483–511.
Lederman, N. G. (2007). Nature of science: Past, present, and future. In S. K. Abell & N. G. Lederman (Eds.), *Handbook of research on science education* (pp. 831–879). Mahwah: Lawrence Erlbaum Associates.
Lee, H., Yen, C.-F., & Aikenhead, G. S. (2012). Indigenous elementary students' science instruction in Taiwan: Indigenous knowledge and western science. *Research in Science Education, 42*, 1183–1199. https://doi.org/10.1007/s11165-011-9240-7.
Longino, H. E. (1990). *Science as social knowledge: Values and objectivity in scientific inquiry*. Princeton: Princeton University Press.
Matthews, M. R. (1994). *Science teaching: The role of history and philosophy of science*. New York: Routledge.
Matthews, M. R. (2012). Changing the focus: From nature of science (NOS) to features of science (FOS). In M. S. Khine (Ed.), *Advances in nature of science research* (pp. 3–26). Dordrecht: Springer.
McComas, W. F. (1998). The principal elements of the nature of science: Dispelling the myths. In W. F. McComas (Ed.), *The nature of science in science education: Rationales and strategies* (pp. 53–70). Dordrecht: Kluwer Academic.
McComas, W. F. (2017). Understanding how science works: The nature of science as the foundation for science teaching and learning. *School Science Review, 98*(365), 71–76.
McComas, W. F. (2020). *Nature of science in science instruction: Rationales and strategies*. Cham: Springer.
McDuffie, T. E. (2001). Scientists – Geeks and nerds? *Science and Children, 38*(8), 16–19.
Morales-Doyle, D. (2018). Students as curriculum critics: Standpoints with respect to relevance, goals, and science. *Journal of Research in Science Teaching, 55*(5), 749–773.
O'Neill, T. B. (2010). Fostering spaces of student ownership in middle school science. *Equity & Excellence in Education, 43*(1), 6–20.
Pedretti, E. (2005). STSE education: Principles and practices. In S. Alsop, L. Bencze, & E. P. Pedretti (Eds.), *Analysing exemplary science teaching: Theoretical lenses and a spectrum of possibilities for practice* (pp. 116–126). London: Open University Press.

Pérez-Garzón, C. A. (2018). Unveiling the meaning of social justice in Colombia. *Mexican Law Review, 10*(2), 27–66.
Proper, H., Wideen, M. F., & Ivany, G. (1988). World view projected by science teachers: A study of classroom dialogue. *Science Education, 72*(5), 547–560.
Ratcliffe, M., & Grace, M. (2003). *Science education for citizenship: Teaching socio-scientific issues*. London: McGraw-Hill Education.
Reiss, M. J. (2003). Science education for social justice. In C. Vincent (Ed.), *Social justice, education and identity* (pp. 163–174). London: Routledge Falmer.
Roberts, D. A. (2007). Scientific literacy/science literacy. In S. K. Abell & N. G. Lederman (Eds.), *Handbook of research on science education* (pp. 729–780). Mahwah: Lawrence Erlbaum Associates.
Rudolph, J. L. (2000). Reconsidering the 'nature of science' as a curriculum component. *Journal of Curriculum Studies, 32*(3), 403–419. https://doi.org/10.1080/002202700182628.
Ryder, J. (2001). Identifying science understanding for functional scientific literacy. *Studies in Science Education, 36*(1), 1–44. https://doi.org/10.1080/03057260108560166.
Sadler, T. D., Barab, S. A., & Scott, B. (2007). What do students gain by engaging in socioscientific inquiry? *Research in Science Education, 37*(4), 371–391.
Santos, W. L. P. (2009). Scientific literacy: A Freirean perspective as a radical view of humanistic science education. *Science Education, 93*, 361–382.
Selby, D. E. (2014). Education for sustainable contraction as appropriate response to global heating. In L. Bencze & S. Alsop (Eds.), *Activist science and technology education* (pp. 165–182). Dordrecht: Springer.
Siegel, H. (1997). Science education: Multicultural and universal. *Interchange, 28*, 97–108. https://doi.org/10.1023/A:1007314420384.
Simonneaux, L. (2014). From promoting the techno-sciences to activism – A variety of objectives involved in the teaching of SSIs. In L. Bencze & S. Alsop (Eds.), *Activist science and technology education* (pp. 99–111). Dordrecht: Springer.
Sjøberg, S. (2000). *Science and scientists: the SAS study*. Retrieved April 4, 2005, from http://folk.uio.no/sveinsj/SASweb.htm.
Sjöström, J., & Eilks, I. (2018). Reconsidering different visions of scientific literacy and science education based on the concept of Bildung. In Y. J. Dori et al. (Eds.), *Cognition, metacognition, and culture in STEM education* (pp. 65–88). Cham: Springer.
Snively, G., & Corsiglia, J. (2001). Discovering indigenous science: Implications for science education. *Science Education, 85*, 6–34. https://doi.org/10.1002/1098-237X(200101)85:13.0.CO;2-R.
Star, S., & Griesimer, J. (1989). Institutional ecology, 'Translations' and boundary objects: Amateurs and professionals in Berkeley's Museum of Vertebrate Zoology, 1907-39. *Social Studies of Science, 19*(3), 387–420.
Stenmark, M. (2004). *How to relate science and religion: A multidimensional model*. Michigan: W. B. Eerdmans Publishing.
Taber, K. S. (2013). Conceptual frameworks, metaphysical commitments and worldviews: The challenge of reflecting the relationships between science and religion in science education. In N. Mansour & R. Wegerif (Eds.), *Science education for diversity. Theory and practice* (pp. 151–177). Dordrecht: Springer.
Varelas, M., Morales-Doyle, D., Raza, S., Segura, D., Canales, K., & Mitchener, C. (2018). Community organizations' programming and the development of community science teachers. *Science Education, 102*(1), 60–84.
Vesterinen, V.-M., Aksela, M., & Lavonen, J. (2013). Quantitative analysis of representations of nature of science in Nordic upper secondary school textbooks using framework of analysis based on philosophy of chemistry. *Science & Education, 22*, 1839–1855.
Villar, P., & Guppy, N. (2015). Gendered science: Representational dynamics in British Columbia science textbooks. *Canadian Journal of Education, 38*(3), 1–24.

Yacoubian, H. A. (2015). A framework for guiding future citizens to think critically about nature of science and socioscientific issues. *Canadian Journal of Science, Mathematics and Technology Education, 15*(3), 248–260.

Yacoubian, H. A. (2018). Scientific literacy for democratic decision-making. *International Journal of Science Education, 40*(3), 308–327.

Yacoubian, H. A., & Bazzul, J. (Eds.). (2015). Rethinking education for citizenship [special issue]. *Canadian Journal of Science, Mathematics, and Technology Education, 15*(3), 211–215.

Yacoubian, H. A., Al-Khatib, L., & Mardirossian, T. (2017). Analysis of the image of scientists portrayed in the Lebanese national science textbooks. *Science & Education, 26*(5), 513–528.

Zacharia, Z., & Barton, A. C. (2004). Urban middle-school students' attitudes toward a defined science. *Science Education, 88*(2), 197–222.

Zajda, J., Majhanovich, S., & Rust, V. (2006). Education and social justice: Issues of liberty and equality in the global culture. In J. Zajda, S. Majhanovich, V. Rust, & E. M. Sabina (Eds.), *Education and social justice* (pp. 1–12). Dordrecht: Springer.

Zeidler, D. L., Sadler, T. D., Simmons, M. L., & Howes, E. V. (2005). Beyond STS: A research-based framework for socioscientific issues education. *Science Education, 89*(3), 357–377.

Chapter 2
From Nature of Science to Social Justice: The Political Power of Epistemic Lessons

Douglas Allchin

2.1 Introduction: From Social Justice to Nature of Science

In pursuing social justice, one usually looks to social workers, charitable programs, or political activists. Not scientists. Still, science can have a significant role in shaping and justifying public policy by documenting injustice and by validating effective solutions. Political rhetoric cannot substitute for the trustworthiness of evidence-based claims. Students should thus learn about the crucial *epistemic* dimension of science. What makes scientific claims reliable? Most current approaches to teaching about how science works, however, are idealized and decontextualized. In this chapter, I describe an alternative approach that incorporates "Whole Science" (Allchin 2011, 2013, 2017a) and conveys fully and concretely the connection between epistemics and science in society. Notably, that includes (as addressed in separate sections below) the roles of science communication, expertise and credibility, uncertainty, and conflicts of interest. Special attention is given to the naturalizing error and to scientific errors rooted in cultural ideology (gender, race and class biases)—and how such errors are mitigated and remedied. That is, students should appreciate how the epistemic practices of science, in conjunction with standard moral principles, can help us expose and resolve the problems that arise from the pursuit of disproportionate privilege, profit or power.

At first, social justice may seem an unlikely topic for a science classroom. Science teachers prepare to teach by learning science, not ethics or politics. They become well versed in scientific concepts and the epistemic tools of empirical investigations, not in justifying moral claims, in methods of discussing economic or ideological values, or in negotiating authority between conflicting interests. Still, plain

D. Allchin (✉)
Minnesota Center for the Philosophy of Science, University of Minnesota, Minneapolis, MN, USA
e-mail: allchindouglas@gmail.com

H. A. Yacoubian, L. Hansson (eds.), *Nature of Science for Social Justice*, Science: Philosophy, History and Education,
https://doi.org/10.1007/978-3-030-47260-3_2

unadorned science can be relevant to fostering social justice in many ways. (By social justice, I mean the disparities in wealth, social benefits, and privileges that result from the exercise of power rather than through equal opportunity and free access to common resources; National Education Association 2017; Center for Economic and Social Justice 2018). For example, DNA evidence can help exonerate persons wrongly convicted of murders and, over many cases, demonstrate systematic racial injustice. Epidemiological patterns can help establish how environmental risks have fallen disproportionately on already impoverished communities, or how workplace safety rules adversely affect certain already disadvantaged populations. Economic analysis can expose the disparities between politicians' claims about tax policy and the ultimate realities about who benefits and who bears the costs. One could easily expand this list to include such issues as equity in access to birth control or other health services; understanding the relationship between lack of economic opportunity and youth gun violence; the role of needle-swap programs in reducing disease transmission among drug-users, or unequal barriers to participation in democratic elections. Trustworthy information and evidence matter. Ironically, perhaps, science can contribute to social justice not through any direct political action, but by providing reliable knowledge that informs arguments used to either justify or challenge the disparities in privilege, profit, and power. *My analysis here builds on this philosophical dimension of science in promoting social justice: through broad epistemic understanding* (see also Kolstø, Chap. 10).[1]

While much science education remains focused on content, or scientific concepts, a growing international consensus has highlighted the role of teaching the *nature of science (NOS)*, or "scientific practices," or how science works (Allchin 2017a; Allchin et al. 2014; Hodson 2008; NGSS Lead States 2013; OECD 2017). *Namely, how does science develop its claims and, more importantly perhaps, how does it establish their reliability, or trustworthiness?* The growing tradition in NOS education forms a foundation here.

Further, NOS is intended to contribute to functional scientific literacy (Kolstø 2001; OECD 2017; Ryder 2001) or what a panel of the U.S. National Science Foundation called "science in the service of the citizen and consumer" (Toumey et al. 2010). Namely, the purpose of NOS instruction is not merely to profile the explanatory power of science, nor strictly to legitimize its cultural authority. Rather, NOS is to aid individuals in a society where public policy and personal decision-making increasingly draw on scientific claims (Rudolph 2005; Rutherford and Ahlgren 1991).[2]

---

[1] My central theme here differs from the social epistemology notion of *epistemic justice* (articulated by Fricker 2007). That is, I do not focus here on how social justice *within* science affects the reliability of its claims. Rather, I am concerned with how the reliability of scientific claims contributes to arguments in a public sphere relevant to social justice.

[2] This widespread institutional orientation thus situates NOS education solidly within basic citizenship goals. Epistemic lessons contribute in personal, social, political and economic contexts precisely because they support the assessment of evidence and arguments, and promote informed decision-making (contrast to Vilanova and Martins, Chap. 7).

Historically, of course, many scientific claims have later proven unreliable. All scientific knowledge is "tentative," or provisional (e.g., McIntrye 2019; Oreskes 2019; Zimring 2019). Errors occur with regularity. However, in some cases the errors have had important cultural or political implications, affecting social justice. For example, in certain episodes, scientists endeavored to portray gender, race, or class disparities as validated by science (see §2.3 and cases in table below, §2.2). Such claims were then used to maintain customs of social privilege, to restrict freedoms, to limit immigration, to hinder social advancement, and/or to deny educational opportunities. Later research exposed the flawed assumptions or methods and other lapses in reasoning. Using history as a guide, it seems vitally important in a social context to know how to identify such erroneous claims when they arise. Understanding how science works means also understanding how or when science does *not* work (Allchin 2012a). When is science vulnerable to bias and to reaching unreliable conclusions, even if apparently supported by some evidence? *Students will ideally learn how both types of claims can develop, and how to differentiate between them—namely, both when to trust scientific claims and, equally, when to doubt them.*

Socially, the authority of science matters, especially in politics. In a sense, science is a form of power. Thus, it should surprise no one that some individuals and monied interests try to secure that authority for themselves, even if their claims do not accord with scientific consensus or are not informed by scientific work (McGarity and Wagner 2008; Mooney 2005b; Nestle 2015, 2018; Oreskes and Conway 2010; Rampton and Stauber 2002). Imitators of science flood print and broadcast media, the Internet, social media, and electronic communications with claims that are deliberately misleading and/or have no scientific merit. Science con-artists and purveyors of fake news are everywhere, vying for advantage through deceit (Allchin 2017a, b, pp. 104–113, 2018a; Goldacre 2010). These misrepresentations present citizen-consumers with additional challenges: interpreting who is a scientific expert and who is not, and evaluating which sources of information are credible, and whose testimony should be trusted. Many issues of social justice now seem to be played out at this level, where non-scientists hope to eclipse the science that would otherwise threaten the profits and privileges they receive from the current power structure. The issue of reliability in science communication adds a significant dimension beyond the standard assessment of scientific evidence and arguments that constitutes most current approaches to NOS. Educators must thus adopt an NOS framework that accommodates these issues at the social level. *We need to shift from nature of science to nature of science-in-society* (Allchin forthcoming; Höttecke and Allchin 2020; Kelly et al. 1993; Raveendran and Chunawala 2013).

Current institutional approaches to NOS are insufficient. Concepts of NOS coalesced in the late 1990s around a set of principles shared across major international curriculum documents, what has come to be known as the "NOS consensus list." Ironically, the "consensus" list no longer enjoys a very wide consensus (Allchin 2017a; Hodson and Wong 2017; see also Bazzul, Chap. 5). The emphasis was on ideas, theory, idealized scientific reasoning, and training future scientists. That is, the view was largely internalist. Accordingly, classrooms tended to adopt decontextualized "blackbox" activities and "cookbook" inquiries (for example, Lederman

Depository 2018) that treated scientific justification as simple and unrealistically formulaic. Ironically, it failed to reflect authentic scientific practice as exhibited in complex historical cases (Allchin 2013, pp. 107–120). This was an impoverished view of the nature of science. In retrospect, the consensus view lacked the contextualization that seems so clearly essential today. NOS needs to address a broader set of questions (see also Bazzul, Chap. 5). In particular, it was not designed to focus on the cultural consequences of science or to develop informed citizens or consumers of science. To do that, one must follow the scientific claims beyond publication in professional journals into society where they are applied and, sadly, sometimes misrepresented. One must focus on the entire reach of science—from test tubes to YouTube, from the lab bench to the judicial bench, from field site to website, from lab book to Facebook—or "*Whole Science*" (Allchin 2010, 2011, 2012b, 2013, 2017a; Höttecke and Allchin 2020).

In the following sections, I elaborate on the relevant dimensions of the more expansive Whole Science approach. I describe a number of epistemic elements that are missing in conventional approaches to NOS, but which are integral to pursuing social justice effectively (see table below, §2.2). These are illustrated with numerous concrete cases, as examples of the kinds of lessons that students might encounter in a classroom transformed to include social justice issues.

2.2 Epistemic Dependence, Expertise and Credibility

Perhaps the most significant socioscientific issue currently is global warming and climate change. Yet many political leaders and media pundits (notably in the U.S.) dismiss the scientific consensus, calling it a hoax, a scam, a fraud (Allchin, 2015). The problem here is not inadequate conceptual understanding of the greenhouse effect. Nor is it failure to appreciate the nature of theories versus laws, or the role of creativity in science (elements of the outmoded NOS "consensus list"). Nor is it about general acceptance of or belief in the authority of science. Rather, it is public contention about what the science legitimately claims. The problem is in *communicating* science and in understanding *scientific expertise and credentials* (Höttecke and Allchin 2020). Who is a credible spokesperson for science? That epistemic challenge is part of the nature of science (Goldman 1999).

No one individual is competent alone to adjudicate all the evidence relevant to climate change (it is far too vast and specialized). We all rely on others for expert knowledge. *Epistemic trust* is essential (Hardwig 1991). One inevitable task of NOS education (for scientific literacy), then, is to teach students how to deal with the second-hand reliance on the knowledge of scientists (Allchin 2017b, pp. 95–103; Gaon and Norris 2001; Norris 1995, 1997; Zemplén 2009). But with all the potential for misinformation, when is trust warranted, and when is it not? As puzzling as it may seem, learning the structure for warranting trust in third-party scientific claims in a public realm is just as important as the original research itself.

Investigating the expertise of climate change naysayers quickly indicates that many of the most prominent voices are not experts at all. Fred Singer, one of the earliest critics, was a nuclear physicist, with no background in atmospheric or climate science. Steve Milloy, frequently featured on Fox News and labeled an "expert," was a lawyer and a lobbyist working for a libertarian think-tank. Indeed, one finds that the whole denial movement has been largely funded and promoted by the fossil fuel industry and other political conservatives (Mooney 2005a; Oreskes and Conway 2010; Union of Concerned Scientists 2007). That sponsored interference is a clue that climate change science is also a significant social justice issue. The eclipse of facts here does not affect the populace uniformly. Those who profit from the carbon-based energy economy continue to benefit, at the expense of increasing the risks and long-term costs for everyone. Those who generate greenhouse gases disproportionately (generally, developed nations) prosper, while the environmental consequences mount globally. Discounting the legitimate science perpetuates and amplifies injustice. In addition, the science can identify who (historically) has generated the greenhouse gases, and thus who may be considered primarily accountable for remedying the situation now. Achieving restorative justice is intimately linked to a scientific analysis of who caused the problem, how they benefitted, and thus who is responsible now, and to what degree. Addressing the justice issue involves, in this case, knowing who is an expert and who is not, and demanding politically that scientific expertise matters.

Expertise does not always align with authority or political leadership. For example, in the early 2000s, while AIDS ravaged South Africa, Dr. Manto Tshabalala-Msimang, the Public Health Minister, adopted a policy that denied the connection between HIV and AIDS. Appealing to anti-Colonial sentiment, she claimed instead that traditional African values and knowledge of nutrition could effectively deal with the "alien" disease (Goldacre 2010; Voude 2007). Yet she was not an expert. Nor did she heed the global consensus of medical science. As a result, hundreds of thousands of people—mostly those already impoverished and modestly educated—died prematurely. All because her power and appeal to cultural values trumped expertise.

As another example of expertise and nationalistic cultural values, consider recent efforts in India to validate Ayurvedic remedies (Kumar 2017). According to the ancient texts, an elixir made of cow urine, dung, milk, yogurt and clarified butter (called anchagavya) is supposed to cure such conditions as diabetes, cancer, schizophrenia, and autism. Testing this proposed medication clinically would certainly exhibit the empirical dimension of science—as dictated in conventional NOS. But current studies are being promoted by nationalists intent on validating those cures, not examining their efficacy objectively. The "science" is expected to lend greater authority to viewing India as a superior culture (Kumar 2019). The political intent, aligned with a presumptive scientific outcome, seems to discount deference to experts. In a similar way, not long ago extreme nationalists suggested that ancient texts provided evidence that Indians once flew interplanetary spacecraft, worked with stem cell therapies, and performed interspecies surgery that yielded a human with the head of an elephant (Desai 2014; Khan 2018; Kumar 2019). Members of the Indian Academy of Sciences were quick to discredit these claims. These cases illustrate the potential for *political conflict of interest* in public scientific claims. Whether the public is susceptible to such claims depends in part on their ability to understand and discern expertise.

Misleading or erroneous support for folk remedies has an additional social consequence. Namely, if ineffective but readily available folk remedies can be misportrayed as effective, then a government need not ensure access to modern—and more costly—health care. The costs of funding health care for the economically disenfranchised would conveniently disappear. As a result, the poor would continue to suffer from illness, while the wealthy paid their way to health, compounding any unjust class disparities that already exist.

Again, ascertaining scientific expertise matters to social justice. That is a dimension of reliability that needs to be added to the NOS curriculum (Table 2.1).

Table 2.1 Features of NOS relevant to social justice, with some example cases

	Example cases
Epistemic dependence; scientific expertise, credibility, and credentials	Purported role of vitamins or nutrition in preventing AIDS (South Africa) (Goldacre 2010)
	Contagiousness of AIDS (USA) (Toumey 1996)
Role of science communication, including conflicts of interest	Nationalistic promotion of Ayurvedic medicines (India)
	Recruiting poorly educated women for cervical cancer research (India)
	Lack of public disclosure of fracking chemicals (USA)
Scientific uncertainty	Dam safety (Uttarakhand, India; Attapeu, Laos; Burmadinho, Brazil)
	False image of uncertainty about safety of fire-retardants or workplace chemicals (USA)
	Precautionary Principle: nuclear power plants (Japan, India, Brazil)
Scientific errors: gender, race and class bias	Historical IQ testing and immigration (USA) (Gould 1983)
	Davenport's historical view of pellagra as genetic, not nutritional (Allchin 2016)
	Historical craniology and women (Fee 1979)
Scientific errors: the naturalizing error (values masquerading as facts)	Genes as determinants affecting cultural potentials or social class
	Natural selection as a socioeconomic process affecting social status and cultural privilege
	Dichotomy of sexes, gender identity, and transgender rights
Sources of research funding and their biases	Agricultural biotechnology as biased towards mechanization & large-scale monocrops
	Ethnobotanical remedies – research & intellectual property vs. availability & distribution
	Energy research on large-scale (industrial) vs. small-scale (household) sources (Terrapon-Pfaff et al. 2014)

2.3 Error, Bias and the Naturalizing Error

One familiar feature of NOS, through almost all characterizations over the last several decades, is (as noted above) that science is "tentative." Namely, scientists revise their claims and theories. Sometimes, that means acknowledging earlier errors or, at least, misleading models. For example, not long ago, the U.S. Preventative Services Task Force revised its recommendations for mammogram tests for breast cancer (Kolata 2009). For most women, it concluded, screening should begin at age 50. Earlier, it was age 40. But was the change justified? This was announced during a time of social concerns about soaring medical costs. So, was this just a way to cut costs—at the unjust expense of women's health? Nature of science was relevant, here—but not merely to admit that science is "tentative" and can change. The specific reasoning mattered. One needed an epistemic analysis. At one level, the lesson of expertise applied here. Yes, the panel members were independent, qualified medical researchers. But how did they justify changing their expert view, then? The new recommendations were based on more data and meta-studies, which provided a better overall view of the benefits and risks of the tests (including harm from added exposure to X-rays). The available evidence changed. We have to be ready to revise our theories and even fundamentally alter our conclusions when that happens (McIntrye 2019; Oreskes 2019; Zimring 2019).

Ironically, in some political contexts, appeals to tentativeness and a "skeptical attitude" can have negative consequences. In many cases in recent history, doubt has been enlisted repeatedly to forestall government policies for protecting human health and the environment. For example, the tobacco industry claimed that there was not sufficient evidence on the effects of secondhand smoke in the 1960s, so (they argued) cigarette sales ought not be regulated. Available evidence was wholly discounted by leveraging an oversimplified NOS concept. The strategy of appeal to tentativeness was used over and over again in the ensuing decades. Industry contended that in the absence of "absolute" proof, informed regulatory policy was not possible, and any action must wait until better knowledge is available. This playbook was echoed in the cases of acid rain; chlorinated fluorocarbons (CFCs) and the ozone layer; DDT use; formaldehyde; flame retardants, hexavalent chromium; vinyl chloride; lead; and ephedra (Kenner 2010; Michaels 2008; Oreskes and Conway 2010). Deeper understanding of NOS is needed, including more nuanced views of "proof" and burden of proof in a policy context.

Genuine errors in science do occur. And they can have a significant impact on social justice, even if only until the errors are identified and remedied. As noted briefly above, historians of science have documented countless major cases involving supposed justification for gender discrimination, racial prejudice, and adverse outcomes for lower classes (Gould 1981, 1983; Schiebinger 1989, 1993). The source of the error may typically be characterized as a collective blind spot by a scientific community without the relevant balance of cultural perspectives. Namely, in the past, male scientists have generally not noticed the flaws of their own gendered assumptions, until an alternate gendered voice emerged to keep their conclusions in check (Fee 1979). Likewise, white Europeans failed to see their assumptions (and thus mistakes) about races and other cultures—until persons from those cultures had standing to challenge their weak evidence in scientific discourse (Barkan 1992). Wealthy individuals have easily overlooked what seems obvious to those without such wealth (Allchin 2017b, pp. 43–59). The general lesson is that scientific evidence is interpreted by scientists, and the cultural perspectives of the scientists can matter. One perspective keeps another in check and accountable to the evidence. Diversity in scientific communities matters—not just on the principles of social justice, but because it is integral *epistemically* to robust reliable outcomes. Philosophers have now articulated more fully the significance of *social epistemology*, at a level above the methods profiled in conventional NOS (Harding 1991, 1998; Longino 1990, 2001; Solomon 2001). The basics of social epistemology are another concept key in a Whole Science approach (Allchin 2013, pp. 107–120), essential to a full NOS curriculum.

One type of scientific error has special significance to social justice: the naturalizing error (Allchin 2008; Allchin and Werth 2017, in press; Raveendran and Chunawala 2015). In these cases, a cultural or political ideology becomes embodied in the scientific conclusions. The value-laden assumptions become inscribed invisibly as unquestionable "facts" of nature. Nature, in turn (due to our native teleological psychology), is viewed as inevitable or unchangeable, even intentional or purposeful. The bias or power structure, a result of social history, thence comes to be regarded (illegitimately) as "natural." Worse, the cultural view seems endorsed by empirical evidence and the authority of science. For example, the conventional stereotyped image of natural selection tends to inappropriately naturalize competition as an integral component of "progress." The scientific concept originated among the Victorian elite, but now seems (with circular reasoning) to implicitly justify open-market views and current economic stratification (Allchin 2017b, pp. 43–59). Also, strict categories of male and female are not warranted biologically, but do help reinforce gendered division of labor and power structures (Allchin 2017b, pp. 114–124). Many views of genetics also portray DNA as destiny, implying that efforts towards social justice are doomed to fail in the context of inherited, "natural" differences (Allchin 2017b, pp. 141–145; Heine 2017; Lewontin et al. 1984). These scientific errors are especially important in education because of the circular link from culture to science to culture again. What appears as scientifically justified may not be, upon closer examination and critical analysis by diverse participants. The solution is not to abandon science (as some contend), but to get the science right.

Scientific errors may seem like the last thing one wants to teach in science, as some admission of its capacity to fail. Yet past errors are also the clues to the methods by which we avoid such errors in the future. Especially contextualized in history, cases or error in science are valuable contributions to healthy epistemic lessons (Table 2.1; Allchin 2012a, 2013, pp. 165–183).

2.4 Uncertainty and the Precautionary Principle

Another challenge for science in social contexts is not susceptibility to error, but *uncertainty*. That is, in some cases, the science is admittedly incomplete. Conclusions are not yet possible, even "tentative" ones. That applies to many contemporary cases of technological risk. Acknowledging the full range of scientific uncertainty matters. Consider the case of installing hydroelectric dams in Uttarakhand province in India in the early 2000s. Construction proceeded heedless of possible adverse effects. That led to disaster in 2013. As a result of heavy rains, several dams failed. Nearby construction debris and mud from unmanaged excavation areas was washed downstream. Villages were wiped out. Over 6000 people died (Joshi 2016; Ministry of Environment and Forests 2014). Here, an appeal by industry to the "tentativeness" of science after the fact may seem disingenuous. The victims of the Uttarakhand disaster were the local residents. Those who benefitted from the dams, by contrast, were the wealthy industrialists and the Indians in other, more prosperous states who drew electrical power (and profit) from the dams. The risk of the projects was not borne by those who benefitted most, but by those with marginal economic status. With a deficit of scientific clarity or openness about the risks—all too obvious now, after the dam failures—the local populations had little political leverage to oppose the dams. Because a fuller respect for environmental science was eclipsed in building the projects, in retrospect the disaster may seem "unexpected": the builders can thus easily frame it as an "accident" triggered by heavy rainfall, a "natural" event for which no one can bear responsibility. Appeal to scientific uncertainty becomes a political escape clause. But many of the inherent risks were known in advance. The disaster could well have been avoided if the dam-makers had fully addressed the environmental risks and concerns of the engineers at the outset. The social injustice in the disaster ultimately resulted from a disregard for "known" science uncertainties. Later, similar events led to major dam collapses in Laos (Ives 2018) and (twice) in Brazil (Douglas 2015; New York Times 2019). Nor is the commercial neglect of safeguards in these episodes that much different from the classic case of building residential communities on top of toxic waste dumps in Love Canal or Times Beach in the U.S. (Newton and Dillingham 1994. pp. 7–28). Science and scientific uncertainty can each be used towards political ends—a key awareness for the scientifically literate citizen, but not found in conventional NOS profiles.

The episode of Uttarakhand dam and related cases underscore the importance of articulating how scientific uncertainty is addressed differently in social versus scientific contexts. Scientists, of course, are typically loathe to advance claims without sufficient evidence. Their principle might be summarized as, "first, publish no wrong." In this case, they could not precisely predict the consequences. That might be an appropriate idealized epistemic posture, aptly reflecting the NOS tenet of "tentativeness." But in a social setting, that posture becomes grossly irresponsible. Policy-makers needed to also consider the ethical dimension of possible environmental consequences, whether fully documented or not. The burden of proof should have been on demonstrating and achieving the absence of significant risk

(Shrader-Frechette 1990). Socially, scientific uncertainty indicates the need for preventative safeguards, using an ethical guiding principle of "first, do no harm." That is the philosophical origin of the Precautionary Principle (Foster, et al. 2000; Harremoës et al. 2001; Ivone 2015; O'Riordan and Cameron 1994; World Commission on the Ethics of Scientific Knowledge and Technology 2005) and its close relation to science. Epistemic and policy postures under scientific uncertainty differ. The relevance of that difference and of the Precautionary Principle is precisely why students need a Whole Science approach to learning NOS (Table 2.1).

2.5 Funding and Conflict of Interest

Bias in science, with corresponding implications for social justice, also occurs in research sponsorship. The growth of knowledge depends on sources of funding. If certain avenues of research or certain investigative problems are privileged, with disproportionate funding, research is led in certain directions at the expense of others (Kitcher 2001). Wealthy interests can thereby influence what science concludes—often in ways that perpetuate that wealth. For example, agricultural biotechnology is based on conceptualizing crops as genes or as individual plants threatened by weeds, pests and limited resources, rather than as a complex interaction of social systems that foster monocropping and large-scale mechanized farming (Allchin 2019; Levidow 1998). That view favors property owners who can increase the productivity of their land and wealthy farmers who can invest in capital equipment. In both ways, viewing biotech as central peripheralizes the roles of laborers and the unequal social distribution of wealth. Biotechnological research yields answers that implicitly reaffirm the interests of the wealthy. Similar biases govern research on marketable pharmaceuticals versus alternative pain treatments (such as acupuncture) that are more labor-intensive and inherently less profitable to business investors. Major research on effective ethnobotanical remedies likewise tends to focus on identifying active ingredients (that can be patented and thus owned as exclusive intellectual property), rather than on analyzing the preparation techniques and therapeutic practices that would generally be more widely accessible and less costly to individuals. What is known scientifically—appearing altogether objective because of a body of evidence—can actually be shaped by funding. The bias in research choices is also a core epistemic concept, again not included in conventional NOS (see also Dagher, Chap. 3).[3]

[3] Here, I underscore the epistemic dimension of funding. That leaves open the cultural question of who funds science, and how. Some advocates (e.g., Kitcher 2001) propose a public, community-based and democratic ideal. Yet the majority of research is funded by commercial interests. Roughly 40% of all geologists are employed by the petroleum industry. Even the majority of government-sponsored research is typically oriented to national defense and the military. The complex mixture of public and private funding renders deeply problematic the question of how one might regulate which research topics are supported.

The cases described thus far should make it clear that the ideals of science profiled in conventional NOS lessons are not always found in the real world. Science underwrites power and scientific claims and authority are often contested. Accordingly, no one should be surprised that monied and ideological interests inevitably endeavor to "bend" science where government regulation of risks to workplace and environmental safety are concerned (McGarity and Wagner 2008; Wagner and Steinzor 2006). Thus, the scientifically literate citizen should always be alert to those with conflicts of interest who may try to distort, discount, or deny good science (Allchin 2017b, pp. 104–113, 2018a). For example, how should one interpret the debates over the safety of nuclear power? Operators of the plant in Fukushima, Japan, defended its safety, back-up systems and response protocols, of course—until the disaster in 2011. Likewise, officials at the Kudankulam plant in Tamil Nadu, India, continue to assert that their design is safe, although six workers were severely injured when a pipe burst in 2014. The plant has been fined for numerous operational violations and has experienced numerous shutdowns for steam leakage and other problems (Economic Times 2017). While the companies should have the best access to useful information about safety, the history of their claims indicates that they cannot be trusted as credible. Even when the Tokyo Electric Power Company first issued its analysis of the causes of the Fukushima "accident," it was sharply criticized for its narrow focus and effort to justify the company's response. Another report followed. Concerns about reliability are more acute in such cases because while the risk probabilities are low, the magnitude and scope of errors is potentially quite large. So citizens need to be educated about the effects of conflicts of interest in those presenting evidence and scientific arguments, not just about weighing whatever evidence is offered to them (see also Benzce and Carter, Chap. 4). However, if social interests endeavor to "bend" science, the appropriate response seems not to disparage all science as inherently flawed, but to "unbend" the problematic cases. One should use a keen understanding of epistemics to leverage awareness, and thereby forestall or remedy any distorting bias (as in the case of social epistemology discussed above, §2.3).

The challenges of conflict of interest extend to research ethics, as well. In another case in India, in 2009 several aid agencies sponsored clinical trials for vaccines against cervical cancer. Many of the patients were from poor tribal populations and were not fully informed about the risks of the study. Nor were the side effects well monitored. The U.S. drug company, Merck, seemed eager to earn approval for—and profit from—mandatory vaccination programs. Eventually, a U.S. researcher revealed Merck's aggressive marketing tactics and its failure to fully disclose risks (Attkisson 2009; Bagla 2013; Chamberlain 2015). Here, the vulnerabilities of the tribal population underscore again how real scientific practices combined with conflict of interest can easily amplify rather than solve social injustice. Of course, historically, other vulnerable groups have been unjustly subjected to the risks in medical investigations. One may consider the cases of prisoners in a malaria study at the Stateville Penitentiary (Comfort 2009); mentally ill patients in studies on a hepatitis vaccine (Robinson and Uhruh 2008); orphans in an interventive experiment on stuttering (Reynolds 2003), prostitutes, prisoners, mental patients and soldiers in a U.S. study of syphilis in Guatemala (Reverby 2012), as well as the more renowned study of Southern blacks in the Tuskegee syphilis experiment (Jones 1981). Funding and conflict of interest are further dimensions that shape the reliability of scientific practice and that should be included in NOS curriculum (Table 2.1).

2.6 Conclusion

In some cases, social justice is served by lessons in plain old scientific content. Biology, for one, can provide a deeper understanding of the properties that are commonly but inappropriately used to "justify" many prejudicial categories. For example, the genetics of skin color, so emblematic of race in the public consciousness, are not closely correlated with either distinct ancestral groups or geographic regions (Allchin 2018b). Indeed, the whole concept of race is biologically unsound. As are the more fundamental widespread beliefs about genes as identity (Allchin 2017a, b, pp. 141–145). Nor are the categories of male and female strictly dichotomous (Allchin 2017b, pp. 117–124). That has implications for the status of transgender individuals and for biases based on gender stereotypes. The presumption that the cultural status quo, with all its political and economic inequities, reflects "survival of the fittest," is based on erroneous understanding of natural selection (Allchin 2017b, pp. 37–64). All are examples of the naturalizing error (§2.3). Science is a potent resource for informing and challenging many of the prejudices that shape social injustice.

In other cases, science can challenge cultural myths about science, scientists, and scientific reasoning that help perpetuate injustice. For example, eyewitness testimony was once considered by most jurists (and juries) as the gold standard for evidence in pursuit of justice. Yet such testimony proves to be strongly biased by preconceptions and memories that have been reconstructed by suggestion. Hence, in a judicial settings, cultural prejudices, rather than be corrected by such testimony, tend to be ironically reinforced. It has taken rigorous science, led largely by Elizabeth Loftus (1996), to begin to remedy the legal perspectives. By the same token, science can also help produce the evidence that exposes injustice. DNA evidence has helped to exonerate over 350 victims of wrongful conviction, over 70% of them originally involving eyewitness misidentification (Innocence Project 2017). Some people seem eager to blame science for social injustice, without considering the many roles of science in actually helping to remedy it. We should recognize that science is not inherently a "weapon of oppression" (see Ogunniyi, Chap. 9) or co-conspirator of coercive capitalism (see Benzce and Carter, Chap. 4), but can sometimes be a tool for liberation and justice. Epistemic lessons can be politically quite powerful.

Achieving social justice often hinges on proper justification of scientific facts in arguments about privilege, profit, and power. Injustices, in many cases, are sustained by appeals to scientific claims that are deliberately misleading or strategically misstated (§2.2). They may exhibit cognitive errors, hide key assumptions, or misrepresent expertise. In policy or economic settings, bogus, distorted or misleading science can shape social privilege, economic advantage, or individual rights. The well informed citizen or consumer, vulnerable to such tactics, should ideally be empowered to defend good science and to expose any flaws or pretenses in unjustified claims. This requires understanding how science works, not just ideally or superficially, but in actual practice. It requires understanding not just how knowledge is produced within a scientific community, but also how it is conveyed through social settings, as well. How does science ultimately justify its claims and how, at other times, does it fail (§2.3)? What are the genuine uncertainties and where is the burden of proof (§2.4)? Who is a credible expert (§2.2)? Who exhibits conflict of interest (§2.5)? A Whole Science approach is needed to replace the current internalist and decontextualized approaches to the nature of science. To contribute to social justice, students need a full understanding of epistemics through lessons in the nature of science.

That approach, in turn, should guide concrete classroom practice. All the examples discussed here (summarized in Table 2.1) epitomize the aim of functional scientific literacy for citizenship. It is not enough to know the scientific concepts, nor simply to be able to reason scientifically about evidence. The role of epistemic dependence (§2.2), cultural bias and error in science (§2.3), uncertainties and the precautionary principle (§2.4), and the potential for conflicts of interest and bias in social arguments appealing to science (§2.5) all underscore the need for more complete understanding of the nature of Whole Science—from test tubes to YouTube, from the lab bench to the judicial bench, from field site to website, from lab book to Facebook. Accordingly, science teachers should actively introduce and discuss appropriate cases, such as those in Table 2.1, in the classroom.

Acknowledgments My deep appreciation to the Editors for their invitation and encouragement in completing this work, to my colleagues in Brazil and India for helping me understand local social justice issues there, and to two anonymous reviewers.

References

Allchin, D. (2008). Naturalizing as an error-type in biology. *Filosofia e História da Biologia, 3*, 95–117.

Allchin, D. (2010). The nature of science: From test tubes to YouTube. *American Biology Teacher, 73*, 591–593.

Allchin, D. (2011). Evaluating knowledge of the nature of (whole) science. *Science Education, 95*, 918–942.

Allchin, D. (2012a). Teaching the nature of science through scientific error. *Science Education, 96*, 904–926.

Allchin, D. (2012b). Towards clarity on whole science and KNOWS. *Science Education, 96*, 693–700.

Allchin, D. (2013). *Teaching the nature of science: Perspectives & resources*. St. Paul: SHiPS Education Press.

Allchin, D. (2015). Global warming: Scam, fraud, or hoax? *American Biology Teacher, 77*, 308–312.
Allchin, D. (2016). Is science self-correcting? *American Biology Teacher, 78*, 695–695.
Allchin, D. (2017a). Beyond the consensus view: Whole science. *Canadian Journal of Science, Mathematics and Technology Education, 17*, 18–26.
Allchin, D. (2017b). *Sacred bovines: The ironies of misplaced assumptions in biology*. New York: Oxford University Press.
Allchin, D. (2018a). Fake news and alternative facts. *American Biology Teacher, 80*, 631–633.
Allchin, D. (2018b). Skin color and the nature of science. *American Biology Teacher, 80*, 163.
Allchin, D. (2019). Science without shiny labs. *American Biology Teacher, 81*, 61–64.
Allchin, D. (Forthcoming). From the nature of science to the nature of science-in-society.
Allchin, D., & Werth, A. (2017). The naturalizing error. *Journal for the General Philosophy of Science, 48*, 3–18.
Allchin, D., & Werth, A. (in press). How we think about human nature: The naturalizing error. *Philosophy of Science*, 87.
Allchin, D., Andersen, H., & Nielsen, K. (2014). Complementary approaches to teaching nature of science: Integrating inquiry, historical cases and contemporary cases in classroom practice. *Science Education, 98*, 461–486.
Attkisson, S. (2009, August 19). Gardasil researcher speaks out. *CBS News*. https://www.cbsnews.com/news/gardasil-researcher-speaks-out.
Bagla, P. (2013, September 9). Indian parliament comes down hard on cervical cancer trial. *Science*. http://www.sciencemag.org/news/2013/09/indian-parliament-comes-down-hard-cervical-cancer-trial.
Barkan, E. (1992). *The retreat from scientific racism*. Cambridge: Cambridge University Press.
Center for Economic and Social Justice. (2018). *Defining economic justice and social justice*. Washington, DC: Author. http://www.cesj.org/learn/definitions/defining-economic-justice-and-social-justice/
Chamberlain, G. (2015. January 13). *Judges demand answers after children die in controversial cancer vaccine trial in India*. http://www.dailymail.co.uk/news/article-2908963/Judges-demand-answers-children-die-controversial-cancer-vaccine-trial-India.html.
Comfort, N. (2009). The prisoner as model organism: Malaria research at Stateville penitentiary. *Studies in History and Philosophy of Biological and Biomedical Sciences, 40*(3), 190–203.
Desai, D. (2014). The car existed in Vedic times and stem cell technology started in the Mahabharata: Gujarat's bizarre school lessons revealed. *Daily Mail India*. http://www.dailymail.co.uk/india-home/indianews/article-2707713/The-car-existed-Vedic-times-stem-cell-technology-started-Mahabharata-Gujarats-bizarre-school-lessons-revealed.html.
Douglas, B. (2015). Brazil's slow-motion environmental catastrophe unfolds. *The Gaurdian*. https://www.theguardian.com/business/2015/nov/13/brazils-slow-motion-environmental-catastrophe-unfolds.
Economic Times. (2017, December 19). *Examine safety of two Kudankulam nuclear plants, shift out new units: Stalin*. https://economictimes.indiatimes.com/news/politics-and-nation/examine-safety-of-two-kudankulam-nuclear-plants-shift-out-new-units-m-k-stalin/articleshow/62139712.cms.
Fee, E. (1979). Nineteenth-century craniology: The study of the female skull. *Bulletin of the History of Medicine, 53*, 415–433.
Foster, K. R., Vecchia, P., & Repacholi, M. H. (2000). Science and the precautionary principle. *Science, 288*, 979–981.
Fricker, M. (2007). *Epistemic injustice: Power and the ethics of knowing*. Oxford: Oxford University Press.
Gaon, S., & Norris, S. P. (2001). The undecidable grounds of scientific expertise: Science education and the limits of intellectual independence. *Journal of Philosophy of Education, 35*, 187–201.
Goldacre, B. (2010). *Bad science: Quacks, hacks, and big pharma flacks*. New York: Faber and Faber.

Goldman, A. I. (1999). *Knowledge in a social world*. Oxford University Press.
Gould, S. J. (1981). *The mismeasure of man*. New York: W.W. Norton.
Gould, S. J. (1983). Science ad Jewish immigration. In *Hen's teeth and horse's toes* (pp. 291–302). New York: W.W. Norton.
Harding, S. (1991). *Whose science? Whose knowledge?* Ithaca: Cornell University Press.
Harding, S. (1998). *Is science multicultural?* Bloomington: Indiana University Press.
Hardwig, J. (1991). The role of trust in knowledge. *Journal of Philosophy, 88*, 693–708.
Harremoës, P., Gee, D., MacGavin, M., Stirling, S., Keys, J., Wynne, B., & Vaz, S. G. (2001). *Late lessons from early warnings: The precautionary principle 1896–2000*. Copenhagen: European Environmental Agency.
Heine, S. J. (2017). *DNA is not destiny: The remarkable, completely misunderstood relationship between you and your genes*. New York: W.W. Norton.
Hodson, D. (2008). *Towards scientific literacy*. Rotterdam: Sense.
Hodson, D., & Wong, S. L. (2017). Going beyond the consensus view: Broadening and enriching the scope of NOS-oriented curricula. *Canadian Journal of Science, Mathematics and Technology Education, 17*(1), 3–17.
Höttecke, D. & Allchin, D. (2020). Re-conceptualizing nature-of-science education in the age of social media. *Science Education, 104*, 641–666.
Innocence Project. (2017). *DNA exonerations in the United States*. New York: Author. https://www.innocenceproject.org/dna-exonerations-in-the-united-states/.
Ives, M. (2018, July 29). Laos dam failure exposes pitfalls of lax regulation. *New York Times*, A7.
Ivone, M. (2015). *Evolution of the precautionary principle*. Saarbrücken: Lambert Academic Publishing.
Jones, J. H. (1981). *Bad blood: The Tuskegee syphilis experiments*. New York: Free Press.
Joshi, H. (2016). *Rage of the river: The untold story of Kedarnath disaster*. Gurgaon: Penguin Random House.
Kelly, G. J., Carlsen, W., & Cunningham, C. (1993). Science education in sociocultural context. *Science Education, 77*, 207–220.
Kenner, R. (2010). *Merchants of doubt* [film]. Sony Pictures.
Khan, R. (2018, April 30). The myth of Vedic scientific achievements. *Rising Kashmir*. http://risingkashmir.com/news/the-myth-of-vedic-scientific-achievements-324736.html.
Kitcher, P. (2001). *Science, truth, and democracy*. New York: Oxford University Press.
Kolata, G. (2009, November 17). Panel urges mammograms at 50, not 40. *New York Times*. https://www.nytimes.com/2009/11/17/health/17cancer.html.
Kolstø, S. D. (2001). Scientific literacy for citizenship: Tools for dealing with the science dimension of controversial socioscientific issues. *Science Education, 85*, 291–300.
Kumar, S. (2017). Critics assail India's attempt to 'validate' folk remedy. *Science, 355*, 898.
Kumar, S. (2019). In India, Hindu pride boosts pseudoscience. *Science, 363*, 679–680.
Lederman Depository. (2018). *SI and NOS activities*. https://science.iit.edu/mathematics-science-education/resources/lederman-depository.
Levidow, L. (1998). Democratizing technology–or technologizing democracy? Regulating biotechnology in Europe. *Technology in Society, 20*, 211–226.
Lewontin, R., Kamin, L., & Rose, S. (1984). *Not in our genes*. New York: Pantheon.
Loftus, E. F. (1996). *Eyewitness testimony*. Cambridge, MA: Harvard University Press.
Longino, H. (1990). *Science as social knowledge*. Princeton: Princeton University Press.
Longino, H. (2001). *The fate of knowledge*. Princeton: Princeton University Press.
McGarity, T. O., & Wagner, W. E. (2008). *Bending science: How special interests corrupt public health research*. Cambridge, MA: Harvard University Press.
McIntrye, L. (2019). *The scientific attitude*. Cambridge, MA: MIT Press.
Michaels, D. (2008). *Doubt is their product: How industry's assault on science threatens your health*. New York: Oxford University Press.

Ministry of Environment and Forests. (2014). *Assessment of environmental degradation and impact of hydroelectric projects during the June 2013 disaster in Uttarakhand.* http://www.indiaenvironmentportal.org.in/files/file/environmental%20degradation%20&%20hydroelectric%20projects.pdf.

Mooney, C. (2005a). Some like it hot. *Mother Jones, 30*(3), 36–94.

Mooney, C. (2005b). *The republican war on science*. New York: MJF Books.

National Education Association. (2017). *Diversity toolkit: Social justice*. Washington, DC: Author. http://www.nea.org/tools/30414.htm.

Nestle, M. (2015). *Soda politics: Taking on big soda (and winning)*. New York: Oxford University Press.

Nestle, M. (2018). *Unsavory truth: How food companies skew the science of what we eat*. New York: Basic Books.

New York Times. (2019, February 9). *Why did the dam in Brazil collapse?* Here's a brief look.

Newton, L. H., & Dillingham, C. K. (1994). *Watersheds classic cases in environmental ethics*. Belmont: Wadsworth.

NGSS Lead States. (2013). *Next generation science standards: For states, by states*. Washington, DC: The National Academies Press.

Norris, S. P. (1995). Learning to live with scientific expertise: Toward a theory of intellectual communalism for guiding science teaching. *Science Education, 79*, 201–217.

Norris, S. P. (1997). Intellectual independence for nonscientists and other content-transcendent goals of science education. *Science Education, 81*, 239–258.

O'Riordan, T., & Cameron, J. (1994). *Interpreting the precautionary principle*. London: Earthscan.

OECD (Organization for Economic Co-operation and Development). (2017). *PISA 2015 assessment and analytical framework: Science, reading, mathematic, financial literacy and collaborative problem solving* (Revised edn). Paris: PISA/OECD Publishing. https://doi.org/10.1787/9789264281820-en.

Oreskes, N. (2019). *Why trust science?* Princeton: Princeton University Press.

Oreskes, N., & Conway, E. M. (2010). *Merchants of doubt: How a handful of scientists obscured the truth on issues from tobacco smoke to global warming*. New York: Bloomsbury.

Rampton, S., & Stauber, J. (2002). *Trust us, we're experts: How industry manipulates science and gambles with your future*. New York: Tarcher/Penguin.

Raveendran, A., & Chunawala, S. (2013). Towards an understanding of socioscientific issues as a means to achieve critical scientific literacy. *epiSTEME* 5. http://dnte.hbcse.tifr.res.in/wp-content/uploads/2018/04/2013_AS_SSI_epi5.pdf.

Raveendran, A., & Chunawala, S. (2015). Values in science: Making sense of biology doctoral students' critical examination of a deterministic claim in a media article. *Science Education, 99*(4), 669–695.

Reverby, S. M. (2012). Ethical failures and history lessons: The U.S. Public Health Service research studies in Tuskegee and Guatemala. *Public Health Reviews, 34*(1), 1–18.

Reynolds, G. (2003, March 16). The stuttering doctor's "monster study." *The New York Times*. https://www.nytimes.com/2003/03/16/magazine/the-stuttering-doctor-s-monster-study.html.

Robinson, W. M., & Uhruh, B. T. (2008). The hepatitis experiments at Willowbrook state school. In E. J. Emanuel, C. Grady, R. A. Crouch, R. K. Lie, F. G. Miller, & D. Wendler (Eds.), *The Oxford textbook of clinical research ethics* (pp. 80–85). Oxford: Oxford University Press.

Rudolph, J. (2005). Inquiry, instrumentalism, and the public understanding of science. *Science Education, 89*, 803–821.

Rutherford, F. J., & Ahlgren, A. (1991). *Science for all Americans*. New York: Oxford University Press.

Ryder, J. (2001). Identifying science understanding for functional scientific literacy. *Studies in Science Education, 36*, 1–44.

Schiebinger, L. (1989). *The mind has no sex? Women in the origins of modern science*. Harvard: Harvard University Press.

Schiebinger, L. (1993). *Nature's body: Gender in the making of modern science*. Boston: Beacon Press.
Shrader-Frechette, K. (1990). Models, scientific method and environmental ethics. In D. Scherer (Ed.), *Upstream/downstream: Issues in environmental ethics* (pp. 90–120). Philadelphia: Temple University Press.
Solomon, M. (2001). *Social empiricism*. Cambridge, MA: MIT Press.
Terrapon-Pfaff, J., Dienst, C., König, J., & Ortiz, W. (2014). A cross-sectional review: Impacts and sustainability of small-scale renewable energy projects in developing countries. *Renewable and Sustainable Energy Reviews, 40*, 1–10.
Toumey, C. (1996). *Conjuring science: Scientific symbol and cultural meanings in American life*. New Brunswick: Rutgers University Press.
Toumey, C., Besley, J., Blanchard, M., Brown, M., Cobb, M., Ecklund, E. H., et al. (2010, October). *Science in the Service of Citizens & consumers: The NSF workshop on public knowledge of science*. Columbia: University of South Carolina Nanocenter.
Union of Concerned Scientists. (2007). *Smoke, mirrors & hot air: How ExxonMobil uses big tobacco's tactics to manufacture uncertainty on climate science*. Cambridge, MA: Author.
Voude, J. R. (2007). *AIDS, South Africa, and the politics of knowledge*. London: Routledge.
Wagner, W. E., & Steinzor, R. (2006). *Rescuing science from politics: Regulation and the distortion of scientific research*. Cambridge: Cambridge University Press.
World Commission on the Ethics of Scientific Knowledge and Technology. (2005). *The precautionary principle*. Paris: UNESCO.
Zemplén, G. Á. (2009). Putting sociology first—Reconsidering the role of the social in "nature of science" education. *Science & Education, 18*, 525–559.
Zimring, J. (2019). *What science is and how it works*. Cambridge: Cambridge University Press.

Chapter 3
Balancing the Epistemic and Social Realms of Science to Promote Nature of Science for Social Justice

Zoubeida R. Dagher

3.1 Introduction

This chapter explores why the majority of science education studies with a nature of science agenda may have contributed scarce insights to those concerned with issues of social justice, and why studies focused on social justice, in turn, have not contributed much to nature of science studies. The chapter begins with an overview of NOS and social justice orientations, describes their connections to science curriculum standards and delineates their characteristics. I propose that nature of science frameworks that are squarely focused on epistemic aspects of scientific knowledge do not possess the conceptual breadth or the analytical capacity to address socio-scientific content adequately. I illustrate how more holistic frameworks like the Family Resemblance Approach to Nature of Science (Erduran and Dagher 2014) can serve as a bridge for connecting "cold" or discipline-based to "warm" or community-based notions of nature of science. The chapter concludes with a discussion that underscores the importance of integrating NOS and social justice orientations for the purpose of improving learning opportunities and outcomes for all students.

3.2 NOS Overview

By virtue of its emphasis, teaching the nature of science in science education has relied on relevant insights gleaned from philosophy of science, history of science, philosophy of the special sciences, sociology of science, cognitive studies of science, and to a much lesser extent feminist critiques of science. Philosophical

Z. R. Dagher (✉)
School of Education, University of Delaware, Newark, DE, USA
e-mail: zoubeida@udel.edu

H. A. Yacoubian, L. Hansson (eds.), *Nature of Science for Social Justice*, Science: Philosophy, History and Education,
https://doi.org/10.1007/978-3-030-47260-3_3

analyses focus on studying the trustworthiness of scientific knowledge, the reliability of scientific theories, and what distinguishes scientific knowledge from mere belief. Insights about the complexity and theory-dependence of observations (Norris 1985), and the myth of the scientific method were examples of two ideas that could impact how science is taught. Studies on how students develop their understanding of the nature of science flourished (e.g. Driver et al. 1996). History of science inspired the development of contextualized science stories that humanize science and facilitate learning (e.g. Allchin 2013; Clough 2011; Conant et al. 1957; Klassen 2007; Matthews 1994, 2004; Stinner and Williams 1993). Historical analyses have pointed out rampant myths in science textbooks and popular accounts of scientists (e.g. Allchin 2003; Dagher and Ford 2005; Milne 1998). Focus on cognitive factors in the development of scientific knowledge sparked another line of research that investigated knowledge development in science and its relevance to improving student learning (Nersessian 1989). Conceptual change theory drew heavily on analogies between the Piagetian learning processes of assimilation and accommodation to Kuhn's account of paradigm change in science (Posner et al. 1982). This association promoted considerable research about the nature of science learning, debates about distinctions between scientists and science learners, and offered research-based strategies for addressing student misconceptions.

A review of empirical studies on NOS in science education by Lederman (1992) traced advocacy for teaching students about nature of science to over 100 years ago in arguments for increased "emphasis on the scientific method and the processes of science" (p. 332). Successive emphases on nature of science took many forms over the years, and during the reforms of the 60's remained closely tied to the scientific method and scientific processes. Also it became closely associated with the elusive goal of scientific literacy, which has been conceptualized in myriad ways over the years (DeBoer 2000; Norris and Phillips 2003; Roberts 2007; Roberts and Bybee 2014). Lederman's (1992) review documents the evolution of thinking of NOS content in science education based on assessment instruments developed over time to assess NOS understanding. Subsequent literature reviews describe salient findings of the accumulating body of research while also noting limited success in attaining major changes in students' and teachers' conceptions of NOS (Lederman 2007; Abd-El-Khalick 2014; Lederman and Lederman 2014).

3.2.1 NOS and Science Education Standards

Concern about teacher and student understanding of nature of science did not remain a mere academic interest, or a focus for curriculum innovation. Given its recognized importance to scientific literacy among key stakeholders, NOS became a staple in science education curriculum policy documents in the United States in the early reforms of the 1990's. Those documents advocated the inclusion of nature of science in K-12 science teaching. *Science for all Americans* (AAAS 1989) dedicated three entire chapters to justify the importance of nature of science, nature of technology, and nature of mathematics. The *National Science Education Standards* (NSES) (NRC 1996) dedicated one out of eight standards to the history and nature of science. This standard focused specifically on three key ideas that were further elaborated: (1) Science is a human endeavor, (2) nature of scientific knowledge, and (3) historical perspectives. Studies of curriculum standards and goals over a period of two decades revealed varying trends in the extent to which NOS is explicitly expressed in several international policy documents (e.g. Dagher 2009; McComas and Olson 1998; Olson 2018).

In the most recent science education reform in the United States, the *Next Generation Science Standards* (NGSS Lead States 2013) did not address nature of science in the same explicit way expressed in the previous reform. Rather than being singled out as one of eight central standards within the NSES document, NOS ideas are discussed in a dedicated appendix (H) that describes eight NOS categories relevant to K-12 education in the NGSS document. The first four NOS categories are associated with science and engineering practices, and the following four are associated with cross-cutting concepts,

1. *Scientific Investigations Use a Variety of Methods*
2. *Scientific Knowledge is Based on Empirical Evidence*
3. *Scientific Knowledge is Open to Revision in Light of New Evidence*
4. *Scientific Models, Laws, Mechanisms, and Theories Explain Natural Phenomena*
5. *Science is a Way of Knowing*
6. *Scientific Knowledge Assumes an Order and Consistency in Natural Systems*
7. *Science is a Human Endeavor*
8. *Science Addresses Questions About the Natural and Material World* (NGSS Lead States 2013, p. 4)

Relegation of the NOS categories to the appendix of the NGSS has received a mixed response. While these categories are broad enough to avoid direct conflict with a variety of NOS perspectives—they are rather too broad to suggest specific actions that explicate or exemplify their implementation. In the minds of the standards' writers, a skillful and explicit articulation of the practices and cross cutting concepts dimensions necessarily involve rich discussion about the nature of science. In the minds of critics, the relegation of these NOS ideas to an appendix amounts to marginalizing them. Visually, they are one step removed from the stated performance expectations, when they should be part and parcel of them. The overriding concern is that, left implicit, these nature of science ideas are likely to be ignored, and not embedded in the standard-based teaching or curriculum. For the NOS ideas to be fully endorsed, McComas (2016) has proposed that they should be part of the NGSS three-dimensional structure and not auxiliary to it, sort of a fourth dimension.

Continued attention to NOS in curriculum and instruction is justified, in part, by the accumulated evidence that document commonly held NOS misconceptions. In a recent article, Clough (2017) samples some of these misconceptions obtained from decades of research on student and teacher understanding of the history and nature of science. They include for example, the ideas that science and scientists "can and should be free from emotions and bias", that "scientific ideas arise from data", and that "scientific models are copies of reality" to name a few (Clough 2017, p. 41).

Why is addressing students' 'misconceptions' about NOS deemed necessary? Because much like science content preconceptions and misconceptions, these ideas are likely to remain undetected and unaddressed by traditional science instruction. Students can easily slip into a state of distrusting science if their perceptions of science are rooted in it as a truthful representation of how things are. If they believe that theories are tentative otherwise they turn into laws, they will appeal to the perceived temporary status of theories to justify why they do not trust them. If they base their ideas of scientific credibility on the outcomes of experimental methods as the most definitive way to adjudicate scientific claims, then they will discount the legitimate contributions of observational and historical methods. If they view science as capable of providing answers for all life's questions, they will be disappointed to find that it cannot answer questions for which it does not have tools to explore. If they think that all scientific-sounding claims should be trusted, then they will not seek to distinguish between trustworthy sources and their impersonators. This is of concern at two levels: (1) distrusting science theories of consequence such as climate change, results in complacence and lack of interest in taking appropriate personal, civic, or political action, and (2) inability to distinguish between scientific and nonscientific claims, makes people vulnerable to accepting and acting on claims that may not be only useless, but downright dangerous.

One of the widely used NOS frameworks in the last two decades is commonly referred to as the consensus view (CV). This view targets seven main NOS ideas that are considered to be appropriate target understandings for the K-12 population (Lederman et al. 2002). These ideas are written in the form of straightforward propositions that can be tailored to science instruction as listed below:

1. Scientific knowledge is tentative
2. Observations and inferences are different
3. Scientific knowledge is theory laden
4. Scientific knowledge involves imagination and creativity
5. Scientific knowledge is socially embedded
6. Scientific laws and theories are different
7. There is no one scientific method

There have been many critiques of this view (e.g. Allchin 2011; Duschl and Grandy 2013; Hodson and Wong 2017; Irzik and Nola 2011, 2014; Matthews 2012; Schizas et al. 2016). Two that are central to the argument in this chapter are that the CV is focused almost exclusively on scientific knowledge and that it provides highly prescriptive statements that do not address several other core epistemic components of science, nor do they account for related social, institutional and societal dimensions.

If the purpose of teaching science is to cultivate more informed citizens, then understanding the nature of scientific knowledge and where this knowledge comes from, what it means, how is it constructed and justified, and how it is being used, is part of what ought to be taught. Because the scientific enterprise is not only defined by its products alone but by its community practices and its institutional and societal interactions, there is a pressing need to represent it in science education in a holistic and nuanced way.

3.3 Social Justice Overview

Approaches for teaching science for social justice derive their theoretical orientations from evolved conceptualizations of Science, Technology, Society (STS) movement (Yager 1996; Aikenhead 1994), culturally relevant pedagogy (Ladson-Billings 1995), and critical race theory (Crenshaw et al. 1995; Ladson-Billings and Tate 1995). These orientations, to varying degrees, encourage science teaching that is youth and community centered and consider the learners' 'funds of knowledge' (Barton and Tan 2009; González et al. 2005) and cultural milieu as central to learning. They challenge assumed deficit models and encourage identifying and connecting with children's experiences in order to make relevant connections with the curriculum (in school settings).

The primary goal of social justice approaches to science teaching is to make learning inherently relevant to the learners' lives and to develop student critical awareness and their sense of agency. For example, Roth's teaching in rural Canada (Roth 2009a), Barton (2003) and Associates' (Tan and Barton 2012) work with afterschool programs, describe how youth engage in scientific thinking in urban settings. The focus in both settings shows strong connections with students and communities, an orientation that characterizes much of the work on science education for social justice. As Roth states, "with a reorientation of science and scientific literacy in and through problematic issues in the lives of people, science educators might actually begin to make inroads into the currently intractable problem of the irrelevance of science in the everyday life of students specifically and all everyday folks more generally" (2009a, p. 2). In other words, science content and practices have most worth when they are made to matter in students' eyes, and when they empower students to make a difference in their lives and their communities. In formal education settings, teachers attuned to equity and social justice tailor the curriculum and, when needed, improvise materials and tools to match the needs and interests of children (see the case of Roth 2009a).

Similar principles apply in informal education settings. Barton (2003) describes how she and her team gave children voice and choice in what they wanted to do in an afterschool program, recounting a variety of projects that empowered the youth and enabled them to understand what is happening in their neighborhoods. Examples of these projects, are using the Urban Heat Index to identify causes and solutions, building a bench to improve the local park, or developing a community garden. Elements of science, technology or engineering design understandings were gained as a result of intentional action on issues of inherent interest to the children. The anticipated end product dictated where the project started and how it developed. That end product was established by the youth, for the youth and their communities. Somehow this seems easier to implement when students are present in afterschool or informal education settings. This is because teachers and students break free from the "fetters" of curriculum standards and traditional disciplinary boundaries and concern themselves with matters that are of pure interest or value to them, and typically those tend to be ones that affect them or their immediate communities (e.g. Barton 2003).

Common across formal teaching contexts in rural areas (e.g. Roth 2009a), after-school programs in urban settings (Barton 2003; Tan and Barton 2012), and summer camp programs (Buxton 2010), are engaged youth developing expertise around topics that matter to them and their communities. The focus is on earning functional knowledge that the youth eventually use in some capacity to effect positive change. Through recounting their own personal experiences, involving family and personal health issues, Roth (2009b) and Barton (2009), describe the laborious translational effort they exercised to *transform* generalized knowledge to personalized knowledge, only then did this knowledge lead to breakthroughs in identifying and managing existing health conditions. In many ways, teaching science for social justice attempts to mediate and accelerate the personalization and appropriation of scientific knowledge.

Developing culturally competent teachers who are capable of understanding the values and cultures of their students is a challenging goal for teacher educators (e.g. Atwater et al. 2014; Cochran-Smith 1995). This is because teaching for social justice requires the ability to see things from the perspective of the other: a positioning that enables teachers in this case to understand, appreciate, and connect instruction to children's backgrounds. Relating empathetically with student communities demands overcoming one's own biases and assumptions about them. Open-mindedness to navigate difference, and a high level of curricular and instructional flexibility are basic ingredients for enhancing learning in urban, rural and other diverse settings. For student-teachers, stretching to establish cultural connections with students whose cultural backgrounds and experiences differed significantly from their own, proved to be very challenging (Larkin 2013).

3.3.1 Social Justice and Science Education Standards

In the United States, both waves of science education reforms in the last two decades have emphasized equity and inclusion, through the expectation that all students should attain the level of competence set out in the standards. The *National Science Education Standards*' commitment to equity is stated early in the document: "The standards apply to all students, regardless of age, gender, cultural or ethnic background, disabilities, aspirations, or interest and motivation in science" (NRC 1996, p. 2) and reiterated later in more detail. It called for fairness in assessment tasks, the importance of accommodating them to students with disabilities or limited language proficiency. It called attention to vetting these tasks against stereotypical perspectives or language or any other features that can negatively interfere with student performance. It called for provision of quality resources, attending to different learning styles and providing necessary adaptations as well as supporting varied opportunities for science learning. In contrast to the reforms of the 90's, the most recent reform (NGSS Lead States 2013) dedicates Appendix (D) to explicating how to make "the Next Generation Science Standards accessible to all students" (NGSS Lead States 2013). Over a span of 21 pages, this Appendix introduces equity issues and describes seven case studies. In a companion link to additional resources, each case study is further elaborated by describing strategies that teachers can use to "ensure that the NGSS are accessible to all students" (see https://www.nextgen-science.org/appendix-d-case-studies). The titles of these cases give an idea of the categories of diverse student groups that are addressed: Economically disadvantaged, race and ethnicity, students with disabilities, English language learners, girls, alternative education, and gifted and talented students.

Critics and supporters of the standards agree that commitment to teaching science to all requires the provision of (1) material resources that are necessary to improve students' access to quality learning tools and environments, (2) human resources in the form of equity-minded teachers who make deliberately inclusive decisions about the what, how and when in their teaching. This involves every detail in the teaching process from getting to know the students and their communities to selecting curriculum materials that connect meaningfully to children's experiences, familiar everyday phenomena, engaging questions and investigations. Rodriguez provides an overview of how commitment to social justice involves attending to the three prongs of engagement, equity and diversity, which, in effect, amount to a fourth dimension that supports the enactment of a more accessible and equitable science curriculum (see Rodriguez 2015, p. 1043).

Teaching science for social justice can be implemented, according to Rodriguez (2015), through deliberate action: strategic choice of curriculum, purposeful use of effective and diverse instructional strategies, offering choices, differentiating and accommodating a variety of needs, inclusion of, or connection to, a broader culture of science and scientists, and attention to students' personal and cultural diversity. In other words, central to the enactment of equity and social justice discourses is a primary focus on students and their communities, their assets, needs, and interests. At a basic level, teaching science for social justice includes providing (1) access to authentic and relevant questions/contexts, (2) rigorous tools and strategies to explore them, and (3) means to empower students to make a difference.

3.4 Bridging NOS and Social Justice Orientations to Science Education

There is a striking similarity in how some NOS and SJ advocates view the status of their orientation in the NGSS (NGSS Lead States 2013), in that their ideas have been partially marginalized. Independent voices from both sides called for their focus to be treated as a fourth dimension and perhaps a central one in the NGSS's narrative (McComas 2016; Rodriguez 2015). At the same time there are notable differences between the two perspectives, which are discussed in the rest of this section.

In examining studies related to both orientations, some distinctions come to the fore. For example, most nature of science studies take place in formal educational settings, whereas most social justice studies are conducted in formal and informal settings. Similar differences can be discerned along various aspects, such as participants, goals, stance towards science, methodologies, connection types, literacy focus and so on. In Table 3.1, I present general observations regarding distinctions between NOS-focused and social justice-focused curriculum or research. These observations describe common patterns not individual cases; they are neither essentialist nor normative. Most importantly, they do not imply that social justice researchers are not in some way addressing or incorporating a particular nature of science view, or that nature of science researchers do not endorse an underlying social justice agenda.

The differences between NOS- and social justice-focused orientations align with the different visions for scientific literacy. Roberts (2007) Vision I of scientific

Table 3.1 Comparison of primary aspects that distinguish mainstream studies in NOS and social justice and a proposed shift

Primary aspects	Nature of science	Social justice	Nature of science for social justice
Participants	General K-16 population	Diverse and underrepresented groups	General and vulnerable populations
Setting	Formal education	Informal and formal education	Formal and informal
Science stance	Knowledge/ standards-based (given)	Personal/community-based (contested terrain and resource)	Improve connections between school and community science
Goal	Enhancement of scientific literacy	Personal and community enhancement	Science literacy in personal, social and political contexts
Methods	Mixed, qualitative, and/or quantitative methods	Participant observation; ethnography, narrative, storytelling	Open, determined by the questions asked
Connection	Cognitive and epistemic	Cultural and personal	Situates the cognitive, epistemic & social dimensions of learning in cultural and personal relevance
Scientific literacy	Vision I (Roberts 2007) Discipline of science is a starting point	Vision II (Roberts 2007) Scientific perspective is situated in societal context	Vision III (Sjöström and Eilks 2018) Critical and transformative learning
Curriculum emphasis (Roberts 1982)	*Structure of science* *Self-as-explainer*	*Everyday coping* *Science, technology, & decisions*	*Combines the four emphases with emphasis on action*

literacy, is mainly focused on supporting student understanding scientific knowledge and processes, it "looks inward at science itself – its products such as laws and theories, and its processes such as hypothesizing and experimenting." Vision II, on the other hand, is primarily focused on science-related situations students are likely to encounter, it "looks outward at situations in which science has a role, such as decision-making about socioscientific issues" (Roberts 2007, p. 9). Using Roberts' framing of these visions, the two orientations appear to subscribe to different visions of scientific literacy with different start and end points. The different visions explains to some extent some of the main observations noted in Table 3.1. Difference in vision emerges from difference in curriculum emphases (Roberts 1982). Two different curriculum emphases seem most closely linked to each of these orientations. Nature of science orientations seem to be mostly aligned with the curriculum emphases of structure of science and self-as-explainer. A structure of science emphasis relates to the nature of science and "how it functions intellectually in its own growth and development" (p. 247), whereas self as explainer focuses on either the institutional or personal engagement in explaining events. Social justice orientations emphasize everyday coping that value "individual and collective understanding of scientific principles, as a means for coping with individual and collective 'problems'" (p. 246). They also support a science, technology and decisions emphasis, which focuses on "the limits of science in coping with practical affairs" (p. 247).

NOS for social justice aligns best with Vision III of scientific literacy. This vision goes beyond mere engagement in socioscientific issues as it combines critical scientific literacy and engaging in socio-political action (Sjöström and Eilks 2018, p. 66). There are two issues that follow in this respect. While it can be argued that NOS frameworks have the potential, in principle, to support a social justice orientation to science teaching or research, I contend that they are not equally equipped to do so. This is because narrowly configured approaches focusing on very specific propositions of scientific knowledge, such as the Consensus View (CV) and its related assessments (e.g. VNOS), address limited epistemic aspects of scientific knowledge-that are inward looking. They do not have built-in tools to interpret scientific knowledge in its social and political dimensions. Consequently, the CV framework "does not capture the flexibility comprised when diverse people come together to wrestle with community problems and democratized forms of science. Sometimes these forms represent cultural understandings not called 'science'" (Mueller 2011, p. 353). In other words, the NOS conceptions and the instruments that measure them do not have the epistemic breadth or agility to connect to emergent ideas that stray from strict disciplinary discourse. In this sense, they are likely to limit what teachers teach and what researchers observe and analyze.

A recognition of the need to capture such discourse was noted in a study that examined the intersection of race, gender and culture with mostly African American elementary school children in two large school districts. In the conclusion, the authors wrote, "in our queries, we focused exclusively on NOS, yet there are many areas to be explored from a more inclusive approach (e.g. scientific inquiry, social and cultural aspects, environmental education, other science content knowledge)" (Walls et al. 2013). I would argue that the issue is not only a function of the exclusive focus on NOS, but of using a limited account of NOS that is well aligned with vision I but is poorly aligned with Vision II and the more advanced Vision III of scientific literacy (see Sjöström and Eilks 2018 for a detailed review on Vision III) which requires awareness and attention to a wider range of social, institutional and political understandings of science.

To conclude, successful integration of nature of science in teaching science for social justice is not likely to occur when using primarily inward looking unidimensional NOS frameworks focused on one aspect of science, namely scientific knowledge. This explains why we seldom see evidence to the contrary in the studies that use these frameworks. Multidimensional frameworks, such as Erduran and Dagher (2014) Family Resemblance Approach, provide the theoretical groundwork that is sympathetic to the goals and concerns of social justice researchers. In the next section, I describe the FRA framework and illustrate how it functions in conjunction with a social justice teaching agenda.

3.4.1 Description of the Family Resemblance Approach

The Family Resemblance Approach (FRA) to nature of science was proposed by Irzik and Nola (2014) to account for perceived shortcomings in the consensus view. Applying Wittgenstein's notion of family resemblance to science, they argued that different branches of the natural sciences share enough common features that when taken together can distinguish science disciplines as members of a family that share common features while maintaining their individual unique characteristics. They proposed eight categories along which such resemblance can be found, and organized these under science as a cognitive-epistemic system and science as a social-institutional system. They identified processes of inquiry, aims and values, methods and methodological rules and scientific knowledge as key components of science as a cognitive-epistemic system, and identified professional activities, scientific ethos, social certification and dissemination of scientific knowledge, and social values under science as a social-institutional system. Erduran and Dagher (2014) saw the potential of this approach for addressing both shared and unique features among the sciences in light of contributions by the philosophies of the special sciences (e.g. biology, chemistry, geology) and insights from the science studies literature. They reconceptualized NOS for science education by adding three NOS categories that are often ignored in science education. These are: social organizations and interactions, financial systems, and political power structures. The 11 categories that comprise this version of the FRA are defined as follows:

Scientific Aims and Values recognize values that include among other things, objectivity, novelty, accuracy, and empirical adequacy.

Scientific Practices involve the understanding that scientists develop through investigations that involve questions about the world and engagement in activities that enable them to understand it better.

Scientific Methods address the idea that scientists use many methods that include, among others, observational and experimental investigations that may or not involve hypothesis testing.

Scientific Knowledge pertains to describing how scientific knowledge is expressed in multiple forms such as theories, laws, models that are used to explain and predict phenomena, how they relate to one another and how they contribute to the growth of knowledge.

Scientific Ethos focuses on understanding norms that are important for producing credible scientific knowledge, such as intellectual honesty, respect for and protection of human subjects, respect for colleagues and the environment.

Social Values include respecting the environment, social utility, and freedom.

Social Certification and Dissemination describes the role of peer review, evaluation and criticism with an understanding that scientific findings get reviewed, criticized and evaluated by peers.

Professional Activities focus on understanding the role of attending professional conferences, presenting findings, writing research proposals, and conducting peer reviews of papers and proposals.

Financial Systems acknowledge the roles of funding priorities, commercial and special interests in enabling, controlling or limiting scientific knowledge.

Political Power Structures address gender issues, colonial interests, ideological influences on scientific knowledge and practices and who benefits from them.

Social Organization and Interactions address the organizational structures and relational transactions within and among scientific communities.

The FRA Wheel (Fig. 3.1), captures Erduran & Dagher's attempt to visually depict all 11 components of the FRA to NOS framework. The circular structure of the wheel and the perforated lines are intended to relay the dynamic and interactive

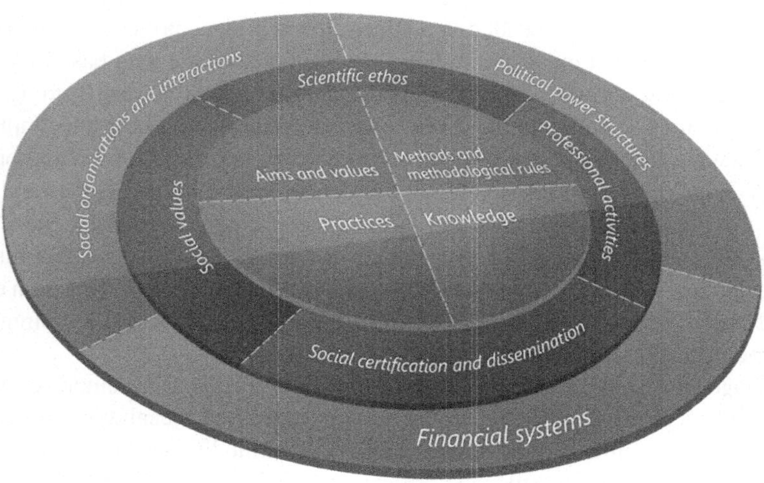

Fig. 3.1 The FRA Wheel representing the 11 NOS categories. (Reprinted from Erduran and Dagher 2014, p. 28)

nature of the Wheel's components. The cognitive-epistemic categories occupy the center of the Wheel, while the social institutional categories occupy the outer two layers. All components interact with each other in a dynamic and fluid way. This representation captures to some extent the complexity of science relative to the larger societal influences within which it operates. This framework, as a whole, emphasizes overarching ideas that view scientific knowledge as part of a larger and more dynamic cognitive-epistemic and social-institutional-political context. The framework's full account (Erduran and Dagher 2014) situates this FRA account in relation to Irzik & Nola's original account, and provides detailed justifications, and pedagogical tools and images that support embedding its components in authentic investigative learning contexts.

Used in conjunction with a Science, Technology, Society and the Environment (STSE) or place-based education where personal, local or global issues become sites for anchoring science explorations, the FRA gives teachers license and tools to embed some of the FRA components in the instructional context. For example, a science project can arise from students' complaints about air quality in their city. These complaints can spark discussions about how to best quantify their perceptions, use measurement tools, identify potential causes, design studies, debate sampling locations and data collection and analysis methods, question the validity of the collected data, locate and question EPA standards, and question policies in light of gathered evidence from data sources and expert opinions. This can lead into further questioning about the source of established EPA standards, and their meaning in contemporary and historical contexts, how they may have changed over time, how do they affect people in different communities, what can be done about them to counter negative implications etc. Answers to such questions could be sought from first hand data, second hand data (archives and data-bases), and by seeking opinions of local experts. The entire investigation could be led by teachers in one school or be part of a school district's project involving teachers and students in several schools. It can also be embedded in a larger citizen science project in which information from different cities and regions is contributed to the data-base by participants of all ages, allowing for a much better understanding of what is going in the school's vicinity and region, the country, and possibly the world (for example, see https://www.citizenscience.gov/air-sensor-toolbox/# and https://scistarter.org/air-quality-citizen-science). The goal is not only to understand the parameters and gravity of the issue but to form evidence-based understandings that enable students to pursue avenues for political action which can involve raising the community's awareness and drafting memos to local politicians to effect change. This is an example of participatory research, leading to learning that is connected to community, and to citizenship and advocacy.

Engagement with socioscientific issues (SSIs) as a context for empowering students and supporting citizenship, helps situate learning in meaningful contexts. In doing so, it is possible to target both visions of scientific literacy by creating understanding while expanding students' critical awareness and acquisition of epistemic and sociopolitical tools to engage and contribute to science-based conversations and related-policies that affect their community and the larger context. The depth of engagement in such contexts can vary on a continuum from extensive to shorter term engagements (Zeidler 2014; Zeidler et al. 2004). Embedding the FRA with SSI cases offers promising possibilities based the preliminary findings of a recent study with seventh grade students (Chaparian 2020). Further integration of FRA and SSI resources with Kolstø's (2001) knowledge-transcendent tools, provides additional means to sort out complexity and attain deeper understanding of these issues.

Framing scientific inquiries in terms of nature of science connections that span a broad range of cognitive, epistemic, social, institutional and political dimensions, provides enhanced possibilities for supporting a social justice orientation. This is because frameworks like the FRA (Erduran and Dagher 2014) encourage not only understanding the practices involved in the generation and validation of scientific knowledge but also encourage critical inquiry. Exploring how scientific aims and values, methods, practices, knowledge, social ethos, political interests, financial considerations affect and are in turn affected by the local or global context, allows students to situate the gained knowledge in its proper perspective.

Recapturing earlier arguments in this paper, most NOS frameworks in science education focus on narrow epistemic aspects of science, select sub-components from the internal part of the FRA Wheel. In contrast, social justice studies stay closer to the outer sphere, where scientific knowledge may be questioned and problematized. Contesting scientific knowledge, who does science, for what purpose, is a shared space where nature of science and social justice goals can meet. It is not the exclusive domain of any field. It just appears to be that way because most nature of science research has concentrated on traditional science content that is defined by curriculum standards, using NOS frameworks that are exclusively focused on characterizing scientific knowledge.

Incorporating the FRA (Erduran and Dagher 2014) in teaching science for social justice opens up discussions along epistemic, cognitive, social, institutional, political or financial aspects of the inquiry whether it is community-based or classroom-based. The outer two layers of the FRA Wheel (Fig. 3.1) go beyond discussing epistemic matters internal to scientific knowledge (considered one of four core components in the Wheel) as in the perennial emphasis on the tentativeness of scientific knowledge or the myth of the scientific method. They legitimize discussion of relevant influences at the intersection of science and society, using historical and contemporary cases. These may relate to debunked scientific theories (phrenology), violation of scientific norms and gender bias (Watson & Crick/Rosalind Franklin); unethical practices resulting in harming vulnerable populations (Tuskegee experiments); exploitation of human subjects (case of Henrietta Lachs); politically/financially/culturally-motivated campaigns challenging the credibility of science (smoking, climate change, asbestos, dioxins, evolution); or showcasing the often ignored role of women and ethnic minority scientists and mathematicians (Pickering Calculators, NASA Computers).

Much in the same way that Pintrich et al. (1993) critiqued conceptual change theory as originally formulated by Posner et al. (1982), for being excessively rational and focused on the cognitive processing of ideas with little attention paid to personal and motivational factors; a similar critique can be applied to nature of science formulations that are too discipline-centered and inward looking to be of value within a social justice agenda. The predominant use of overly epistemic and narrowly-focused 'cold' NOS formulations has not contributed to crossing the apparent NOS and social justice divide. Moving beyond these cold formulations requires the adoption of warm NOS conceptions that are cultivated at the intersection of disciplinary knowledge *and* personal-social-emotional relevance.

Promoting nature of science for everyday living and for social justice, rests on developing a critical stance towards science, equipping students with tools that prepare them to be thoughtful producers and consumers of knowledge. This can be promoted through teachers' posing reflective questions along the lines Allchin (2013) proposed in the preface to his book:

> what claims are reliable, and why? Which experts can you trust, especially when they seem to disagree? Do the circumstances reflect a warranted change in scientific consensus? …. What assumptions may have been made and how they may have biased the conclusions? Who sponsored the research, and what are the affiliations and interests of the researchers? Where does verifiable information end and value judgments begin?

While most nature of science researchers agree that such queries make excellent reflective questions, I suggest that addressing them can be supported more systematically by using NOS frameworks that include social and political dimensions.

The FRA's components involve features that appeal to the human side of science. These features provide tools and spaces to ask questions about the what, the who, the why, and the goals of both science and science education. Whether extending the science connections from standards-inspired phenomena to the neighborhood/community, or using neighborhood/community-inspired questions to make sense of standards-based concepts, the FRA framework can be used as a tool to interrogate the science learned in both frames of reference. The categories of the FRA offer grounds for a measured dose of skepticism and curiosity moving questions from the "what" to the "how and why" and "for what purpose".

3.5 Implications

In this chapter, I have outlined differences between typical NOS and social justice approaches to science education, and proposed a vision that reconciles and invites more interaction between them. I have argued that holistic NOS frameworks such as the FRA (Erduran and Dagher 2014) contain the structural potential and flexibility to guide metacognitive reflection within an equity and social justice agenda. I also argued that NOS frameworks that are highly focused on a set of epistemic statements are too limited in scope to support NOS for social justice. Finally, I called for using "warm" and contextualized NOS conceptions that support scientific inquiry for social justice.

Barton reminds us that "contrary to the goals of the national reform initiatives, which urge teachers to work with students to help them feel like members of a scientific community through engaging in the activities of scientists, the youth [they worked with] did not feel as if they were members of a scientific community unless their scientific activities actually contributed to their lives or their neighborhoods" (Barton 2003, p. 154). Social justice approaches typically capitalize on students' lived experiences, and develop scientific practices and knowledge that empower them to raise questions and find answers that are of consequence to their community. Holistic nature of science perspectives cast a robust metacognitive layer that can potentially drive and enhance the knowledge building process that characterize robust learning communities.

Teaching NOS for social justice demands a distinctive orientation to science curriculum and instruction. This orientation involves active cross-linking of standards-based science with community-based issues. It also involves a re-positioning of metacognitive NOS questions that may cause us, teachers and teacher educators, to question our goals: What is science? What are we teaching science for? Who are we teaching it to, or with? Why should they care? What aspects of it are we missing? What functional aspects of that knowledge can provide hope and agency to students? Such questions go to the heart of choices we make about curriculum, instruction and research. Students have much to gain from linking nature of science and social justice orientations to science teaching. We owe it to them to make it happen.

References

Abd-El-Khalick, F. (2014). The evolving landscape related to assessment of the nature of science. In N. G. Lederman & S. K. Abell (Eds.), *Handbook of research on science education, Volume II* (pp. 621–650). New York: Routledge/Taylor & Francis.

Aikenhead, G. (1994). Consequence to learning science through STS: A research perspective. In J. Solomon & G. Aikenhead (Eds.), *STS education: International perspectives on reform* (pp. 169–186). New York: Teachers' College.

Allchin, D. (2003). Science myth-conceptions. *Science Education, 87*, 329–351.

Allchin, D. (2011). Evaluating the knowledge of nature of (whole) science. *Science Education, 95*(3), 518–542.

Allchin, D. (2013). *Teaching the nature of science: Perspectives and resources*. St Paul: SHiPS Education Press.

American Association for the Advancement of Science. (1989). *Science for all Americans*. Washington, DC: American Association for the Advancement of Science.

Atwater, M., Russell, M., & Butler, M. (Eds.). (2014). *Multicultural science education: Preparing teachers for equity and social justice*. Dordrecht: Springer. https://doi.org/10.1007/978-94-007-7651-7.

Barton, A. C. (2003). *Teaching science for social justice*. New York: Teachers College.

Barton, A. C. (2009). Mothering and scientific literacy: Challenging truth-making and authority through counterstory. In W.-M. Roth (Ed.), *Science education from people for people: Taking a stand(point)* (pp. 134–145). New York: Routledge.

Barton, A. C., & Tan, E. (2009). Funds of knowledge and discourses and hybrid space. *Journal of Research in Science Teaching, 46*(1), 50–73.

Buxton, C. (2010). Social problem solving through science: An approach to critical, place-based, science teaching and learning. *Equity & Excellence in Education, 43*(1), 120–135. https://doi.org/10.1080/10665680903408932.

Chaparian, S. (2020). *Changes in grade 7 learners' NOS understandings and argumentation skills after engaging in reflective discussions following alternative information evaluation in the context of socio-scientific controversial issues*. Unpublished master's thesis. American University of Beirut, Beirut, Lebanon.

Clough, M. P. (2011). The story behind the science: Bringing science and scientists to life in post-secondary education. *Science & Education, 20*(7), 701–717.

Clough, M. P. (2017). History and nature of science in science education. In K. S. Taber & B. Akpan (Eds.), *Science education: An international course companion* (pp. 39–51). Dordrecht: Springer.

Cochran-Smith, M. (1995). Color blindedness and basket making are not the answers: Confronting the dilemmas of race, culture, and language diversity in teacher education. *American Educational Research Journal, 32*(3), 493–522.

Conant, J., Nash, L., & Roller, D. (Eds.). (1957). *Harvard case histories in experimental science* (Vols. I & II). Cambridge, MA: Harvard University press.

Crenshaw, K., Gotanda, N., Peller, G., & Thomas, K. (Eds.). (1995). *Critical race theory: The key writings that formed the movement*. New York: The New Press.

Dagher, Z. R. (2009). Epistemology of science in curriculum standards of four Arab countries. In S. BouJaoude & Z. Dagher (Eds.), *The world of science education: Arab states* (pp. 41–60). Rotterdam: Sense Publishers.

Dagher, Z. R., & Ford, D. (2005). How are scientists portrayed in children's science biographies. *Science & Education, 14*, 377–393.

DeBoer, G. (2000). Scientific literacy: Another look at its historical and contemporary meanings and its relationship to science education reform. *Journal of Research in Science Teaching, 37*(6), 582–601.

Driver, R., Leach, J., Millar, R., & Scott, P. (1996). *Young people's images of science*. Philadelphia: Open University Press.

Duschl, R., & Grandy, R. (2013). Two views about explicitly teaching nature of science. *Science & Education, 22*, 2109–2139.

Erduran, S., & Dagher, Z. R. (2014). *Reconceptualizing nature of science for science education: Scientific knowledge, practices, and other family categories*. Dordrecht: Springer.

González, N., Moll, L., & Amanti, C. (Eds.). (2005). *Funds of knowledge: Theorizing practices in household, communities, and classrooms*. Mahwah: Lawrence Erlbaum Associates.

Hodson, D., & Wong, S. L. (2017). Going beyond the consensus view: Broadening and enriching the scope of NOS-oriented curricula. *Canadian Journal of Science, Mathematics and Technology Education, 17*(1), 3–17.

Irzik, G., & Nola, R. (2011). A family resemblance approach to the nature of science. *Science & Education, 20*, 591–607.

Irzik, G., & Nola, R. (2014). New directions for nature of science research. In M. Matthews (Ed.), *International handbook for research in history, philosophy, and science teaching* (pp. 999–1021). Dordrecht: Springer.

Klassen, S. (2007). The application of historical narrative in science learning: The Atlantic cable story. *Science & Education, 16*(3–5), 335–352.

Kolstø, S. D. (2001). Scientific literacy for citizenship: Tools for dealing with the science dimension of controversial scientific issues. *Science Education, 85*, 291–310.

Ladson-Billings, G. (1995). Toward a theory of culturally relevant pedagogy. *American Educational Research Journal, 32*(3), 465–491.

Ladson-Billings, G., & Tate, W. (1995). Toward a critical race theory of education. *Teachers College Record, 97*(1), 47–68.

Larkin, D. (2013). *Deep knowledge: Learning to teach science for understanding and equity*. New York: Teachers College Press.

Lederman, N. G. (1992). Students' and teachers' conceptions of the nature of science: A review of the research. *Journal of Research in Science Teaching, 29*(4), 331–359.

Lederman, N. G. (2007). Nature of science: Past, present, and future. In S. K. Abell & N. G. Lederman (Eds.), *Handbook of research on science education – Volume II* (pp. 831–880). New York: Routledge/Taylor & Francis.

Lederman, N. G., & Lederman, J. S. (2014). Research on teaching and learning the nature of science. In N. G. Lederman & S. K. Abell (Eds.), *Handbook of research on science education, Volume II* (pp. 600–620). Mahwah: Lawrence Erlbaum Associates.

Lederman, N. G., Abd-El-Khalick, F., Bell, R., & Schwartz, R. (2002). Views of nature of science questionnaire (VNOS): Toward valid and meaningful assessment of learners conception of nature of science. *Journal of Research in Science Teaching, 39*(6), 497–521.

Matthews, M. (1994). *Science teaching: The role of history and philosophy of science*. New York: Routledge.

Matthews, M. (2004). The pendulum: Its place in science, culture, and pedagogy. *Science & Education, 13*(4–5), 261–277.

Matthews, M. (2012). Changing the focus: From nature of science (NOS) to feature of science (FOS). In M. S. Khine (Ed.), *Advances in nature of science research: Concepts and methodologies* (pp. 3–26). Dordrecht: Springer.

McComas, W. (2016, April). *The nature of science (NOS) and the Next Generation Science Standards: The role of cross cutting concepts*. Paper presented at the NARST conference in Baltimore, MD.

McComas, W. F., & Olson, J. (1998). The nature of science in international science education standards documents. In W. F. McComas (Ed.), *Nature of science in science education: Rationales and strategies* (pp. 41–52). Dordrecht: Kluwer Academic Publishers.

Milne, C. (1998). Philosophically correct science stories? Examining the implications of heroic science stories for school science. *Journal of Research in Science Teaching, 35*, 175–187.

Mueller, M. (2011). Ecojustice in science education: Leaving the classroom. *Cultural Studies of Science Education, 11*(6), 351–360.

National Research Council. (1996). *National science education standards*. Washington, DC: National Academies Press.

Nersessian, N. (1989). Conceptual change in science and in science education. *Synthese, 80*(1), 163–183.

NGSS Lead States. (2013). *Next Generation Science Standards: For states, by states*. Washington, DC: National Academies Press.

Norris, S. P. (1985). The philosophical basis of observation in science and science education. *Journal of Research in Science Teaching, 22*(9), 817–833.

Norris, S. P., & Phillips, L. M. (2003). How literacy in its fundamental sense is central to scientific literacy. *Science Education, 87*, 224–240.

Olson, J. (2018). The inclusion of the nature of science in nine recent international science education standards documents. *Science & Education, 27*, 637–660.

Pintrich, P., Marx, R., & Boyle, R. (1993). Beyond cold conceptual change: The role of motivational beliefs and classroom contextual factors in the process of conceptual change. *Review of Educational Research, 63*(2), 167–199.

Posner, G. J., Strike, K. A., Hewson, P. W., & Gertzog, W. A. (1982). Accommodation of a scientific conception: Towards a theory of conceptual change. *Science Education, 66*(2), 211–227.

Roberts, D. (1982). Developing the concept of "curriculum emphases" in science education. *Science Education, 66*(2), 243–260.

Roberts, D. (2007). Scientific literacy/science literacy. In S. K. Abell & N. G. Lederman (Eds.), *Handbook of research on science education* (pp. 729–780). Mahwah: Lawrence Erlbaum Associates.

Roberts, D., & Bybee, R. (2014). Science literacy, scientific literacy, and science education. In N. G. Lederman & S. K. Abell (Eds.), *Handbook of research on science education, Volume II* (pp. 545–558). New York: Routledge/Taylor & Francis.

Rodriguez, A. (2015). What about a dimension of engagement, equity, and diversity practices? A critique of the Next Generation Science Standards. *Journal of Research in Science Teaching, 52*, 1031–1051.
Roth, W.-M. (Ed.). (2009a). *Science education from people for people: Taking a stand(point).* New York: Routledge.
Roth, W.-M. (2009b). Living with chronic illness: An institutional ethnography of (medical) science and scientific literacy in everyday life. In W.-M. Roth (Ed.), *Science education from people for people: Taking a stand(point)* (pp. 146–171). New York: Routledge.
Schizas, D., Psillos, D., & Stamou, G. (2016). Nature of science or nature of the sciences. *Science Education, 100*(4), 706–733.
Sjöström, J., & Eilks, I. (2018). Reconsidering different visions of scientific literacy and science education based on the concept of *bildung*. In Y. J. Dori, Z. R. Mevarech, & D. Baker (Eds.), *Cognition, metacognition, and culture in STEM* (pp. 65–88). Dordrecht: Springer.
Stinner, A., & Williams, H. (1993). Conceptual change, history, and science stories. *Interchange, 24*(1&2), 87–103.
Tan, E., & Barton, A. C. (2012). *Empowering science and mathematics education in urban schools*. Chicago: University of Chicago.
Walls, L., Buck, G., & Akerson, V. (2013). Race, culture, gender, and the nature of science in elementary settings. In J. Bianchini, V. L. Akerson, A. Calabrese Barton, O. Lee, & A. Rodriguez (Eds.), *Moving the equity agenda forward: Equity research, practice, and policy in science education* (pp. 132–150). Dordrecht: Springer.
Yager, R. (1996). *Science/technology/society as reform in science education*. Albany: SUNY.
Zeidler, D. (2014). Socioscientific issues as a curriculum emphasis. In N. G. Lederman & S. K. Abell (Eds.), *Handbook of research on science education, Volume II* (pp. 697–726). New York: Routledge/Taylor & Francis.
Zeidler, D., Sadler, T., Simmons, M., & Howes, E. (2004). Beyond STS: A research-based framework for socioscientific issues education. *International Journal of Science Education, 89*, 357–377.

Chapter 4
Capitalism, Nature of Science and Science Education: Interrogating and Mitigating Threats to Social Justice

J. Lawrence Bencze and Lynette C. Carter

4.1 Introduction

For at least the last century, there have been recommendations for teaching students about the 'nature of science' (NoS) (Lederman and Lederman 2014). Through such education, students may learn, for instance, about attributes of practitioners, characteristics of their approaches, significance of products of their work and how products of their work are used. Broadly, teaching about NoS may draw from 'science and technology studies' (STS)—using concepts, approaches, etc. from such fields as history, philosophy and sociology (Hodson 2008). Educating students about the nature of science has been justified from several perspectives. It may, for example, assist people in judging relative merits of different societal knowledge systems—including those that some scholars consider outside boundaries of what may be considered 'science' (e.g., Matthews 2017). Students also may become less dependent on authority figures, such as teachers and others, in making judgements about knowledge claims—intellectual and cultural independence that appears helpful for effective participation by all citizens in democratic decision making (Osborne et al. 2003).

Despite long-standing, justifiable, attempts to encourage teachers and others to help students to develop deep and complex conceptions of the nature of science, many scholars in this field suggest that such education often has been compromised. As Clough (2018) recently said, "little of what is known [about NoS] is widely implemented in science classrooms" (p. 3).

J. L. Bencze (✉)
Department of Curriculum, Teaching & Learning, OISE, University of Toronto, Toronto, ON, Canada
e-mail: larry.bencze@utoronto.ca

L. C. Carter
Australian Catholic University, Melbourne, VIC, Australia

H. A. Yacoubian, L. Hansson (eds.), *Nature of Science for Social Justice*, Science: Philosophy, History and Education,
https://doi.org/10.1007/978-3-030-47260-3_4

An aspect of NoS that appears to be frequently compromised in school science pertains to influences of powerful pro-capitalist individuals (e.g., financiers) and groups (e.g., corporations, governments) on fields of science and technology (and many other entities) that seem to compromise wellbeing of many individuals, societies and environments (e.g., Carter 2008). Analyses by, for example, Darren Hoeg and Larry Bencze (2017), Annette Gough (2015) and Dana Zeidler (2016) of the latest curriculum in the USA, for instance, suggest that it prioritizes teaching and learning of 'products' (e.g., laws, theories & innovations) and skills (e.g., for experimentation) of 'STEM' (Science, Technology, Engineering & Mathematics) fields, largely omitting or misrepresenting problematic influences of pro-capitalist entities on such fields. Compromises to more holistic and critical education about science and technology appear to represent threats to *social justice* of students. Although a contentious term that either, for example, recommends opportunities for or equal access to resources enabling parity of participation in social (and economic) life (Fraser 2008), students learning chemistry, for example, not enlightened about injustices of labour in extraction, manufacturing and disposal relating to elements like aluminum are being given an unjust education (beyond injustices to labourers in such contexts) (Levinson 2014).

In this chapter, after elaboration of potential and realized injustices linked to influences of powerful people and groups on fields of science and technology and on other biotic and abiotic entities, perspectives and practices in science education and science teacher education are discussed that may contribute to increases in social justice and environmental health.

4.2 Pro-Capitalist Science and Technology and Relevant Education

4.2.1 A Networked World

Traditionally, science has been taught in isolation from other subjects, a priority partly fed by claims that its focus on abstract, decontextualized knowledge makes it fundamental to progress in related fields like technology and engineering that, perhaps, have more direct impacts on societies (Ziman 1984). Since about the early 1970s, however, educational researchers, book publishers, school systems and others claim that it should be taught in relationships with many other disciplines. In this period, there have been, for example, numerous efforts around the world to integrate or interrelate fields of science and technology and societies and environments (STSE) (Pedretti and Nazir 2011). Such more holistic, socially and environmentally integrated conceptions of the nature of science and technology has much evidential and theoretical support. Based, for example, on *actor-network theory* (ANT), a set of ontological perspectives that largely developed from studies of scientists and their work, all living, non-living and symbolic (semiotic) entities ('actants')—including fields of science and technology and related disciplines (e.g., mathematics)—are intertwined in a global network of dynamic, reciprocating, relationships (Latour 2005).

In learning about global dynamic relationships involving fields of science and technology, a particularly essential aspect of them that students should consider is that they are *imbalanced*—in that some actants appear more influential than others. In *Pasteurization of France*, for instance, Latour (1988[1984]) claimed that it was not until cooperative associations among various actants—such as laws, technologies, transportation mechanisms and public education—were established that pasteurization was widely-practised. Therefore, within global networks, there appear to be coordinated 'sub-networks' that consist of living, nonliving and symbolic entities aligned (more or less) to support particular purposes. Michel Foucault (2008), in his analyses of power, named such cooperating sets of actants *dispositifs*. In this light, one can ask about dispositifs that appear to overly-influence fields of S&T in ways that may contribute to various personal, social and environmental injustices.

4.2.2 Capitalist Dispositifs

Considerable evidence and argument suggest that most actants around the world are greatly *orchestrated* to 'perform' interests of relatively few pro-capitalist individuals and groups. Although financial and other elite have long had significant influences over large fractions of societies, their dominance appears to have dramatically escalated since about the 1970s. Largely in response to decreases in their shares of wealth, in part due to costs of infrastructure re-building and social programming after the second world war, economic elite, government officials, military leaders and others worked to develop *neoliberal* socio-economic systems. These, generally, appear to have assimilated governments, corporations, financiers, think tanks, transnational trade organizations, banks, transportation networks, trade agreements, perspectives and practices of large fractions of societies and many more entities into a 'super-dispositif' promoting policies and practices like competition, individualism, cost externalizations, business de-regulation, tax reductions and avoidance, privatization, etc. that, ultimately, seems to funnel wealth and wellbeing towards socio-economic elite (Springer et al. 2016). This neoliberal dispositif appears to have, indeed, worked well, dramatically increasing differences between super-rich and most other people on the planet—wealth disparities that, moreover, are predicted to dramatically increase, despite such interruptions as the 2008 global financial crisis (Piketty 2014).

Although capitalism appears to have gained much of its influence and resiliency through *global* infiltration—while often somewhat customized to local traditions (Roudometof 2016)—of pro-capitalist perspectives (e.g., individual competitiveness) and practices (business de-regulation), there are now numerous cases around the world, such as in the Austria, Hungary, Turkey, the USA and elsewhere, where extreme nationalist sentiments appear to have taken hold (Pelinka 2013). On the one hand, while such isolationist tendencies may weaken capitalist networks, there are suggestions that they may be interpreted as adaptations that, ultimately, continue to protect or, indeed, augment capitalist perspectives and practices. In her analyses of right-wing populism in the USA, for instance, Naomi Klein (2017) suggests that such nationalist sentiments may represent examples of *disaster capitalism*; that is, 'capitalizing' on citizens' states of shock, allowing capitalists to further implement their perspectives and practices. Apparently after considerable destabilization of societies associated with neoliberalism, such as in terms of job losses (e.g., via moving them to jurisdictions with cheaper labour costs and lax environmental regulation) and/or precarity (e.g., part-time, on-call, with no/few benefits), populist leaders can portray themselves as saviours of the people against the 'evil' political class. They can then declare a *state of exception* (to normal relationships between governors and governed [Agamben 2004]) and enact policies supporting capitalists, that, ironically, further impoverish many citizens and compromise environments (Bauman 2017).

4.2.3 Hyper-capitalist Science and Technology

With little doubt, key agents in capitalists' programme of wealth concentration, apparently regardless of globalizing or localizing tendencies, are many fields of science and technology (S&T). Laws and theories about phenomena and inventions and innovations are, of course, useful for thoughts and actions—all of which can be commodified and marketed with help of fields of science and technology that have some form of financial arrangements with or influences by for-profit entities (Mirowski 2011). Influences of capitalists on fields of science and technology likely can be understood from many perspectives. A schema presented by Wolff-Michael Roth (2001), however, an adaptation of which is provided in Fig. 4.1, may be helpful (Bencze and Carter 2015). This schema was meant to depict 'science' as (often a series of) translations from phenomena of the World into Signs to represent them and depict 'technology' (and engineering) as (often a series of) translations from Signs into phenomena of the World. Because such translations are reciprocal (World and Sign co-affecting each other), the model might best be thought of as

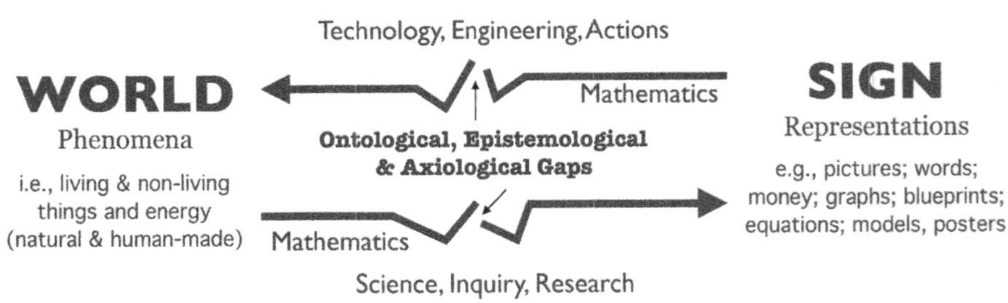

Fig. 4.1 Capitalist-friendly technoscience/STEM

representing *technoscience* (Sismondo 2008). Since both fields often use mathematics, however, the schema also may represent 'STEM' fields.

A key aspect of capitalist influences on science and technology (or STEM) fields that may compromise social justice pertains to 'gaps' (inefficiencies) in translational processes. *Ontological* gaps occur because composition of two entities involved are different, such as with difficulties of a relatively fixed two-dimensional map (Sign) being used to represent a three-dimensional (and changing) space of land. Such inefficiencies seem, to some extent, unavoidable. *Ideological* gaps, on the other hand, involve value-laden (axiological) choices that can vary—depending on individuals' or groups' ideological or ethical positions. A particular problem in this regard is capitalism's uses of STEM fields for promotion of *subterfuge* often associated with repeating cycles of consumption and disposal of material commodities (Leonard 2010) and, more recently, abstractions such as speculation on future for-profit ventures (McMurtry 2013). Subterfuge can occur, for example, when technology designers and/or advertisers promote *hyperreal* (dissociated from entities they are to represent) constructions (Signs) of phenomena (World), such as 'sexiness,' 'efficiency' or 'upper class' associated with automobiles (Baudrillard 1998). Such manipulation of thought and action may represent a form of *biopolitics* (Lemke 2011), in which powerful actors subjectify (e.g., influence character formation of) populations of living things (Foucault 2008). Such management of thought, identity and action can be particularly problematic when idealizations—like sexiness, etc.—distract consumers from noticing or acknowledging compromises to phenomena being purchased. A common technique in this regard is, apparently, *punctualization* (Callon 1990); that is, making an entity (e.g., commodity) appear (with Signs) simpler than it may be in 'reality.' As suggested by Clayton Pierce (2013), for example, genetically modified (GM) salmon can be seen in significantly reductionist ways as major food sources compared to wild salmon, perhaps distracting customers from noticing possibly-problematic actants in larger, more complex (*de-punctualized*), networks—such as government regulation policies (by, for example, the FDA [Federal Drug & Food Administration]) favouring corporations over consumers and sea lice pests that can harm both GM and wild salmon.

On the one hand, there is much to celebrate about societal contributions of fields of science and technology. Humans enjoy longer lifespans, for instance, through numerous developments in medical and agricultural fields. Nevertheless, in our current era of the so-called *Anthropocene*, Earth systems appear to be rapidly degrading in terms, for example, of out-of-control climate change and devastating species losses related, for example, to extensive habitat destruction (Steffen et al. 2018). An international panel of scientists recently suggested that existence of about a million species are currently under extreme threat (IPBES 2019). Meanwhile, advent of right-wing populism as a form of disaster capitalism, as discussed above, may involve manipulation of perceptions—such as delegitimizing climate science (Klein 2017)—while attempting, perhaps with some resistance, to further deregulate industries that may contribute to problems like climate change that often disproportionately harm more vulnerable societal members (McCarthy et al. 2018). Similarly, those people electing to use apparently relatively free social media services, such as

Facebook™ and Google™, may find that—as part of *surveillance capitalism* (Zuboff 2019)—their 'voluntarily'-provided information is mined using complex computer and algorithmic systems to manipulate their thoughts and actions, such as in terms of promotion of certain shopping and electoral choices.

4.3 Towards Socially-Just 'NoS' Education

4.3.1 Preamble

Given problematic effects of *hegemonic* influences of capitalists on fields of science and technology and on corresponding educational initiatives, such as STEM education, it seems difficult to envisage acceptance of more de-punctualized and problematized conceptions of the nature of science and technology. Regarding rapid emergence of various right-wing populist movements, Naomi Klein (2017) suggests that many counter-revolutionary groups have achieved some successes by not just critiquing their opposition but by also providing supporters with visions of different, perhaps more 'progressive,' societies. This recommendation seems aligned with Thomas Kuhn's (1970) suggestion that revolutionary change thrives on existence of alternative paradigms (if only emergent ones). Accordingly, in the two sections below, we provide narratives of approaches to school science and science teacher education, respectively, that we feel may provide educators and others with possibilities for education about the nature of science and technology that prioritize social justice.

4.3.2 Ecojustice Through Education for Critical and Altruistic Civic Actions (Larry Bencze)

For at least the last two decades, I have been convinced that school science (and, now, STEM education) has been used as a mechanism serving economic elite for selecting and educating a small group of potential *knowledge producers* (e.g., scientists & engineers) who can develop and manage mechanisms of production and consumption and a relatively large mass of *knowledge consumers*; that is, citizens who will dutifully follow labour instructions and enthusiastically purchase commodities (Bencze 2001). More recently, I also have become increasingly convinced that many governments have tended to mainly support capitalists, often sacrificing wellbeing of most individuals, societies and environments (Bencze 2008). Accordingly, since 2006, I have worked with graduate students in collaborations with educators in many different contexts (e.g., schools and after-school clubs) to create and evaluate teaching and learning approaches and resources that may help all students to develop more 'realistic' (*de-punctualized* and *problematized*) conceptions of relationships among fields of science, technology, societies and environments (STSE) and to develop expertise, confidence and motivation to design and carry out well-informed actions to address harms they perceive in such relationships (Bencze 2017). In addition to being informed by students' previous experiences and learning, we have encouraged and enabled them to base actions on their research (secondary and primary), which can be highly motivating, and on negotiations they have with peers. Overall, we have been working to encourage and enable learners to self-direct research-informed and negotiated action (RiNA) projects to address harms students perceive in STSE relationships.

Although there may be some exceptions, our research suggests that most students benefit from teacher-led teaching and learning activities that may help them develop sufficient expertise, confidence and motivation for *eventually* self-directing RiNA projects to address harms in STSE relationships concerning them. Working with teachers, graduate students and others, we have found that the schemas like that in Fig. 4.2 have helped many students achieve such outcomes. Called 'STEPWISE' (Science & Technology Education Promoting Wellbeing for Individuals, Societies & Environments), this schema suggests that teachers guide students—in a *stepwise* fashion—through a series of constructivism-informed 'apprenticeship' lessons and activities before asking students to self-direct RiNA projects on topics of their choice.

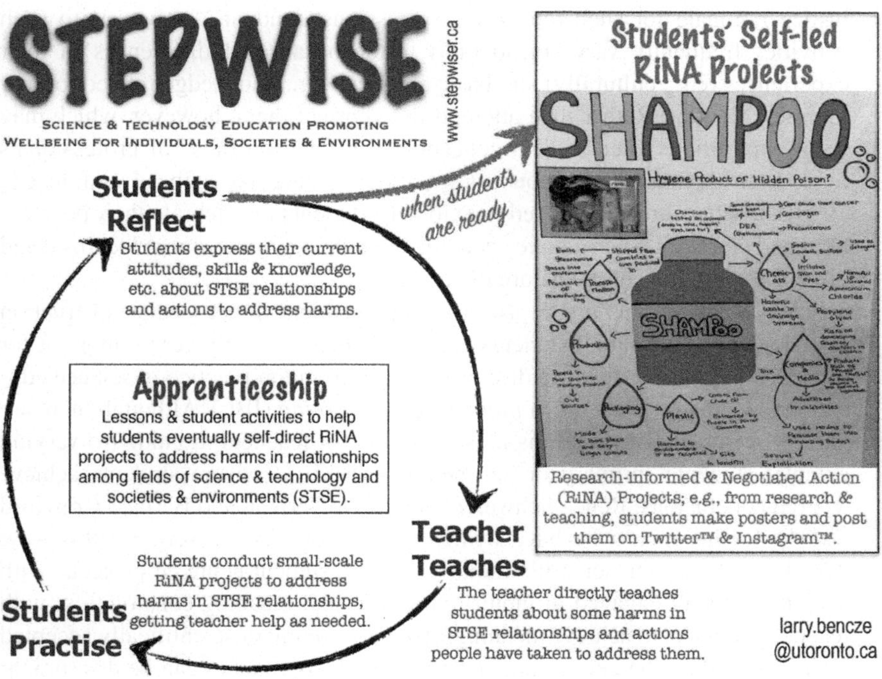

Fig. 4.2 STEPWISE pedagogical framework

The STEPWISE framework was designed specifically to help students achieve 'expectations' for learning specified in Ontario science curricula (e.g., MoE 2008), which are divided into three learning domains; that is, (i) STSE education (e.g., influences of capitalists on scientists), (ii) Skills education (e.g., experiment design) and (iii) 'Products' education (e.g., understanding laws, theories and inventions/innovations). It may be adapted, however, to align with curricula in other jurisdictions. Some suggestions, with examples, for each of the three 'phases' of apprenticeships lessons and activities to address such expectations are provided in the sub-sections below:

- Students Reflect: Activities in this phase are intended to stimulate students to reflect on their prior experiences (including their education) and to *express* their personal attitudes, skills and knowledge ('ASK') regarding a science-related topic (usually specified in curricula). For stimuli, it has been helpful, for example, to show students different products of S&T fields, such as: hamburgers and French fries, cell phones, drugs, clothing fashions, cosmetics, weapons, etc. Students can say or write about what they like and dislike about these, and discuss people and groups who may support (e.g., companies & advertisers) or critique (e.g., citizen activists, some government representatives) such products. Such reflection activities are based on foundational constructivist learning theory (Osborne and Wittrock 1985), which suggests that expressing existing ASK can help learners to become more self-aware of them—and, later, when confronted with conflicting ASK, consider changing them. There also are, however, social justice rationale for such reflections/expressions. Students' responses to such activities frequently vary considerably, often because of differences in their experiences (e.g., culturally) and basic abilities (e.g., knowledge related to their families' wealth). Rather than attempting to correct them, however, which may seem appropriate when student conceptions differ from those of professionals (e.g., as with many Indigenous views about science [Aikenhead and Jegede 1999]), teachers are encouraged here to help students to celebrate their perspectives and practices. Such a policy can, for example, legitimize more localized and diverse conceptions of the nature of science;
- Teacher Teaches: This phase prioritizes planned and direct teacher instruction aimed at ensuring that all students learn ASK that are likely very important for their lives, but are difficult to discover. A very common practice in science education is to engage students in *inquiry-based learning* (IBL). Although there are various forms of IBL, it seems most common for teachers to guide (to varying extents) students' primary or secondary research to ensure students achieve desired conclusions. Reneé Schwartz, Norman Lederman and Barbara Crawford (2004), for instance, who have written much about IBL, describe it this way: "Within a classroom, scientific inquiry involves student-centered projects, with students actively engaged in inquiry processes and meaning construction, with teacher guidance, to achieve meaningful understanding of scientifically accepted ideas targeted by the curriculum" (p. 612). Such approaches can be discriminatory for many students, however, who often lack sufficient intelligence and/or

cultural capital to discover expected ASK (Bourdieu 1986). It also can suggest to them that science inquiry is a relatively flawless process, leading to well-accepted claims—placing perhaps unwarranted trust in fields of science and technology and their products and services, a form of manipulation by private sector members (albeit indirectly) that may result in increased consumerism (Bencze and Alsop 2009).

- To overcome problems of discrimination and manipulation like those noted above, it seems particularly important that students are taught about problematic exercise of power in STSE relationships and what people might do to overcome such situations. In our work, teaching students about actor-network theory, how to develop actor-network maps to depict STSE relationships involving particular commodities, etc., concentrations of network power in dispositifs and subterfuge in World ←→ Sign relationships (focusing on ideological gaps [Fig. 4.1]) have been very helpful (Bencze and Krstovic 2017a). Students may be taught, for instance, how governments and transnational trade agreements often allow companies to add sugars/sweeteners, salts, fats, chemical colourings, flavours and preservatives to foods and that research suggests are linked to human illnesses, like heart disease, diabetes and cancer. The teacher also could show them a video that describes how other students researched food industry problems and developed and carried-out a campaign to educate citizens about possible harms from manufactured foods. Such teaching about problematic power relations seems particularly necessary in light, for example, of evidence and arguments to suggest that capitalists have paid scientists, journalists and others to discredit science research findings that would, if made more public, discourage people from consuming products like cigarettes, pesticides, petroleum and by-products, nuclear energy, etc. (Oreskes and Conway 2010).
- Although it is important, as stressed above, for teachers to directly teach important ASK, such teaching needs to be complemented with activities in which students apply—and, essentially, *evaluate*—newly-taught attitudes, skills, and knowledge. Student learning appears to be deepest and most committed when they have increased control over translations in both directions of the schema in Fig. 4.1 (Wenger 1998). This may be the case when students use new ASK to: i) evaluate World ➤ Sign translations from previous experiences; and, ii) develop plans for possible actions (Sign ➤ World). An excellent choice in this regard is to engage students in STSE *case methods,* such as questions and suggestions (methods) surrounding a documentary (case) depicting citizens' efforts to eliminate (what they determined to be) toxic dust (containing many heavy metals, such as lead and cobalt) dispersal from the local port onto their neighbourhood (Bencze and Pouliot 2017).
- <u>Students Practise</u>: To further deepen students' expertise, confidence and motivation for them, students are asked (in this phase) to develop and implement small-scale RiNA projects to address harms they determine in STSE relationships—obtaining help from the teacher, as needed and/or requested by them. Typically, the teacher will encourage small groups of students to choose an STSE issue/problem to explore and then ask them to complete a RiNA project to

address it—providing them with deadlines for separate parts (e.g., issue/problem; research methods; actions). For some students apparently needing them, teachers can provide ideas, stimuli for exploration, etc. These can range from single words or phrases, such as *climate breakdown*, *cell phones*, *surveillance*, *species losses*, etc., through to lists of questions (e.g., 'To what extent should governments regulate international transportation of potentially harmful substances such as petroleum?') and on to more information-rich stimuli like our 'multi-actant documentaries' of possible STSE relationships (tinyurl.com/y97254t3).

At the end of one cycle of apprenticeship lessons and activities like that outlined above, the teachers may ask students to reflect again on their conceptions of the nature of STSE relationships and RiNA projects—claims that students can use for planning and conducting future such projects. The teacher may then decide either to provide at least one more such cycle of lessons and activities or, if students seem ready, ask students to self-direct RiNA projects.

Once the teacher believes that most (if not all) students have sufficient expertise, confidence and motivation, students should be asked to *self*-direct RiNA projects to address harms they perceive in STSE relationships. Typically, this means that the teacher will provide students with a formal assignment—often with a broad description of projects, deadlines for smaller parts of them (e.g., topics, methods, results, actions, etc.) and an assessment/evaluation scheme. As a culminating event, teachers also could ask students to give presentations about their projects in public fora (e.g., an 'STSE-Action Fair'). Although these culminating projects should be largely student-directed and open-ended, teachers find that many students appreciate receiving lists of possible project topics (e.g., as brief descriptions of controversies surrounding S&T products). In formal school contexts, 'guidance' also often occurs in terms of assessment/evaluation forms linked to such projects—the extent of guidance varying, depending on perceptions of students' needs.

The STEPWISE project seems to have helped many students to critically analyze STSE relationships to identify power-related harms for individuals, societies and/or environments and also develop and implement personal and social actions intended to increase social justice and environmental sustainability. Examples of such projects appear in three special issues of the *Journal for Activist Science & Technology Education* (JASTE) (goo.gl/N00b3s; bit.ly/2JGIgtf) edited by science teachers and featuring reports of RiNA projects written by students. Rich descriptions of RiNA projects and related pedagogical practices also are provided in the 'STEPWISE' edited book (Bencze 2017). Commonly used by students as actions are educational (and 'activist') videos, like that provided here: tinyurl.com/yagkgmue. Such projects represent *potential* for changes in phenomena (potential Sign ➛ World translations, Fig. 4.1). Less common in school science are actions that may *realize* changes to phenomena; such as technology/engineering products. After apprenticeship lessons and activities like that depicted in Fig. 4.2, however, students were able to at least design engineering products that they believed may improve social justice and/or environmental sustainability (Bencze and Krstovic 2017a). For example, as

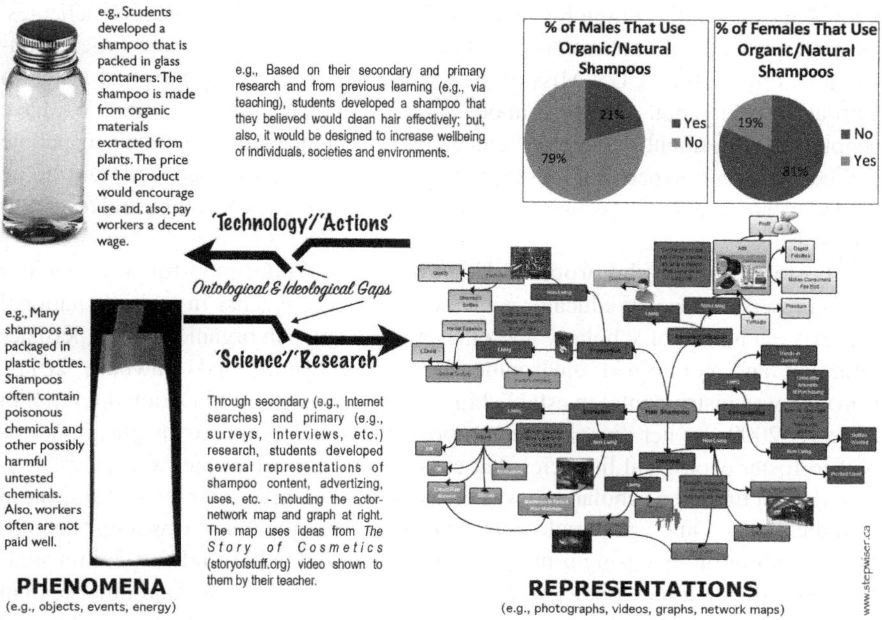

Fig. 4.3 Students' ANT-based RiNA

indicated in Fig. 4.3, a group of students investigated actor-network relations involving hair shampoo and, in light of related education they had received from their teacher and through their secondary and primary research about shampoos, developed an organic shampoo (as a 'technology') that could clean hair effectively; but, as well, would use recyclable glass containers and pay workers a living wage.

4.3.3 The Nature of Science (Education): Science Teacher Educators Working Together for Social and Ecojustice (Lyn Carter)

One alternative to pro-capitalist STEM education has centred on professional collaborations of my colleagues and our shared approach to preparation of our preservice science teachers. Over the past two decades, Drs. Caroline Smith, Mellita Jones, Jenny Martin and Carolina Castano Rodriquez and I have made a somewhat unique science education team at the Melbourne and Ballarat campuses of the Australian Catholic University (ACU). ACU is a unique multi-campus public university across five states and territories in Australia. On the Melbourne and Ballarat campuses (hereafter ACU Vic, both campuses being located in Victoria state), we have worked with broader sociocultural and political conceptualisations of science education and the nature of science, rather than the mastery of reductive science knowledge and skills commonly associated with narrow STEM education schemes. Despite pressures of curriculum and standards frameworks of various types, mandated or otherwise, ACU's Mission Statement has enabled a conceptual space to embrace fundamental questions about human experience and meaning. We have adapted the statement's focus on 'enhancing the dignity and wellbeing of people and communities, especially those most marginalised or disadvantaged' and 'to be guided by social and eco-justice principles' in developing our teaching and researching projects.

My story begins with Caroline Smith, a science and education-for-sustainability (EfS) preservice teacher educator, a classroom science teacher in multiple national contexts, an author of scholarly and other manuscripts, an organic farmer, permaculturalist and committed environmental activist (Smith and Dawborn 2011). Caroline was instrumental in establishing ACU Vic's science education direction in the early 2000s, as her diverse interests provided her with unique insights into how best to foster ecological literacies that are essential for life in the twenty-first century. At that time, my scholarship was concerned with impacts on science education of radical social and epistemological injustices consequent upon twenty-first century globalisation. Focusing principally on educational policy and curriculum studies, and utilising a methodology of critical philosophical inquiry and other textual analyses, I examined ways in which postcolonialism, indigenous knowledge and ecological sustainability could act as counter discourses to globalisation and resource new approaches to teaching and learning in science (Carter 2008). As colleagues for more than a decade and a half, our work coalesced into a shared vision that believed science education should not only work towards a deeper understanding of our planetary systems, but also towards explicit goals of creating a more just, equitable and sustainable world (Carter and Smith 2003).

During the early and mid-2000s, Caroline, Mellita, and I implemented student units (courses) whose organising framework drew from literatures of science studies, nature of science, cultural diversity and sustainability science to depict development of science as cultural stories reiterating themes of recognition, difference and localism. In a departure from what would be regarded as typical science content, our major core unit began by exploring cosmologies from various cultures to show that human societies have always tried to understand and shape their world; sciences and technologies are as old as humanity, and that there are as many sciences as there are contexts/cultures. Western science could, thus, be understood as a particular form of localised ethnoscience, shaped by, and reproductive of, political, economic, cultural and social forces of the times. Through its epistemological robustness, reliability and usefulness, Western science was shown to have transcended its immediate determinants, eclipsing other ways of knowing and ensuring its universal acceptance as *the* powerful way of understanding our world. Reviewing precepts of energy and matter conceptually, and within their historical context, as necessary precursors for potent technologies of the nineteenth, twentieth and twenty-first centuries, our unit explored how Western science has been responsible for much human flourishing. But enmeshed as it is in the global capitalist progress paradigm, Western science was also shown to have been co-productive of hegemonic interests resulting in much ecological devastation and many forms of imperialism. This 'warts and all' approach to teaching about science, and its nature as Western science, at the same time as developing its concepts and skills, was our attempt at working within politics of the practical (see Carter 2008).

Caroline's departure (upon retirement) from ACU in 2010, coupled with growth in the university, enabled new colleagues Carolina Castano Rodriquez and Jenny Martin to continue evolution of our sociocultural agenda while STEM education increased popularity in Australia and overseas. Carolina's experience in South America using empathy with animals as a way of mitigating violence within disadvantaged communities brought a new perspective to our work (Castano 2012). A committed environmentalist and outdoor educator, Jenny's interest in student agency and a unique methodological approach from discursive psychology, along with Mellita's strength in reflective pedagogies (Jones and Ryan 2014), added further insights. Our newly minted team was just as like-minded in its desire for science education to promote eco-social justice rather than corporatised/neoliberal agendas. Our collective scholarship somehow seemed to coalesce around interests in facilitating sociopolitical activism, both our own and that of our preservice teachers.

While Jenny (with a little help from me) continued her investigation of student agency (Martin and Carter 2015), Carolina, Mellita and I worked with transformative learning theory (TL), attractive for its focus on promoting action. First described by Jack Mezirow in the late 1970s, TL argues that critical reflection and emancipatory education practices (which was perhaps where our earlier emphases lay) are necessary but not sufficient conditions for transformation. Individuals must experience their own *conflict/disorienting dilemmas/triggering event* to make learning transform into action (Cranton 2006). Accordingly, we developed and implemented an elective unit for preservice teachers to explore whether TL could become pedagogical for socio-political activism within science education (Carter et al. 2018). Challenged with the proposition that 'any sort of egg/chicken consumption contributes to animal cruelty,' designed to create required disorientation or conflict, our results showed that preservice teachers' reflections on what supported their assumptions were critical to generating awareness of their own choices and actions.

Our efforts, of course, continue and are, as always, a work in progress. More recently under Carolina's leadership, we have begun exploring ethics of care (EoC) as an approach to science education jointly developed by Carol Gilligan (1982), Nel Noddings (1992) and other feminist scholars. More recently, posthuman and new material feminism has embraced care ethics within relational ontologies as pedagogical 'intra-actions' (after Barad 2007) (see, for example, Romano 2018). In our practice, EoC furthers our focus on action, as it shifts the moral/ethical question from 'what is just?' to 'how to respond,' while it works to enhance positive relations and recognition of affective needs. We have already completed a small EoC in science education research project at an outer suburban Melbourne primary school with low socio-economic students, typically with first generation migrant and refugee backgrounds (Carter et al. 2019). With a focus on development of collective practices and participants' personal senses of science education, we found invention and construction of new tools and patterns of practice philosophically grounded in an EoC. Our team has taken some of these insights into further developing and implementing EoC into our preservice science education units.

It is an exciting time to be involved in initial science teacher education. There is much that can be done to broaden perspectives on the nature of science and science education for future generations of teachers. It is only to be hoped that enthusiasm and passion with which our preservice teachers embrace these ideas are not to be completely trammeled under the weight of neoliberal compliance that abounds within schools. There lies the next challenge.

4.4 Summary and Ways Forward

Given complicity of most people in the world in helping—like, for example, engineers and scientists, factory labourers, financiers, corporations, advertisers, consumers, lawyers, bankers, science education scholars and myriad more—to create personal, social and environmental harms, there seems to be ample evidence to claim we are experiencing the *Anthropocene* epoch; that is, a period of human dominance, typically seen in a negative light, of most Earth systems (Steffen et al. 2018). On the other hand, in light of Foucault's (2008) concept of *dispositif* formation, there may be certain actants able to orchestrate many living, nonliving and symbolic actants in ways mostly co-supporting aims of few powerful actants. In that vein, there is much evidence and argument to suggest that the most influential force organizing actants around the world is *capitalism* (McMurtry 2013). This seems clear when we note that about 26 billionaires have total wealth equivalent to half the world's population (about 3.8 billion people) (Oxfam 2019). Therefore, it may be more appropriate to consider that many or most of our problems are due to hegemony of the current *Capitalocene* (Moore 2016). Regardless of ultimate causes, however, it is also apparent to many that masses of citizens need to become much more active in helping to address problems like the climate and ecological crises—for at least the reason that governments have tended to be greatly complicit in facilitating systems engendering personal, social and environmental harms.

With little doubt, although generally helpful in many ways, fields of science and technology (and related fields, like engineering and mathematics) often serve as major instruments of wealth concentration, largely at expense of most living and non-living entities on Earth. Accordingly, science education—which has, traditionally, played major roles in selecting and educating potential scientists, engineers and other knowledge generators—must focus on helping to create citizens willing and able to critically evaluate systems of power and, where they perceive injustices, harms, etc., develop and carry out actions that they feel may improve situations. In this chapter, therefore, we have provided narratives of programmes in school science and science teacher education, respectively, that we believe may provide inspiration and practical suggestions for generation of more critical and action-oriented community members. We must, however, temper that claim by noting that the two contexts of our narratives are relatively non-representative of global realities and, indeed, are perhaps somewhat *colonialist*—as they are situated in two 'developed' British Commonwealth countries. Lessons learned from such narrow contexts must, therefore, be relatively broad and 'elastic'—amenable to considerable adaptation to local situations.

Broadly, as suggested in this chapter, a major mechanism in science education benefiting (if not used by) capitalists is to present the nature of science and technology (NoST) as a 'Trojan horse'; that is, an object of desire that harbours dangers. More specifically, NoST is often presented, to varying extents, as *punctualized*; that is, through relatively intense foci on well-established 'products'—such as laws, theories and inventions/innovations—and skills (e.g., experimentation) of S&T, avoiding discussions about their associations with broader sets of actants, such as those involving private sector interests. In doing so, NoST may be seen by students and others as idealized and, by extension, capitalists/capitalism may be interpreted similarly—not so much in need of critical analyses and oppositional actions. At risk of over-simplification, it seems that science/STEM education tends to prioritize presentation of what Steve Fuller (1993) called 'High Church' inquiries and conceptions of science and technology that often are found in *Science and Technology Studies* (STS); that is, relatively reductionist, logical foci on knowledge systems, answering questions like, To what extent is there a 'scientific method?,' What characteristics make science knowledge claims credible?, and In what ways might new S&T disciplines emerge? Ironically, according to Steve Breyman and co-authors (2017), although STS began after world war two with concerns around nuclear warfare, acid rain and toxic chemicals in foods, a period Fuller (1993) referred to as focusing on 'Low Church' studies, STS has evolved to more greatly prioritize High Church forms. The same can be said, it seems, for science/STEM education—presenting S&T fields in *High Church* ways, relatively *punctualized* and *de-problematized*, that are likely to be unjust for many students (Levinson 2018).

In light of evidence and arguments made above about ways in which capitalist influences on fields of science and technology have adversely affected social justice and environmental wellbeing, etc., it seems clear that students in democratic societies need to be educated about such potentially problematic relationships and, where they determine potential injustices, harms, etc., be given education that enables them to develop and implement actions that they believe appropriate (Hodson 2011). In other words, students need more 'Low Church' education about the nature of science and technology (and related fields) so that they may use such more de-punctualized and problematizing education in ways they see fit. Having made such a claim, it may, however, be prudent to stress—as did Sergio Sismondo (2008)—that students may benefit from education that balances Low and High Church STS, a tack that may be more just, not overly prioritizing findings from the two STS 'camps.'

Science/STEM education that provides emphases on Low and High Church STS perspectives may, as well, provide students with what Jesper Sjöström and co-authors (2017) call 'Vision III' forms of science literacy. They suggest, as echoed by scholars like Hoeg and Bencze (2017), Ralph Levinson (2018) and Dana Zeidler (2014), that much science/STEM education prioritizes Doug Roberts' (2007)

Visions I and, to some extent, Vision II forms of science literacy. The former focuses more on reductionist and de-problematized knowledge and skills aimed mainly at potential professionals (e.g., scientists and engineers) while the latter provides education for other members of societies, paying some attention to, for example, societal relationships with professionals, but not strongly critiquing or challenging them. Instead, they suggest emphases must be placed on "[politicized] science education aiming at emancipation and societal participation, and includes aspects like socio- and eco-justice" (Sjöström et al. 2017, p. 182). In making such a recommendation, they also note that STEPWISE-informed practices (and, likely, those highlighted by Lyn Carter in this chapter) appear to prioritize such critical and action-oriented literacy.

Those wanting to promote more Vision III—perhaps in conjunction with Vision I and II—forms of science literacy, which may involve presenting students with more Low Church, along with High Church, STS perspectives about the nature of science and technology, may want to keep in mind that such a programme is unlikely to be easy. As argued here, 'the' capitalist dispositif appears *hegemonic*—continuously acting to minimize perspectives and practices, such as more Low Church STS perspectives, that may challenge its dominance. Indeed, over a decade of action research using the STEPWISE schema to encourage and enable students to critique fields of science and technology and develop and carry out actions to address injustices perceived by them, it was very much apparent that implementation was restricted to relatively rare contexts in which a supportive dispositif existed—including cooperation among actants like, congruent government-sanctioned curriculum objectives (MoE 2008); colleagues and administrators amenable to critical, reflective, practice; and, a teacher adhering to Low Church STS perspectives (i.e., *Naturalist-Antirealist* NoS views [Loving 1991]) (Bencze and Krstovic 2017b). Accordingly, as suggested by Peter Evans (2012), perhaps a reasonable tack for promoting (or 'mobilizing') Low Church STS conceptions in more science/STEM education contexts is for scholars, teachers, administrators and others to actively work to form such a supportive dispositif. Indeed, in the study cited above about apparent toxic dust accumulation in neighbourhoods from the city's docklands, citizens appeared to have some successes in reducing dust depositions through their efforts over a few years to rally a range of living (e.g., fellow citizens), nonliving (e.g., an interactive website) and symbolic (e.g., valid community-generated data) actants to form a relatively effective activist dispositif (Bencze and Pouliot 2017).

In eyes of activists, individuals/groups promoting economic growth with less than desirable attention to wellbeing of many individuals, societies and/or environments may be considered oppressors (Freire 1970). Educating students about potentially problematic power relations, as seems applicable to pro-capitalist dispositifs described here, may, therefore, represent a kind of *conscientization*—a critical consciousness about a (and/or one's own) social milieu (Freire 1970). At the same time, educators in democracies may not want to be guilty of oppression, in the sense of providing students with mis-translations of STS representations like those suggested here—presenting pro-capitalist individuals/groups in unrealistically bad lights. It seems that no educator can avoid ontological gaps and also, likely, ideological gaps (see Fig. 4.1). Accordingly, as Paulo Freire (1970) recommended, to be free of potential oppressors (including science education scholars), citizens need to be given full control over 'praxis'; that is, critical, reflective, practice. Levinson (2010) echoes this call in his discussion of possible citizenship roles in the context of socioscientific issues education when he suggests that education to promote 'Dissent and Conflict' (critical and activist) citizenship should be accompanied by 'Praxis,' in which citizens continually work to be critical of all perspectives and practices as they negotiate being in the world.

References

Agamben, G. (2004). *State of exception* (K. Attell, Trans.). Chicago: University of Chicago Press. https://doi.org/10.7208/chicago/9780226009261.001.0001.

Aikenhead, G. S., & Jegede, O. J. (1999). Cross-cultural science education: A cognitive explanation of a cultural phenomenon. *Journal of Research in Science Teaching, 36*(3), 269–287. https://doi.org/10.1002/(SICI)1098-2736(199903)36:3%3C269::AID-TEA3%3E3.0.CO;2-T.

Barad, K. (2007). *Meeting the universe halfway: Quantum physics and the entanglement of matter and meaning*. Durham: Duke University Press. https://doi.org/10.1215/9780822388128.

Baudrillard, J. (1998). *The consumer society: Myths and structures*. London: Sage. https://doi.org/10.4135/9781526401502.

Bauman, Z. (2017). *Retrotopia*. Cambridge, UK: Polity Press.

Bencze, J. L. (2001). Subverting corporatism in school science. *Canadian Journal of Science, Mathematics and Technology Education, 1*(3), 349–355. https://doi.org/10.1080/14926150109556475.

Bencze, J. L. (2008). Private profit, science and science education: Critical problems and possibilities for action. *Canadian Journal of Science, Mathematics & Technology Education, 8*(4), 297–312. https://doi.org/10.1080/14926150802506290.

Bencze, J. L. (Ed.). (2017). *Science & technology education promoting wellbeing for individuals, societies & environments*. Dordrecht: Springer. https://doi.org/10.1007/978-3-319-55505-8.

Bencze, J. L., & Alsop, S. (2009). A critical and creative inquiry into school science inquiry. In W.-M. Roth & K. Tobin (Eds.), *The world of science education: North America* (pp. 27–47). Rotterdam: Sense. https://doi.org/10.1163/9789087907471_005.

Bencze, J. L., & Carter, L. (2015). Capitalists' profitable virtual worlds: Roles for science & technology education. In P. P. Trifonas (Ed.), International handbook of semiotics (, Vol. 1 & 2, pp. 1197–1212). Dordrecht: Springer. https://doi.org/10.1007/978-94-017-9404-6_57.

Bencze, L., & Krstovic, M. (2017a). Science students' ethical technology designs as solutions to socio-scientific problems. In J. L. Bencze (Ed.), *Science & technology education promoting wellbeing for individuals, societies & environments* (pp. 201–226). Dordrecht: Springer. https://doi.org/10.1007/978-3-319-55505-8_10.

Bencze, L., & Krstovic, M. (2017b). Resisting the Borg: Science teaching for common wellbeing. In J. L. Bencze (Ed.), *Science & technology education promoting wellbeing for individuals, societies & environments* (pp. 227–276). Dordrecht: Springer. https://doi.org/10.1007/978-3-319-55505-8_11.

Bencze, L., & Pouliot, C. (2017). Battle of the bands: Toxic dust, active citizenship and science education. In J. L. Bencze (Ed.), *Science & technology education promoting wellbeing for individuals, societies & environments* (pp. 381–404). Dordrecht: Springer. https://doi.org/10.1007/978-3-319-55505-8_17.

Bourdieu, P. (1986). The forms of capital. In J. G. Richardson (Ed.), *The handbook of theory: Research for the sociology of education* (pp. 241–258). New York: Greenwood Press.

Breyman, S., Campbell, N., Eubanks, V., & Kinchy, A. (2017). STS and social movements: Pasts and futures. In U. Felt, F. Rayvon, C. A. Miller, & L. Smith-Doerr (Eds.), *The handbook of science and technology studies* (4th ed., pp. 289–317). Cambridge, MA: MIT Press.

Callon, M. (1990). Techno-economic networks and irreversibility. *The Sociological Review, 38*(1), 132–161. https://doi.org/10.1111/j.1467-954X.1990.tb03351.x.

Carter, L. (2008). Globalisation and science education: The implications for science in the new economy. *Journal of Research in Science Teaching, 45*(5), 617–633. https://doi.org/10.1002/tea.20189.

Carter, L., & Smith, C. (2003). Revisioning science education from a science studies and futures perspective. *Journal of Futures Studies, 7*(4), 45–54.

Carter, L., Castano, C., & Jones, M. (2018). Sociopolitical activism and transformative learning: Expanding the discourse about what counts in science education. In L. Bryan & K. Tobin (Eds.), *13 questions: Reframing education's conversation: Science* (pp. 437–451). New York: Peter Lang.

Carter, L., Castano Rodriguez, C., & Martin, J. (2019). Embedding ethics of care into primary science pedagogy: Reflections on our criticality. In J. Bazzul & C. Siri (Eds.), *Critical voices in science education research: Narratives of hope and struggle* (pp. 59–72). Dordrecht: Springer.

Castano, C. (2012). Fostering compassionate attitudes and the amelioration of aggression through a science class. *Journal of Research in Science Teaching, 49*(8), 961–986.

Clough, M. P. (2018). Teaching and learning about the nature of science. *Science & Education, 27*(1–2), 1–5. https://doi.org/10.1007/s11191-018-9964-0.

Cranton, P. (2006). *Understanding and promoting transformative learning: A guide for educators of adults* (2nd ed.). San Francisco: Jossey-Bass.

Evans, P. (2012). Counter-hegemonic globalization. In G. Ritzer (Ed.), *The Wiley-Blackwell encyclopedia of globalization* (pp. 1–7). Chichester: Wiley-Blackwell. https://doi.org/10.1002/9780470670590.wbeog114.

Foucault, M. (2008). *The birth of biopolitics: Lectures at the Collége de France, 1978–1979* (M. Senellart, Ed.). New York: Palgrave Macmillan.

Fraser, N. (2008). Abnormal justice. *Critical Inquiry, 34*(3), 393–422. https://doi.org/10.1086/589478.

Freire, P. (1970). *Pedagogy of the oppressed*. New York: Continuum.

Fuller, S. (1993). *Philosophy, rhetoric, and the end of knowledge: The coming of science and technology studies*. Madison: University of Wisconsin Press.

Gilligan, C. (1982). *In a different voice*. Cambridge, MA: Harvard University Press.

Gough, A. (2015). STEM policy and science education: Scientistic curriculum and sociopolitical silences. *Cultural Studies of Science Education, 10*(2), 445–458. https://doi.org/10.1007/s11422-014-9590-3.

Hodson, D. (2008). *Towards scientific literacy: A teacher's guide to the history, philosophy and sociology of science*. Rotterdam: Sense.

Hodson, D. (2011). *Looking to the future: Building a curriculum for social activism*. Rotterdam: Sense. https://doi.org/10.1007/978-94-6091-472-0.

Hoeg, D., & Bencze, L. (2017). Values underpinning STEM education in the USA: An analysis of the Next Generation Science Standards. *Science Education, 101*(2), 278–301. https://doi.org/10.1002/sce.21260.

Intergovernmental Science-Policy Platform on Biodiversity and Ecosystem Services [IPBES]. (2019). *Summary for policymakers of the global assessment report on biodiversity and ecosystem services* (Unedited advance version). Bonn: IPBES.

Jones, M., & Ryan, J. (2014). Learning in the practicum: Engaging pre-service teachers in reflective practice in the online space. *Asia-Pacific Journal of Teacher Education, 42*(2), 132–146. https://doi.org/10.1080/1359866x.2014.892058.

Klein, N. (2017). *No is not enough: Resisting Trump's shock politics and winning the world we need*. Chicago: Haymarket.

Kuhn, T. S. (1970). *The structure of scientific revolutions* (2nd ed.). Chicago: University of Chicago Press.

Latour, B. (1988[1984]). *The Pasteurization of France* (A. Sheridan & J. Law, Trans.). Cambridge, MA: Harvard University Press.

Latour, B. (2005). *Reassembling the social: An introduction to actor-network-theory*. Oxford: Oxford University Press.

Lederman, N. G., & Lederman, J. S. (2014). Research on teaching and learning of nature of science. In N. G. Lederman & S. K. Abell (Eds.), *Handbook of research on science education, Volume II* (pp. 600–620). New York: Routledge. https://doi.org/10.4324/9780203097267.ch30.

Lemke, T. (2011). *Biopolitics: An advanced introduction*. New York: New York University Press. https://doi.org/10.22439/fs.v0i14.3906.

Leonard, A. (2010). *The story of stuff: How our obsession with stuff is trashing the planet, our communities, and our health – and a vision for change*. New York: Free Press.

Levinson, R. (2010). Science education and democratic participation: An uneasy congruence? *Studies in Science Education, 46*(1), 69–119. https://doi.org/10.1080/03057260903562433.

Levinson, R. (2014). Undermining neo-liberal orthodoxies in school science: Telling the story of aluminium. In L. Bencze & S. Alsop (Eds.), *Activist science and technology education* (pp. 381–397). Dordrecht: Springer.

Levinson, R. (2018). Realizing the school science curriculum. *The Curriculum Journal, 29*(4), 522–537. https://doi.org/10.1080/09585176.2018.1504314.

Loving, C. C. (1991). The scientific theory profile: A philosophy of science model for science teachers. *Journal of Research in Science Teaching, 28*(9), 823–838. https://doi.org/10.1002/tea.3660280908.

Martin, J., & Carter, L. (2015). Preservice teacher agency concerning education for sustainability (EfS): A discursive psychological approach. *Journal of Research in Science Teaching, 52*(4), 560–573. https://doi.org/10.1002/tea.21217.

Matthews, M. R. (2017). Feng Shui: Educational responsibilities and opportunities. In M. R. Matthews (Ed.), *History, philosophy and science teaching* (pp. 3–41). Dordrecht: Springer. https://doi.org/10.1007/978-3-319-62616-1_1.

McCarthy, C., Noll, B. D., Weaver, S. L., Kovaks, K., & Wagner, F. (2018). The burden of unburdening: Administrative law of deregulation. *Environmental Law Reporter, 48*(9), 10767–10779.

McMurtry, J. (2013). *The cancer stage of capitalism: From crisis to cure*. London: Pluto. https://doi.org/10.2307/j.ctt183p2k8.

Ministry of Education [MoE]. (2008). *The Ontario curriculum, grades 9 and 10: Science*. Toronto: Queen's Printer for Ontario.

Mirowski, P. (2011). *Science-mart: Privatizing American science*. Cambridge, MA: Harvard University Press. https://doi.org/10.4159/harvard.9780674061132.

Moore, J. W. (2016). *Anthropocene or capitalocene?: Nature, history, and the crisis of capitalism*. Oakland, CA: PM Press.

Noddings, N. (1992). *The challenge to care in schools: An alternative approach to education*. New York: Teachers College Press.

Oreskes, N., & Conway, E. (2010). *Merchants of doubt*. London: Bloomsbury Press.

Osborne, R., & Wittrock, M. (1985). The generative learning model and its implications for science education. *Studies in Science Education, 12*, 59–87. https://doi.org/10.1080/03057268508559923.

Osborne, J., Collins, S., Ratcliffe, M., Millar, R., & Duschl, R. (2003). What "ideas-about-science" should be taught in school science?: A Delphi study of the expert community. *Journal of Research in Science Teaching, 40*(7), 692–720. https://doi.org/10.1002/tea.10105.

Oxfam. (2019). *Pubic good or private wealth?* Oxford: Oxfam International.

Pedretti, E., & Nazir, J. (2011). Currents in STSE education: Mapping a complex field, 40 years on. *Science Education, 95*(4), 601–626. https://doi.org/10.1002/sce.20435.

Pelinka, A. (2013). Right wing populism: Concept and typology. In R. Wodak, B. Mral, & M. Khosravi Nik (Eds.), *Right-wing populism in Europe politics and discourse* (pp. 3–22). London: Bloomsbury. https://doi.org/10.5040/9781472544940.ch-001.

Pierce, C. (2013). *Education in the age of biocapitalism: Optimizing educational life for a flat world*. New York: Palgrave Macmillan. https://doi.org/10.1057/9781137027832.

Piketty, T. (2014). *Capital in the twenty-first century*. Cambridge, MA: Cambridge University Press. https://doi.org/10.4159/9780674982918.

Roberts, D. A. (2007). Scientific literacy/science literacy. In S. K. Abell & N. G. Lederman (Eds.), *Handbook of research on science education* (pp. 729–780). Mahwah: Lawrence Erlbaum Associates.

Romano, N. (2018). Aesthetic wit(h)nessing and the political ethics of care: Generating solidarity and trust in pedagogical encounters. Paper presented at the annual conference of the *American Education Research Association*, New York, April 13–17, 2018.

Roth, W.-M. (2001). Learning science through technological design. *Journal of Research in Science Teaching, 38*(7), 768–790. https://doi.org/10.1002/tea.1031.

Roudometof, V. (2016). *Glocalization: A critical introduction*. London: Routledge. https://doi.org/10.4324/9781315858296.

Schwartz, R. S., Lederman, N. G., & Crawford, B. A. (2004). Developing views of nature of science in an authentic context: An explicit approach to bridging the gap between nature of science and scientific inquiry. *Science Education, 88*(4), 610–645. https://doi.org/10.1002/sce.10128.

Sismondo, S. (2008). Science and technology studies and an engaged program. In E. J. Hackett, O. Amsterdamska, M. Lynch, & J. Wajcman (Eds.), *The handbook of science and technology studies* (3rd ed., pp. 13–31). Cambridge, MA: The MIT Press. https://doi.org/10.4324/9781315640051-72.

Sjöström, J., Frerichs, N., Zin, V. G., & Eilks, I. (2017). Use of the concept of Bildung in the international science education literature, its potential, and implications for teaching and learning. *Studies in Science Education, 53*(2), 165–192. https://doi.org/10.1080/03057267.2017.1384649.

Smith, C., & Dawborn, K. (2011). *Permaculture pioneers: Stories from the new frontier*. Daylesford: Holmgren Design Service.

Springer, S., Birch, K., & MacLeavy, J. (Eds.). (2016). *The handbook of neoliberalism*. New York: Routledge. https://doi.org/10.4324/9781315730660.

Steffen, W., Rockström, J., Richardson, K., Lenton, T. M., Folke, C., Liverman, D., Summerhayes, C. P., Barnosky, A. D., Cornell, S. E., Crucifix, M., Donges, J. F., Fetzer, I., Lade, S. J., Scheffer, M., Winkelmann, R., & Schellnhuber, H. J. (2018). Trajectories of the earth system in the Anthropocene. *Proceedings of the National Academy of Sciences of the United States of America, 115*(33), 8252–8259. https://doi.org/10.1073/pnas.1810141115.

Wenger, E. (1998). *Communities of practice*. Cambridge, UK: Cambridge University Press.

Zeidler, D. L. (2014). Socioscientific issues as a curriculum emphasis: Theory, research and practice. In N. G. Lederman & S. K. Abell (Eds.), *Handbook of research on science education, Volume II* (pp. 697–726). New York: Routledge. https://doi.org/10.4324/9780203097267.ch34.

Zeidler, D. L. (2016). STEM education: A deficit framework for the twenty first century? A sociocultural socioscientific response. *Cultural Studies of Science Education, 11*(1), 11–26. https://doi.org/10.1007/s11422-014-9578-z.

Ziman, J. (1984). *An introduction to science studies: The philosophical and social aspects of science and technology*. Cambridge, UK: CUP. https://doi.org/10.1017/cbo9780511608360.

Zuboff, S. (2019). *The age of surveillance capitalism: The fight for a human future at the new frontier of power*. New York: PublicAffairs.

Chapter 5
Political Entanglement and the Changing Nature of Science

Jesse Bazzul

5.1 Beginning with a Reminder...

I would like to begin with a reminder that the research leading to the current literature on the 'nature of science' is by no means a neutral field of scholarship in science education (more specifically US-based science education). It has a detailed history, and can be situated within multiple historical, political, and social contexts that intimately shape the 'nature' of educational research (Alsop and Gardner 2017). This means that an educator's orientation to science and its entanglements (a word intended to provoke your imagination) is largely sociohistorical, and by extension political. In other words, it is very much dependent on the collective narratives, material circumstances, disciplinary/ideological landscapes, and relationships that surround science knowledges and practices. In North America for example, the relevance and importance of science and science education is intricately linked to the production of human capital (Hoeg and Bencze 2017; Bazzul 2017c; Carter 2017; Pierce 2012), and the perceived need for competitive industrialized states to promote research innovation (Ziman 2000; Mirowski 2011). Given this context, a relevant critical question is how state-approved science education works to constitute relationships to others and the world. Fundamentally challenging the very nature of the nature of science is a vital activity for critical science educators, who see science and education as crucial sites of sociopolitical engagement.

This chapter advocates that education about the nature of science include a politics of knowledge by recognizing that science is inextricably entangled with different disciplines, ways of knowing, and sociohistorical and political contexts. In times of ecological crisis engaging transdisciplinarity and politics as crucial parts of

J. Bazzul (✉)
Department of Education, Maynooth University, Maynooth, Ireland
e-mail: jesse.bazzul@mu.ie

H. A. Yacoubian, L. Hansson (eds.), *Nature of Science for Social Justice*,
Science: Philosophy, History and Education,
https://doi.org/10.1007/978-3-030-47260-3_5

science is increasingly becoming an ethical imperative for multispecies[1] survival. I will develop this premise by first discussing the need to open up the discipline of science education, which includes the nature of science, to other ways of knowing. Then I will discuss how science is entangled in a politics of knowledge due to its role in modern governance and the exercising of (bio)power. Overall, I take a basic critical stance to the nature of science by viewing it as an area of inquiry into what science entails, and could possibly entail, in all of its connections and embodiments historically, politically, epistemologically, ontologically, socially, culturally, ecologically, aesthetically, as well as its transdisciplinarity. A general critical stance toward the nature of science and science education writ large, however, only partially addresses the larger and more urgent problem regarding science, education and the creation of socially and ecological just futures. Along with this critical stance a pressing concern for educators must include how communities of humans and their 'more-than-human'[2] kin, might create and preserve communal ways of living together on a damaged planet. It seems more and more relevant for science educators today to move away from nature of science as merely a distillation of what is most characteristic of science, and the idea that a handful of educational researchers, or research paradigms, can definitively outline what the fields of philosophy, history of science, and science studies combined could not – science's nature. The nature of science must be *left open* to multiple forms of ongoing thought and inquiry. Think about it for a minute: does it make sense that one of the most consequential cultural developments of human history could be distilled so easily? It is not reasonable to also think that science, art, music, history, or economy change substantially over time?

My point is not just that maintaining a *boxed-in* view of science is unrealistic, but also that such a detached and discrete perspective of science is not a good way to orient scientific knowledge and practice during ecologically precarious times (COVID-19 being an example most people could identify with) when transdisciplinarity is becoming necessary for survival. It is not that looking at what makes Modern Western Science different than other ways of knowing is useless. Rather, it is to say that finding new ways to live together in an age of rapid climate change, extinctions, and the growing distance between the Global North and South[3] means recasting science as co-extensive with diverse ecologies, culture, history, and politics. The relationships of science to everything else have always been messy (Harding 1998), and will continue to get messier as human activities such as the burning of fossil fuels and the unchecked killing of the planet's species continues to

[1] Multispecies is an inclusive term that simply means the consideration of more than one species.

[2] The more-than-human is a fluid term that refers to entities other than human in ways that abolish a hierarchical order (plants, rocks, insects, humans, soil, crustaceans, etc.). In my view, the 'more-than' aspect is a reminder that inherent to abolishing this hierarchy is a radical relationality that sees any human meaning, value, or self-understanding as fundamentally dependent on relationships where humans are just one part.

[3] The Global South and North should be understood as a sociopolitical divide along the lines of exclusions, colonialisms, racisms, and socio-economic disparity.

lead to an irreversible geological age some scientists and scholars are now calling the Anthropocene[4] (Lewis and Maslin 2015). It may now be time for science educators to question the idea of always orienting to the nature of science through mainstream educational literature. Should it not also be 'mainstream' to consider scholarship across the academy in the humanities, sciences, arts, and social sciences? If the current nature of science literature purposely ignores a wide range of contexts and literatures, it risks being painfully out of touch.

To summarize, for the nature of science to be relevant to our planet's current ecological and social moment educators should keep in mind that:

1. Science involves ever changing sets of practices and knowledges that are fully entangled with other ways of knowing, histories, social contexts, economies, and ecological relationships.
2. Characterizations of science as separate from other ways of knowing, while useful in highlighting differences, may hinder the search for different ecologically and socially just ways of living on a rapidly deteriorating planet.
3. In order for science and science education to be relevant to surviving mass extinctions, climate change, and widening poverty a more radical view of science's ethico-political potential(s) must emerge.

The growing antiquation of current NOS research, policy, and instructional approaches related to nature of science, is not because many of these paradigms originated in the 1990's, but because they do not take seriously enough the pressing wicked ethical, social, and existential problems of the twenty-first century (Carter 2011). The nature of science will seem more relevant to educators and students when it facilitates collective ways of living with the beings that share our planet. Indeed, this entire volume by Hagop Yacoubian and Lena Hansson (2020) is a testament to the idea that nature of science scholarship is most relevant when it is employed as/for justice education (see also Yacoubian 2015).

To summarize, the nature of science as it has already been articulated in the literature has a history, however due to the fact that we live in times of ecological crisis the nature of science needs to be expanded to allow for a politics of knowledge and a serious and multifaceted engagement with transdisciplinarity.

5.2 Keeping the Nature of Science Discussions Going

I'd like to begin this chapter by building on a pluralistic notion of nature of science based on what long standing thinkers on the subject were already saying in a special issue of the *Canadian Journal of Science, Mathematics, and Technology Education*. The issue was organized around Derek Hodson and Siu Ling Wong's (2017) critique

[4]The Anthropocene is a label used to describe our current geological moment where human activity will come to mark the geological and biological course of the planet for millions of years.

of the 'consensus view' of nature of science, and included six responses to their critique (Bazzul 2017a).[5] I view this special issue's value as a movement toward plurality in relation to science and its nature, and in this way one of many points of departure toward a more open and multiplicitous science.

It would be helpful to present some of the key arguments about the nature of science arising from the special issue (as I see them of course). First, while some might argue otherwise, the 'consensus' view of nature of science – the idea that there is more or less a final agreement about what science entails, how it is done, or is/isn't' – can no longer boast a relevant consensus (Allchin 2017). Second, any description of science and its nature is sociohistorically, politically and materially situated. In this vein, Steve Alsop and Sam Gardner (2017) argue that educators, scholars, and scientists leave room for engaging *the particular*, or those aspects of science that don't fit neatly into boxes. One reason being is that any consensus is historically situated – or more figuratively, one tracing of a dynamic map. Another way to attend to the particular in science and science education is to focus on the intricacies of specific local entanglements and the ethical and political possibilities that emerge. Third, both large and small scale scientific enterprise literally function as active transdisciplinary entanglements, for example in the geopolitical and technology-rich agricultural sciences, such that there is no nature of science distinct from science-in-the-making (Simonneaux 2017). Fourth, science circulates multiple narratives generated by a wide range of communities that draw from many different ways of thinking and seeing the world. This includes those important narratives that describe the overall purposes of science – narratives that are most often not conceived or decided upon by scientists. One of the most important (and 'non-postmodern') points of Jean Francois Lyotard's *The Postmodern Condition* is that scientific knowledges have powerful legitimating processes nonetheless that cannot legitimate the narratives 'about science'. No matter what, science's social, political, cultural importance, and ethos, is decided elsewhere. Fifth, any ethos or set of practices related to science are always coextensive with power relations (Bazzul 2014; Dagher and Erduran 2017). It is this point in particular that this chapter argues should be made explicit when constructing or learning about science's nature. Sixth, ways of knowing and understanding phenomena that are not deemed 'scientific' by rigid disciplinary boundaries are still absolutely fundamental to science practices and knowledge production (Berkovitz 2017). And for educators interested in the relationship between science,culture, politics, spirituality, ethics and history, this is a pretty big understatement. From this vantage point we can see the consensus view of nature of science (see also Lederman et al. 2002), along with the moves toward pluralism outlined in the paragraph above as constellation points in an evolving network of practice. It is now time for another paradigm shift for science that the field of education may be well positioned to help bring about. Science educators, as part of a larger community of caring and ethical beings, need to engage

[5] Fouad Abd-el-Khalick (2012) articulates support for this consensus, and in a way that highlights the tensions that arise when this consensus is potentially undermined.

and respond to ecologically and socially urgent contexts and imperatives in the times called the Anthropocene.

Science education might realize its potential for social and environmental transformation through philosopher Jacques Rancière's (Rancière and Corcoran 2010) theorization of politics, which involves *dissensus* in the name of equality (equality not as a simplistic excuse for ignoring difference, but as a radical democratic principle). The basic idea of Rancière's politics is that the dominant social order casts the world in ways that exclude the marginalized, which includes the more-than-human (or nonhuman). Politics is a process of disruption of what is sensible/thinkable/doable in order to realize new forms of equality. While science education can cultivate a disinterest in politics, certainly it has done this, it can alternatively be part of forms of dissensus that help make those who have been marginalized and oppressed come to count equally. Since the narratives that guide and inform science education are fluid and not decided beforehand, there is no fundamental reason why science education cannot be a force for justice in the name of equality (Tolbert and Bazzul 2017).

What makes political dissensus an interesting addition to nature of science discussions is its opposition to consensus. It may well be that vigorous debates about the nature of science become a vital part of political struggles to combat climate change, environmental racisms, destructive resource extraction, extinctions and widening social inequality. This makes sense because science has already come to play a large role in both the emergence and mitigation of these phenomena. While consensus building has an important place in the normative functioning of institutions, knowledge production, and everyday cultural life, science educators might also ask what a particular consensus excludes. What relationships, interests, or beings are not being counted equally in the social/natural world? Considering much of the planet's wildlife is rapidly disappearing, and humans are exploiting all life for the benefit of a small few ('the 1%'), can scientists and science educators ethically *not* take a strong political stance? Can science educators afford not to expand the purview, powers and reach of science (and its nature)? Accepting entanglement means there is no logical sequence for the work of multispecies justice, precisely because the work emerges in relationships. An educational community can begin by looking at what is important to their own community, make connections, and look outward. It should not look the same each time. The uses, practices, and implications of science, whilst having historical, practical, technical and contextual similarities will also contain differences, which is a major implication of Douglas Allchin's (2012) historical work.

To summarize the main points in this introduction, while some science education scholars have already tried to expand understandings of the nature of science to include considerations of broad sociopolitical contexts, I suggest that a more concentrated engagement with plurality and politics is imperative for multi-species flourishing and survival. In the next section I take a closer look at politics, power, and dissensus as contexts for science, science education and the nature of science.

5.3 Science Education, Modern Governance, and a Case for Political Dissensus

> There is no model of truth that does not refer back to a kind of power, or no knowledge or even science that does not express or imply, in an act, power that is to be exerted. All knowledge runs from a visible element to an articulable one, and vice versa; yet there is no such thing as a common totalizing form, not even a conformity, or bi-univocal correspondence. (Deleuze, in *Foucault* 1988 p. 39).

Critics of continental philosophy and staunch scientific realists may not like the epigraph to this section. Afterall, it was not written by a scientist or science educator! However, if educators cannot listen to a multiplicity of voices, I am not sure what hope there is for the future. Educational communities who want to work toward environmental and social justice should consider that the nature of science is completely coextensive with politics (exactly how should remain a contested terrain of argument), which means that science needs to be conceived and practiced as a politicized activity involving an intricately entangled transdisciplinary set of practices. While many science educators might argue that science and politics have connections, and that other knowledges are helpful to scientific practice, much of the literature and discourses of science education may not put these in relation to the current sociopolitical and ecological moment in which we are currently living.

One reason for this is that the education of scientists, and science educators for that matter, often does not include an examination of how science practices, institutions, and knowledges are necessary for the governance of both human and more-than-human bodies. However, both science and education are *biopolitical*[6] sites of engagement – where dissensus and disagreement become ways to engage distributions of power and technologies of social control. Engaging politically through science education involves aesthetic shifts – allowing something different to be sensed, seen or done. These shifts necessitate a focus on difference and relationality, and this includes the affective-political dimensions of science education (Kayumova and Tippins 2016), which are key to finding new ethical forms of communal living.

A more relational and politically relevant view of the science will involve letting go of the need for absolute descriptions of science, or what science educators like Maria Wallace (2018), following Gilles Deleuze (1994), call a *majority*[7] view of science: overarching abstractions that mask complex relationships between

[6] In short, biopolitics refers to the way modern forms of governance constitute, the conduct, values, and social life of human beings (and their more-than-human kin). Science discourses, which purport a certain amount of objectivity, are power-infused discourses that accompany the exercising of power in modernity (Bazzul 2014).

[7] Although there is not enough space to give a thorough explanation of Deleuze's view of majority/minority, majority views think with concepts already outlined while minority views attempt to make a flight away from these categories. Science, to some extent, struggles with a majority/minority dilemma as scientists attempt to work within paradigms of already accepted concepts, whilst simultaneously attempting to develop knowledges and practices that break free from accepted ways of understanding the world.

phenomena (e.g. between humans and the more-than-human). 'Majority' views are perhaps unavoidable because they serve multiple functions, but they are always incomplete. Therefore alongside majority views of science, educators who care about seeking and nurturing complex relationships with the world must also seek what might be called a *minority* view: an escape from being locked into rigid views of the world and how to engage it. Science educators need space to dream differently in-community with our more-than-human kin (Bazzul et al. 2018). This will involve not just an epistemological shift, but an ontological one. To pursue science's ecological and social entanglements, educators must also seek to *become-minor* as far they are open to complex ecological, technical, historical, social, political, ethical, and relational context they (or others) may not yet understand very well. A minority view finds itself in the *middle of something*, it does not claim to be the ending or a superior view of our shared world.

The epigraph to this chapter suggests that there are always multiple capillaries of power running through relationships of knowledge and the social order. A basic lesson of philosopher Michel Foucault's work is that objective knowledge – of which science is a huge part – is co-extensive and entwined with power relations and modern governance. If this is true, then science's integral role in governance and disciplinary power must be engaged by a wide variety of educators, scientists, and community groups. In order to understand how science and the exercising of power are co-extensive educators need to engage social theory; and Foucault's work is an effective theoretical guide to the functioning of power. One important aspect of power is that it is not just exercised 'from above' through institutions, capital, or the interests of a ruling class. It is also exercised 'from below' at the level of conduct and bodies (Foucault 1978). Think about this: how does scientific knowledge bring people to understand themselves as healthy individuals, a biological species, or a labouring subject in an economy? Modern science is not only linked to economic interests, but helps form the 'grid of intelligibility' by which people make sense of their lives. To put it in aesthetic terms, scientific knowledge helps cast what is visible and sensible, and in this way helps keep particular configurations of power, for better or for worse, in place. However, power relationships are not straightforward. It is not just a matter of determining ideological or economic agendas within a scientific project, although this is a worthwhile endeavour. An important question for science educators to ask is: What do our pedagogies allow science students and teachers to see and do? What does it enable, and how might it be disabling? If seeing and doing things differently for multispecies survival is an obvious ethical imperative for the Anthropocene, science educators will increasingly find themselves engaging contexts of political disagreement and dissensus.

There is a tendency for science educators to get hung up on epistemology. Indeed part of the 'science wars' of the 1990's consisted of 'battles' over epistemology and the limits of science. These debates pitted 'scientific realists' against 'postmodernists' in ways that oversimplified both perspectives. While a range of views exists about how much science is influenced by power, ideology, and technologies of governance, etc., it is undeniable that relationships between knowledge and power exist. These relationships do not necessarily preclude any specific epistemology, no matter what people on either side of the ‘science wars’ may say. Maintaining that there is a relationship between knowledge and power, does not mean things cannot be ‘true’. Foucault (2003) is clear on this: ‘truth’, regardless of whether something is *actually* true or not, has a relation to power. This view is quite compatible with many epistemological positions, including realisms (e.g. critical realisms, scientific realisms), which are perhaps the most appealing epistemic outlooks for science educators. Exploration of science’s political entanglements is not any sort of assessment of science’s ‘veracity’, or the ‘scientificity’ of its claims. Again, this is why different descriptions and deployments of science will involve many ‘minor inquiries’, rather than one overarching truth about knowledge, science, politics and the social world.

Science’s entanglement with politics, a vital part of its nature in precarious times, may be thought of as the exercising of power over life, what many critical scholars refer to as biopolitics. Biopolitics is wide-ranging term used to characterize the exercising of (bio)power, and also resistance to this (bio)power (Hardt and Negri 2000), in the lives of both human and non-human life, however I use the term here mostly as it relates to human bodies (see Thomas Lemke’s (2011) comprehensive overview of biopolitics).[8] It is helpful to think about power relations using Michel Foucault’s (1982) work on subjectivity and governmentality, where knowledge/science, along with the various techniques of subjectification (the constituting of individuals as governed subjects), circulate the effects of power. Essentially, power flows along multiple capillaries and this directionality is always two fold. Like the flow of electricity, the exercising of power always meets with resistance. If there is no resistance, there is no exercising of power. This means a situation of absolute domination and docility, political or otherwise, ceases to be a power relationship. The exercising of freedom and ethical reflexivity naturally becomes very important within this conception of power, especially when thinking about how education brings students into spaces where they must consider how to lead ethical lives (Bazzul 2017b).

Scientific knowledges and discourses, including discourses of science education, carry the effects of this power to a greater extent than other discourses because they purport to objectivite. In the case of science education this objectivity is doubled, as the very fact that state institutions have sanctioned curriculum resources and policy also adds an objective quality. The exercising of power, is used to constitute subjectivity by working to ‘conduct the conduct’ of individuals; however this power is both ‘accepted’ and resisted. Therefore power works to actively constitute the subjectivities of individuals ‘from above’, and individuals collectively function to constitute the institutions, organizations and strategies of power ‘from below’. Figure 5.1 is a simple diagram that helps illustrate these basic attributes of power. The diagram is not meant to define or represent power, but to provoke educators to think about it.

[8] Biopolitics is also a useful frame for understanding modern science and ethics (Bazzul 2015a), as well as the production of labour and science education (Bazzul 2017c).

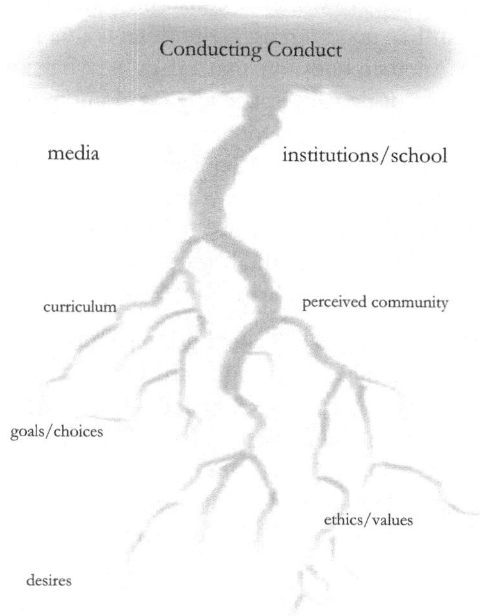

Fig. 5.1 Illustration of power based on the work of Michel Foucault (1978, 1982)

So how is thinking about power important to science education? If we take a slightly broader perspective, all ecologically dangerous and socially unjust conditions, from colonialisms to capitalist economic systems, involve the widespread exercising of (bio)power, which is subsequently met with (bio)political resistance from below. Sandra Harding's (2008) book, *Sciences from Below: Feminisms, Colonialities, and Modernities* documents some of the ways activists in India have helped shaped agriculture science and science policy, as well as how citizen groups have altered the direction of medical research (e.g. autism). Science research and science education are both entangled with the exercising of biopower and the (bio) political resistance to this power. The way science-related ethical issues are introduced and taken up in science classes, can be attached to interests that forward a particular agenda or configuration of power such as neoliberalisms and/or 'naturally' associated a scientific mindset with the Global North (see Bazzul 2015a). Modern governance takes both the population and the individual as objects of intervention in terms of 'who to let live' and 'who to let die' (Bazzul 2014; Foucault 1978). Historically, the co-development of the biological concepts of a 'milieu' and population genetics with practices of modern governance, from census statistics to eugenic programs, demonstrates a parallel story of the co-emergence of scientific knowledges and forms of social control (Rose 2009; Murphy 2017; Canguilhem 2001; Pierce 2012). The interests of the state and corporations, more and more eclipse the needs of communities. Science needs to play a role in larger cultural, geological, ecological, and political 'plays'; therefore imagination is absolutely necessary. How science constitutes students and teachers as 'subjects of knowledge' is still an underrepresented research area in science education (Blades 1997).

Another question that arises is what kinds of scientific discourses carry the effects of power? Practically speaking, it is helpful to look to those discourses that attempt to intervene in how individuals come to understand their own conduct, values and relationships. Science discourses that constitute the activities, purposes and ethical considerations of life, such as those of biology, tend to lend themselves to social, political, and institutional prerogatives. However, educators should also examine sciences like physics and geology, because they orient humans to things like resources and land. Other helpful considerations for educators include how:

- the biological sciences bring people to see their bodies and the bodies of others (as racialized or genetically disposed); and how these bodies should be organized and cared for, which includes how people come to work on ourselves.
- science raises questions of collective existence, for example the appropriateness of technologies, resource development, ecological relations between various populations (human and otherwise).
- science knowledges and discourses help delineate 'how' one might think and act rationally, and ethically. What does it mean to be a science person? Says who? What else might be considered in this view?

The point is precisely for educators to discuss how science is deeply interwoven with governance and politics. This is because engagement with political entanglement, and not a blind faith in technology, is vital for our survival. This aspect of science must be seen as vital to its nature. The warrant for science education should be ecological and social justice for survival. Ajay Sharma (2012) argues as much when he argues to position Climate Change, which is a political issue as well as an ecological issue, as the central guiding phenomenon of science education. In light of this pressing new warrant for science education it would be unethical to ignore the ecological-political landscape affecting communities globally. Science educators and students need to *get real*. Yet, justice oriented scholarship in science education is often 'domesticated' or co-opted for mainstream science education, by clambering to the self-interested directives of governments and corporations (Tolbert et al. 2018).

To summarize, science is integral to modern forms of political power and governance. Due to the relationships between knowledge, power and institutions, science education that endeavours to teach for justice must not ignore this aspect of science's nature. Not only do truth discourses exercise power (independent of their actual truth value), schools that disseminate such discourses work to shape the conduct of subjects. Science education, due to its specific relationship to truth discourses, both official discourses of government and science knowledge, must pay extra attention to its own political entanglement.

5.4 Science Education and Political Entanglement

In my own teaching the Anthropocene, along with its more descriptive alternatives such as the capitalocene (Moore 2017) and the chthulucene (Haraway 2016), has become central to why students should learn about science in the first place. The Anthropocene is one potential context through which ways of knowing the world become instantaneously connected (Moore 2017); and make no mistake; facing the Anthropocene is an ethical-political task. This task should be seen as less of a choice, and more a necessity. The Anthropocene is a time when modern binaries of cultural/natural, geologic/geographic, political/scientific, and human/more-than-human are quickly dissolving. Big agriculture's razing of large forests, European colonialism and the resulting displacements and genocides, industrial-military nuclear testing, and global capitalism's disregard for life have permanently marked the geological record for millions of years to come.

It is therefore unethical for science educators and researchers to somehow render the sociopolitical/geological/environmental as secondary or tertiary aspects of the nature of scientific endeavour. Jay Lemke (2011) maintains that science education is distinctly conservative in character, and that not engaging issues of pressing concern for students is a form of *collaborationism*. Simply put, a liberal frame might be inadequate for mediating our current position (Lloro-bidart 2015). Denying the sociopolitical realities that form the backdrop of science, and vice-versa, is becoming extremely difficult to maintain. According to physicist and philosopher Karen Barad (2007), 'the scientific' must now contend with the fact that it is always already sociohistorical; and likewise the 'sociohistorical' must contend with the fact that is always already biological and geological. The Anthropocene sets the stage for exploring an entangled science, one that is more in-tune with ecological necessity. If science knowledges are integral to the exercising of (bio)power, this means that science and science education are also sites of political engagement. This is where an openness to multiplicity and the freedom to dream become vitally important.

A more radical way of looking at politics can be found in Jacques Rancière's (Rancière and Corcoran 2010) vision of politics as dissensus, disruption of the status quo, the domain of what is sensible and thinkable, in the name of equality (as outlined briefly above). Elsewhere, I have suggested that science education communities align their pedagogical approaches with social movements, such as the Indigenous grassroots movement "Idle No More", in order to commit science education to bringing about more ecologically and socially just futures (Bazzul 2015b). Of course, engaging social movements in the classroom can be difficult and sometimes risky.[9] It is much safer to engage more traditional citizenship education and non-controversial ethical issues related to science. However, traditional forms of

[9] For example, teacher colleagues in Saskatoon, Canada created a disturbance when they took their eco-justice students to support and an Indigenous rights sit-in (MacPherson 2016). There was also many that showed solidarity for the teachers, including many members of our faculty of education at the University of Regina (see open letter)

citizenship education, and 'safe-to-discuss' ethical issues very often tacitly assume a particular social, political, and historical backdrop that remains unchallenged. Citizenship education that embraces Rancière's vision of dissensus means teaching less content and more about the necessity of dissent; not for its own sake (who would want to dissent for no good reason?) but where the equality of communities, including nonhuman life, is concerned. Educational theorist Gert Biesta (2011) argues for this form of citizenship education in his figure of the *Ignorant Citizen*. This form of citizen knows less about what the state would prefer them to do (or be) as citizens, and more about disruptive actions.

Instilling an element of radical democracy also means examining how we as science educators run our own classes. Rancière argues that education for emancipation requires that educators let go of holding their explanations as forms that students must simply reproduce. Such education stultifies the creative powers of students. A basic equality runs through education communities when everyone is given the equal ability to offer a different interpretation, creation, or result (Rancière 1991). Science educators might say: 'Wait a minute! There are some works/knowledges that are more correct or better!'. Yes it is certainly true, in many ways, some works are better than others – the history of literature tells us this much. It is also true that some works fit better with traditional paradigms of established science. However, the epistemic dimensions of knowledge and science deserve much more attention from educators. Following Lyotard's (1984) *The postmodern condition: a report on knowledge*, science has a more or less egalitarian form of knowledge transfer, because there is an assumed equality between the sender and the addressee of knowledge (Bazzul 2013). As science educators we can nurture this radical equality inherent in science communication in our classrooms. At the same time, however, science students and teachers also offer unique work around the narratives, practices, contexts, and knowledges that surround and infuse science – and these are infinite in scope. Science classrooms may therefore be unique in their *dual-capability* to render an equality in communication, both at the level of communication of science among equals who engage in practices of verification and communication; and also in the validation of different situated ideas around the contexts, meanings, applications, and knowledges of science. In summary, equality manifests in science classrooms through a simultaneous recognition of equality in communication and also in the diversity of the speaker in terms of how they are embedded in the entanglements of science and life.

Although criticised for not being 'activist' enough, The US-based "March for Science" demonstration, was a careful stand taken by concerned scientists and members of the US public, along with their global allies, to defund health and climate change research by the Environmental Protection Agency and the National Institute of Health (St. Fleur 2017). Some criticized the March for only being a response to a shift in the original status quo (e.g. science is funded and therefore scientists often did not see the need to get involved in issues outside their particular research purview). Would scientists be silent if a sufficient amount of funding and a modicum of relevant policy were made available? Was the professional (largely white male) scientists' unique position as knowledge producers and experts simply being threatened (and they were responding only to a loss of power)? While these questions are worth engaging, there was, however, a significant rupture in how scientists viewed their entanglement with the social and political world. At a basic level, there was a realization that it is the sociopolitical backdrop that determines what science gets done, and more importantly if it gets done at all!

As stated above a dissensus view challenges the taken-for-granted notion of consensus, which, as I suggest, might include 'consensus' around what constitutes science and its nature. This is not to negate consensus building as a key process to the everyday work in which communities engage. Research into socioscientific issues (Sadler 2011; Zeidler and Nichols 2009) is exemplary for classroom teachers and students to learn the difficult work of coming to a community consensus around socioscientific issues – whether these activists create a breach in the domain of the sensible in the name of equality is however another question and would depend on context. This is because consensus building often happens after certain considerations, positions, and actors/actants are already legimated – thereby concluding or bypassing the process of politics of equal inclusion (if we continue to use Rancière's definition). Politics enters the field when occluded groups, their claims, and the basis for these claims attempt to achieve equal consideration by breaking with what was hitherto sensible, visible or thinkable. How are decisions made about science related issues concerning how communities live? How are the ideological and material frames of these questions already decided beforehand? Whose ideas and viewpoints are never considered? How is power exercised through science so as to render certain actions commonsensical? Disagreement achieves politics when it causes a reshaping of what can sensed or thought in the name of equality (Rancière and Corcoran 2010). Case in point, scientific consensus on climate change has not been enough for governments and corporations to take significant action. Simply put, science and science education must disrupt the status quo, and realize its many political entanglements as a fundamental part of its practice.

In summary, science education might align itself with political movements so that it is a force for survival and justice for diverse communities. "Staying neutral" might be a tragically unethical goal for science education. Staying oppressively disengaged will not achieve the kind of ecologically and socially relevant science 99.9% of people in the world desire.

5.5 Conclusion

It may be much more important for science educators to dwell on the precarity of our current global moment that renders all (singular) personal and collective strategies, green consumerism, personal responsibility, pipeline protests, as inadequate responses to the onset of the Anthropocene (Shotwell 2016). In these times, science, and what we think it is (its nature), needs to be part of a revolution in thinking and material organization on planet earth. And like geographer and ecologist Jason Moore (2015, p. 2) warns: "Modes of thought are tenacious: they are no easier to transcend than the "modes of production" they reflect and help shape". Indeed educators will teach and research in increasingly dire times that make older contexts for constructing knowledges and pedagogies seem embarrassingly antiquated to our current moment (can we not see that just with the events of 2020). What the onset of the Anthropocene shows is that radical ontological shifts in our thinking are immanent, both in terms of the ways diverse educators are coming together to challenge categories and knowledges that position our shared worlds as something to be consumed by capitalism and ongoing forms of colonialism.

Nature of science for justice must radically open up what science means, and give a big space for politics in increasingly intricate ways. I consider the following summary points essential to this endeavour.

- Any nature of science research, like any history or philosophy of science research, is socially, culturally, politically, historically, and even materially situated – and therefore open to interpretation and criticism. Always.
- In light of the situated nature of science, a move toward plurality, transdisciplinary and relationality makes more sense in twenty-first century contexts and the need to search for communal life with our more-than-human kin.
- Scientific knowledge is heavily entangled in modern forms of governance of individuals and populations. This manifests in the exercising of power to conduct the conduct of individuals (biopower), and also resistance to this power (biopolitics).
- Citizen roles/activities as usual, while useful in broaching ethical topics, often leave the social fabric, what is thinkable and doable, in place.
- Engaging politics, as forms of dissensus in the name of (multispecies) equality, is a necessary way for science to contribute to just futures and realize its vast entanglement in the world, as part of its nature.
- Science classrooms are very fitting spaces to engage questions of collective existence because of the influence and input science has over an ethics and politics of life itself. Science education's potential for cultivating values of equality may be underestimated.

So where to begin recontextualizing an entangled and pluralistic nature of science? It could simply begin where you are: what is culturally, materially, historically and politically entangled and urgent. Why not? Who says? Overall, it should look messy, with connections leading disagreement and even conflict: but also to exciting and productive questions and projects. Innovative work already underway can be seen in the way nondominant knowledges and communities are beginning to be engaged in science classrooms (Bang et al. 2012). Engaging marginalized communities includes engaging gender diversity and sexuality (Fifield and Letts 2019), Indigenous knowledges and methodologies (Cajete 2000); and communities of colour experiencing long-standing poverty and environmental racism (Morales-Doyle 2017; Mensah 2011; Kayumova et al. 2019). It also means engaging other ways of expansively thinking about our shared world such as arts and aesthetics and science fiction (Gough 2017); or Black and Indigenous futurisms (Okorafor 2018; Dillon 2016). My hope is that educators move in solidarity to bring about the shifts required to deal with collective living in the Anthropocene. Now is the time to expect more from the sciences and from the potentially dynamic field of science education.

References

Abd-El-Khalick, F. (2012). Examining the sources for our understandings about science: Enduring conflations and critical issues in research on nature of science in science education. *International Journal of Science Education, 34*, 353–374.

Allchin, D. (2017). Beyond the consensus view: Whole science. *Canadian Journal of Science, Mathematics and Technology Education, 17*(1), 18–26.

Alsop, S., & Gardner, S. (2017). Opening the black box of NOS: Or knowing how to go on with science education, Wittgenstein, and STS in a precarious world. *Canadian Journal of Science, Mathematics and Technology Education, 17*(1), 27–36.

Bang, M., Warren, B., Rosebery, A. S., & Medin, D. (2012). Desettling expectations in science education. *Human Development, 55*(5–6), 302–318.

Barad, K. (2007). *Meeting the universe halfway: Quantum physics and the entanglement of matter and meaning*. Durham, USA: Duke University Press.

Bazzul, J. (2013). Emancipating subjects in science education: Taking a lesson from Patti lather and Jacques Rancière. *Cultural Studies of Science Education, 8*(1), 245–251.

Bazzul, J. (2014). Science education as a site for biopolitical engagement and the reworking of subjectivities: Theoretical considerations and possibilities for research. In *Activist science and technology education* (pp. 37–53). Dordrecht: Springer.

Bazzul, J. (2015a). Tracing "ethical subjectivities" in science education: How biology textbooks can frame ethico-political choices for students. *Research in Science Education, 45*(1), 23–40.

Bazzul, J. (2015b). Towards a politicized notion of citizenship for science education: Engaging the social through dissensus. *Canadian Journal of Science, Mathematics and Technology Education, 15*(3), 221–233.

Bazzul, J. (2017a). From orthodoxy to plurality in the nature of science (NOS) and science education: A metacommentary. *Canadian Journal of Science, Mathematics and Technology Education, 17*(1), 66–71.

Bazzul, J. (2017b). The 'subject of ethics' and educational research OR ethics or politics? Yes please! *Educational Philosophy and Theory, 49*(10), 995–1005.

Bazzul, J. (2017c). Biopolitics and the 'subject of labor in science education. *Cultural Studies of Science Education, 12*(4), 873–887.

Bazzul, J., Wallace, M. F., & Higgins, M. (2018). Dreaming and immanence: Rejecting the dogmatic image of thought in science education. *Cultural Studies of Science Education*, 1–13.

Berkovitz, J. (2017). Some reflections on "going beyond the consensus view" of the nature of science in K–12 science education. *Canadian Journal of Science, Mathematics and Technology Education, 17*(1), 37–45

Biesta, G. (2011). The ignorant citizen: Mouffe, Rancière, and the subject of democratic education. *Studies in Philosophy and Education, 30*(2), 141–153.

Blades, D. W. (1997). *Procedures of power and curriculum change: Foucault and the quest for possibilities in science education*. New York: Peter Lang.

Cajete, G. (2000). *Native science: Natural Laws of interdependence*. Santa Fe: Clear Light Publishers.

Canguilhem, G. (2001). The living and its milieu. *Grey Room*, 7–31.

Carter, L. (2011). *Gathering in threads in the insensible global world: The wicked problem of globalisation and science education*.

Carter, L. (2017). Neoliberalism and STEM education: Some Australian policy discourse. *Canadian Journal of Science, Mathematics and Technology Education, 17*(4), 247–257.

Dagher, Z. R., & Erduran, S. (2017). Abandoning patchwork approaches to nature of science in science education. *Canadian Journal of Science, Mathematics and Technology Education, 17*(1), 46–52.

Deleuze, G. (1988). *Foucault*. Minneapolis, MN: University of Minnesota Press.

Deleuze, G. (1994). *Difference and repetition*. Columbia University Press.

Dillon, G. L. (2016). *Introduction: Indigenous futurisms, Bimaashi Biidaas Mose, flying and walking towards you.*

Fifield, S., & Letts, W. (2019). Prolegomenon: Queer theories and STEM education. In W. Letts & S. Fifield (Eds.), *STEM of desire* (pp. 3–40). New York: Brill Sense.

Foucault, M. (1978). *The history of sexuality: An introduction (Vol. 1).* (R. Hurley, Trans.). New York: Pantheon.

Foucault, M. (1982). The subject and power. *Critical Inquiry, 8*(4), 777–795.

Foucault, M. (2003). Ethics of a concerned self. In P. Rabinow & N. Rose (Eds.), *The essential Foucault, selections from essential works of Foucault, 1954–1984* (pp. 25–42). New York: New Press.

Gough, N. (2017). Specifying a curriculum for biopolitical critical literacy in science teacher education: Exploring roles for science fiction. *Cultural Studies of Science Education, 12*(4), 769–794.

Haraway, D. J. (2016). *Staying with the trouble: Making kin in the Chthulucene*. Durham, NC: Duke University Press.

Harding, S. G. (1998). *Is science multicultural?: Postcolonialisms, feminisms, and epistemologies.* Indianapolis: Indiana University Press.

Harding, S. (2008). *Sciences from below: Feminisms, postcolonialities, and modernities*. Durham, NC: Duke University Press.

Hardt, M., & Negri, A. (2000). *Empire*. Cambridge, MA: Harvard University Press.

Hodson, D., & Wong, S. L. (2017). Going beyond the consensus view: Broadening and enriching the scope of NOS-oriented curricula. *Canadian Journal of Science, Mathematics and Technology Education, 17*(1), 3–17.

Hoeg, D., & Bencze, L. (2017). Rising against a gathering storm: A biopolitical analysis of citizenship in STEM policy. *Cultural Studies of Science Education, 12*(4), 843–861.

Kayumova, S., & Tippins, D. (2016). Toward re-thinking science education in terms of affective practices: Reflections from the field. *Cultural Studies of Science Education, 11*(3), 567–575. https://doi.org/10.1007/s11422-015-9695-3.

Kayumova, S., McGuire, C. J., & Cardello, S. (2019). From empowerment to response-ability: Rethinking socio-spatial, environmental justice, and nature-culture binaries in the context of STEM education. *Cultural Studies of Science Education, 14*(1), 205–229.

Lederman, N., Abd-El-Khalick, F., Bell, R., & Schwartz, R. (2002). View of nature of science questionnaire: Toward valid and meaningful assessment of learners' conceptions of nature of science. *Journal of Research in Science Teaching, 39*, 497–521.

Lemke, J. (2011). The secret identity of science education: Masculine and politically conservative? *Cultural Studies of Science Education, 6*(2), 287–292.

Lemke, T., Casper, M. J., & Moore, L. J. (2011). *Biopolitics: An advanced introduction*. New York: NYU Press.

Lewis, S., & Maslin, M. (2015, March). Defining the Anthropocene. *Nature, 519*, 171–180. https://doi.org/10.1038/nature14258.

Lloro-Bidart, T. (2015). A political ecology of education in/for the Anthropocene. *Environment and Society, 9*(1), 128–148.

Lyotard, J. F. (1984). *The postmodern condition: A report on knowledge* (Vol. 10). Minneapolis, MN: University of Minnesota Press.

MacPherson, A. (2016, October 31). Student participation in anti-pipeline rally a 'significant concern': Greater Saskatoon Catholic Schools. *Saskatoon Star Phoenix*. Retrieved From: http://thestarphoenix.com/news/local-news/student-participation-in-anti-pipeline-rally

Mensah, F. M. (2011). A case for culturally relevant teaching in science education and lessons learned for teacher education. *The Journal of Negro Education*, 296–309.
Mirowski, P. (2011). *Science-mart*. Harvard University Press.
Moore, J. W. (2015). *Capitalism in the Web of Life: Ecology and the Accumulation of Capital*. New York: Verso Books.
Moore, J. W. (2017). The capitalocene, part I: On the nature and origins of our ecological crisis. *The Journal of Peasant Studies, 44*(3), 594–630.
Morales-Doyle, D. (2017). Justice-centered science pedagogy: A catalyst for academic achievement and social transformation. *Science Education, 101*(6), 1034–1060.
Murphy, M. (2017). *The economization of life*. Durham, NC: Duke University Press.
Okorafor, N. (2018). *Forthcoming. Binti: The complete trilogy*. New York: Penguin Random House.
Pierce, C. (2012). Education in the age of biocapitalism: Optimizing educational life for a flat world. In *New York*. Palgrave: Macmillan.
Rancière, J. (1991). *The ignorant schoolmaster: Five lessons in intellectual emancipation*. Stanford, CA: Stanford University Press.
Rancière, J., & Corcoran, S. (2010). *Dissensus: On politics and aesthetics*. London: Continuum.
Rose, N. (2009). *The politics of life itself: Biomedicine, power, and subjectivity in the twenty-first century*. Princeton University Press.
Sadler, T. (2011). *Socio-scientific issues in the classroom*. Dordrecht: Springer.
Sharma, A. (2012). Global climate change: What has science education got to do with it? *Science & Education, 21*(1), 33–53.
Shotwell, A. (2016). *Against purity: Living ethically in compromised times*. Minneapolis, MN: University of Minnesota Press.
Simonneaux, L. (2017). Au-delà de la polémique, compléter l'approche macro consensuelle de la NOS avec l'approche micro de la recherche en train de se faire [Beyond the controversy, complementing the consensus-based macro approach of NOS with a micro approach based on research currently underway]. *Canadian Journal of Science, Mathematics and Technology Education, 17*(1), 58–65.
St. Fleur, N. (2017, April 22). Scientists, feeling under Siege, March against Trump policies. Retrieved From: https://www.nytimes.com/2017/04/22/science/march-for-science.html
Tolbert, S., & Bazzul, J. (2017). Toward the sociopolitical in science education. *Cultural Studies of Science Education, 12*(2), 321–330.
Tolbert, S., Schindel, A., & Rodriguez, A. J. (2018). Relevance and relational responsibility in justice-oriented science education research. *Science Education, 102*(4), 796–819.
Wallace, M. F. (2018). The paradox of un/making science people: Practicing ethico-political hesitations in science education. *Cultural Studies of Science Education*, 1–12.
Yacoubian, H. A. (2015). A framework for guiding future citizens to think critically about nature of science and socioscientific issues. *Canadian Journal of Science, Mathematics and Technology Education, 15*(3), 248–260.
Yacoubian, H. A., & Hansson, L. (2020). *Nature of science for social justice*. Dordrecht: Springer.
Zeidler, D. L., & Nichols, B. H. (2009). Socioscientific issues: Theory and practice. *Journal of Elementary Science Education, 21*(2), 49.
Ziman, J. (2000). *Real science. What it is, and what it means*. Cambridge, UK: Cambridge University Press.

Chapter 6
Does Research on Nature of Science and Social Justice Intersect? Exploring Theoretical and Practical Convergence for Science Education

Sibel Erduran, Ebru Kaya, and Lucy Avraamidou

6.1 Introduction

The rise in inequality in the distribution of income among people is well-documented and displays the characteristics of a trend, having affected large numbers of countries, from the poorest to the most affluent (United Nations 2006). The inequality gap between the richest and poorest countries, measured in terms of national per capita income, is growing as well. Concurrently, new socio-political realities caused by the massive migration of refugees to Europe and the urgency for including refugee children into society. In 2017 UNHCR registered 172,301 sea arrivals of refugees and migrants, mainly from Nigeria and the Syrian Arab Republic, to Europe. In the first 6 months of 2017, 16% of all arrivals were children, 72% of which were unaccompanied and separated children (UNHCR et al. 2017). UNICEF reports show that children are increasingly showing signs of deep psychological trauma as a result of their suffering and displacement and are excluded from the communities they now live. These new sociopolitical realities and the rise in poverty in all its manifestations, along with the increase in the numbers of refugees, displaced persons and other victims of circumstance and abuse, represent sufficient evidence for a judgment of persistent, if not growing, injustice in the world. Addressing such

S. Erduran (✉)
Department of Education, University of Oxford, Oxford, UK
e-mail: sibel.erduran@education.ox.ac.uk

E. Kaya
Department of Mathematics and Science Education, Bogazici University, Istanbul, Turkey
e-mail: ebru.kaya@boun.edu.tr

L. Avraamidou
Institute for Science Education and Communication, University of Groningen, Groningen, The Netherlands
e-mail: L.Avraamidou@rug.nl

H. A. Yacoubian, L. Hansson (eds.), *Nature of Science for Social Justice*, Science: Philosophy, History and Education,
https://doi.org/10.1007/978-3-030-47260-3_6

injustice at different levels and areas creates new challenges for education in general and science education in particular, that centre around goals related to social justice. The theme of social justice as an intellectual theme is relatively new:

> None of history's great philosophers—not Plato or Aristotle, or Confucius or Averroes, or even Rousseau or Kant—saw the need to consider justice or the redress of injustices from a social perspective. The concept first surfaced in Western thought and political language in the wake of the industrial revolution and the parallel development of the socialist doctrine. It emerged as an expression of protest against what was perceived as the capitalist exploitation of labour and as a focal point for the development of measures to improve the human condition. It was born as a revolutionary slogan embodying the ideals of progress and fraternity. Following the revolutions that shook Europe in the mid-1800s, social justice became a rallying cry for progressive thinkers and political activists. Proudhon, notably, identified justice with social justice, and social justice with respect for human dignity. (United Nations 2006, pp. 11–12)

From 2006 until today, quite a few researchers in various disciplines such as education, sociology, social psychology and gender studies have engaged with research that aims to promote goals related to social justice. Situated within socio-cultural research traditions, social justice has been a concern for science educators for more than a decade now (e.g. Calabrese-Barton and Upadhyay 2010; Reeve and Sharkawy 2014). Researchers interested in promoting social justice in the context of science education have suggested various programs, frameworks and interventions that aim to provide equal opportunities for science learning to all students regardless of their gender, race, culture, ability, language, and religion. This broad conceptualization of social justice is consistent with the accounts presented in this chapter.

A review of literature indicates there can be at least two senses of inclusion of social justice in science education. First, social justice can be conceived as an overarching goal and a vision for schooling and curriculum. In this case, education can serve the function of promoting and maintaining social justice. A second sense of inclusion of social justice in science education is more directly related to learning goals and outcomes. Here the emphasis would be on the equipment of students with habits of mind and values that would ensure that they contribute to social justice themselves as active citizens. Enhancement of students' understanding of social justice is thought to contribute to good citizenship. In what follows, we argue that contextualising social justice in science education remains challenging for teachers as social justice is not conventionally a common feature of science teaching and learning. This is partly due to the fact that missing remains a conceptualization of school science that explicilty addresses social justice. As a matter of fact, the way that science is conceptualised in school science does not tend to lend itself to invite discussion of social justice issues.

There is a vast body of work on nature of science (NOS) (Erduran and Dagher 2014a; Erduran and Kaya 2019; Lederman et al. 2002) which concerns understanding of and about science. Understanding NOS is thought to contribute to scientific literacy as well as citizenship. Providing equal access to opportunities for scientific literacy and the development of skills needed for active citizenship is at the heart of such account of social justice. Although social justice and NOS literatures might share similar themes such as citizenship goals, the precise intersection of these literatures has not been investigated. There are other accounts in science education such as argumentation and deliberative democracy which raise similar issues in relation to synthesis of conventionally disparate areas of research (Erduran and Kaya 2016). In this chapter, we review some selected perspectives on social justice and NOS literatures leading to a synthesis of common themes that can potentially serve both the purposes of learning about NOS while at the same time advance goals related to social justice through science instruction. We focus on definitions of social justice offered by Rawls (1985) and Miller (2001), and map some of their characterisations to the framework on social aspects of NOS defined by Erduran and Dagher (2014). We thus contribute to NOS literature by drawing on theories from political philosophy, an area that has not been previously explored explicitly in the NOS literature. Our goal is to explore how social justice might be fostered through NOS instruction. In so doing, we trace the potential overlap of social justice and NOS concepts, and draw out example recommendations for curriculum policy as well as teaching and learning.

6.2 Theoretical Framework

In this section we describe how we conceptualize social justice by discussing some influential theories and offering definitions of key constructs that have been used in contemporary research in science education. Subsequently, we discuss how a particular approach to characterising NOS (i.e. the Family Resemblance Approach) can be used in science education to promote goals related to social justice. Recommendations are provided for curriculum policy as well as teaching and learning.

6.2.1 Theories of Social Justice

Social justice is generally equated with the notion of equality or equal opportunity in society. Although equality is undeniably part of social justice, the meaning of social justice is actually much broader (Scherlen and Robinson 2008). Further, "equal opportunity" and similar phrases such as "personal responsibility" have been used to diminish the prospective for realizing social justice by justifying enormous inequalities in modern society (Berry 2005). The most recent theories of and scholarly statements about "social justice" illustrate the complex nature of this theoretical construct. Two prominent accounts of social justice are based on Rawls (1985) and Miller (2001). While neither of these theories can be considered an exhaustive treatment of the subject matter, each offers a complex theory of social justice that illustrates its broad meaning. Both conceptions of social justice are similar, so there is significant overlap between the main ideas.

We chose to focus on Rawls' (1985) and Miller's (2001) accounts of social justice because they offer broad conceptualizations of social justice that provide the basis of articulating goals related to social justice in education, especially in identifying how education can tackle equal access to learning opportunities. According to Rawls, social justice is about assuring the protection of equal access to liberties, rights, and opportunities, as well as taking care of the least advantaged members of society. Rawls posits that rational, free people will agree to play by the rules under fair conditions and that this agreement is necessary to assure social justice because public support is critical to the acceptance of the rules of the game (Rawls 1985, pp. 27–28). These rules or principles *"specify the basic rights and duties to be assigned by the main political and social institutions, and they regulate the division of benefits arising from social cooperation and allot the burdens necessary to sustain it"* (Rawls 1985, p. 7). It is important to note that Rawls' theory is one of domestic justice (principles that apply to the basic structures of society) and not of local justice (principles that apply to institutions and associations in society) or global justice (principles applying to international law) (Rawls 1985, pp. 11–12).

Miller's (2001) account of social justice, on the other hand, deals with the distribution of good (advantages) and bad (disadvantages) in society, and more specifically with how the 'good' should be distributed within society. Further, social justice is concerned with the ways that resources are allocated to people by social institutions (Miller 2001, p. 11). Need is a claim that one is lacking the basic necessities and is being harmed or is in danger of being harmed and/or that one's capacity to function is being impeded (Miller 2001, p. 207–210). Desert is a claim that one has earned reward based on performance, that superior performance should attract superior recognition (Miller 2001, p. 134–141). Equality refers to the social ideal that society regards and treats its citizens as equals, and that benefits such as certain rights should be distributed equally (Miller 2001, p. 232). Furthermore, Miller explains three basic modes of human relationships which are solidaristic community, instrumental associations, and citizenship. The solidaristic community exists when people share a common identity as members of a relatively stable group with a common ethos. Instrumental associations is about how people relate to one another in a utilitarian manner; each has aims and purposes that can best be realized by collaboration with others. The citizenship is about members of a political society in modern liberal democracies who are related not just through their communities and their instrumental associations but also as fellow citizens. The main concepts in Rawl's and Miller's accounts are summarised in Table 6.1.

It should be noted that some researchers in science education (e.g. Bencze and Alsop 2014) emphasize not only the need for providing all students with equal opportunities to learning science despite their differences but also the need for political activism towards such a goal. As such, social justice has become not only an educational goal but also a political activity. In order to better understand science as a political activity, and to implement teaching practices and curricular that aim to promote social justice, some theoretical constructs can serve as tools to conceptualize social justice through a political stance. For example, Kayumova et al. (2018) summarize through a review of contemporary literature conceptualizations of the

Table 6.1 Key concepts from Rawls and Miller

Rawls (1985)	**Miller** (2001)
Equal liberties	Need
Equal opportunity	Desert
Difference principle	Equality
Fair share of benefits to the least advantaged members of society	Solidaristic community
Opportunities for healthy and fulfilling lives	Instrumental associations
Freedom	Citizenship
Human rights	Formal equality
	Accuracy
	Publicity
	Dignity
	General ethos of the community
	Human rights
	Freedom

constructs related to social justice in light of their criticism of current reform recommendations in Europe and the US as being a-theoretical and a-political. These constructs are not meant to serve as a framework of social justice. Rather, they are meant to serve as operational definitions of constructs that have been used in contemporary research that addresses social justice in the context of science education. These are the following: diversity, equity, identity, and creativity. Diversity is used to hightlight the differences among individuals. Equity is used to refer to addressing barriers to equal access. Identity is used to conceptualize the process of learning and development. Finally, creativity elaborated further below is an alternative construct to innovation. It is socially-just. In particular, the further unpacking of these constructs provides a framework through which to investigate social justice concepts:

Diversity: refers to the inclusion of different types of people, with unique characteristics that might influence science teaching and learning: ethnicity/race, gender and socioeconomic status/social class, dis/ability, linguistic, sexuality, gender identity, political, religious, geographical origins, age etc.

Equity: refers to broadening participation, achievement, and/or access and an examination of issues of power and equity within the structural, cultural, and curricular organization of science education, teaching and learning.

Identity: refers to how one views him/herself and how he/she is viewed by others, and can be better understood as a process of identity construction through social participation and lived experience.

Creativity: an expanded view of innovation that both challenges existing scientific epistemologies and centers addressing global challenges from a critical and socially just perspectives, and which goes beyond a traditional view of innovation and scientific entrepreneurialism that only serves to reproduce inequalities.

These overarching costructs provided by Kayumova et al. (2018) can provide us with the lenses and tools to engage in discussions related to notions of what constitutes truth, knowledge, and power, and to study science learning in the context of bigger questions related to social justice.

6.2.2 Nature of Science in Science Education

Nature of science (NOS) is a significant area of research in science education (e.g. Allchin 2011; Erduran and Dagher 2014; Irzik and Nola 2014; Lederman et al. 2002). Different accounts of NOS have emphasised the social aspects of science in various ways. For example, Erduran and Dagher (2014) provided a comparative overview of NOS from the consensus view (Lederman et al. 2002), Features of Science (FOS) (Matthews 2012) and Family Resemblance Approach (FRA) (Irzik and Nola 2014) perspectives (see Table 6.2). Although all these approaches have reference to the social contexts of science, FOS and FRA make explicit reference to social values where concepts related to social justice are likely to reside. For example, Irzik and Nola (2014) refer to social values in a fairly broad sense. Erduran and Dagher (2014), on the other hand, provide further categories including financial systems, political power structures, and social organisations and interactions. The latter categories provide some nuance through which social justice concepts can be explored. For instance, the category of political power structures inherently addresses power relations that are conventionally at the root of social inequality (United Nations 2006).

Erduran and Dagher (2014) discuss NOS from a "Family Resemblance Approach" (FRA) (see Fig. 6.1) which provides an account of NOS based on epistemic, cognitive and social-institutional aspects of science. The framework is based on Wittgenstein's family resemblance idea which was adapted to NOS by philosophers of science Irzik and Nola. A description of the FRA is available in Irzik and Nola (2014). Essentially, the idea of a family resemblance implies that the various sciences are akin to members of a biological family that share certain characteristics although they also possess some differences. For example, while all sciences might rely on evidence, the precise articulation of what counts as evidence in astronomy versus chemistry can be fairly nuanced. In many instances of astronomical investigations, the evidence is historical in nature, based on the time it takes for celestial bodies to be observed given the distance it takes light to travel to earth. However, in a chemistry investigation for example, it is possible to manipulate materials and collect data that are represented at this point in time.

There is now a growing body of research focusing on FRA in science education (e.g. Cheung 2020; Couso and Simarro 2020; Erduran et al. 2019; Park et al. 2020). FRA-based NOS covers a range of aspects of science including aims and values, methods, practices, knowledge as well as social-institutional dimensions of science. As such, FRA is consistent with other frameworks arguing for an inclusive and holistic characterisation of nature of science (e.g. Allchin 2011). Furthermore,

Table 6.2 Comparative overview of Nature of Science (NOS) perspectives (From Erduran and Dagher 2014, p. 26)

NOS Consensus View	Features of Science Approach	Family Resemblance Approach
?	Lists:	Includes scientific aims and values that subsume rationality and theory choice as an aim and value
	Theory choice and rationality which involve a set of aims and values	
?	Lists practices that include:	Includes nature of scientific practices pertaining to observation, experimentation, classification and so on.
	Experimentation	
	Idealization	
	Technology	
	Explanation	
	Mathematization	
Focuses on the idea that scientists use many methods: No one scientific method	?	Methodologies and methodological rules
Distinguishes between	Includes	Scientific knowledge: Epistemic-cognitive aspects of models, theories, laws and explanations and aspects pertaining to them such as knowledge revision
Scientific theories and laws	Models	
Observations from inferences		
Focuses on tentativeness		
Highlights cultural embeddedness	Includes	The expanded social context recognizes cultural embeddedness and societal and religious values.
	Values and socio-scientific issues	
	Worldviews and religion	
Includes	?	Creativity is a psychological component that characterizes aims and methods, practices, and scientific knowledge. It in implicit in the FRA.
Creativity		
?	Includes the following philosophical positions:	The FRA does not make a commitment to any of these positions. In this sense, it is philosophically neutral.
	Realism	
	Constructivism	
	Feminism	

FRA is a framework that accommodates for domain-general as well as domain-characteristics of science highlighting both what is universal across science disciplines and what is particular. The key components of the FRA include cognitive and epistemic dimensions of science including the aims and values of science as well as the social-institutional dimensions of science which are the social certification and dissemination, social ethos, social values, professional activities, social organisations and interactions, financial systems and political power structures. However, research evidence points to the fact that these social-institutional dimensions of science are absent from the curriculum. As shown by studies in numerous

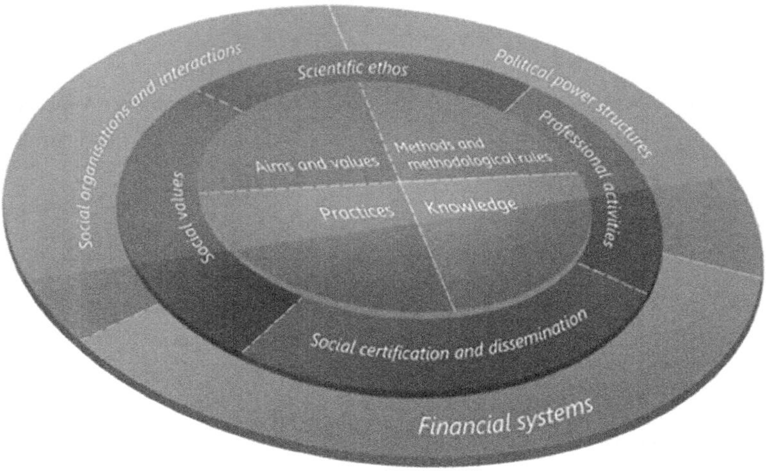

Fig. 6.1 FRA wheel: Science as a cognitive-epistemic and social-institutional system. (Erduran and Dagher 2014, p. 28)

national contexts including Taiwan (Yeh et al. 2019) and Turkey (Kaya and Erduran 2016), curriculum documents tend to contain statements that identify science as a cognitive-epistemic system and they underemphasize science as a social-institutional system.

In developing the social-institutional component of NOS in their framework, Erduran and Dagher (2014) highlight that science involves individual scientists working in social groups in social institutions, exercising social values and activities. The inclusion of the social dimension of science in science education is warranted for various reasons. First, the ways in which scientists organize science socially might have relevance for how science learning environments can be structured. In other words, students may benefit from acquiring the social aspects of scientific communities, and the inclusion of social features of science in the classroom may facilitate students' learning of science. Second, the particular social values and norms that dominate communities of scientists could be considered as potential learning outcomes for students. What this means essentially is that educating students in science goes beyond merely addressing cognitive and epistemic aspects of science to including the social dimension of science.

Hence, understanding science in its entirety suggests that students learn about the social norms that scientists work by. Without the inclusion of the social context of science in science education, students are bound to have limited understanding of how the scientific enterprise works, and how the social structures, relationships and issues guide the advacement of science. Erduran and Dagher (2014) argue that categories of science as a social-institutional system can be visualized in terms of (a) the core features of professional activities, scientific ethos, social certification and dissemination and social values, and (b) the broader features of political power structures, financial systems and social organizations and interactions. The latter features are referred as broad because finance, politics and institutions are integral components of the larger society in which science, like other organized human activity, is being practised. In reality, however, all categories of this system are interactive, hence the porous boundaries that are symbolically represented in the Fig. 6.1.

Social certification and dissemination refer to the social mechanisms through which scientists review, evaluate and validate scientific knowledge for instance through peer review systems of journals. Scientific ethos refers to the norms that scientists employ in their work as well as in interaction with colleagues. Social values refer to specific values such as freedom, respect for the environment, and social utility. Professional activities is about how scientists engage in professional settings such as attending conferences and doing publication reviews. Social organisations and interactions refer to how science is arranged in institutional settings such as universities and research institutes. Financial systems are defined as the underlying financial dimensions of science including the funding mechanisms. Political power structures are the dynamics of power that exist between scientists and within science cultures. Social certification and dissemination, scientific ethos, social values, scientists' professional activities, social interactions, financial systems and political power structures are all key constructs in conceptualizing and implementing curricula that promote goals related to social justice. In what follows, we elaborate on these constructs, through a discussion that cuts across conceptualisation of social justice and NOS.

6.3 Intersection of Social Justice and Nature of Science

In this section we provide educational examples for an inclusive agenda that promotes the teaching and learning of NOS and social justice concurrently and in ways in which goals related to social justice can be achieved through understanding NOS. The first example focuses on the formulation of curriculum standards that serve the purposes of both NOS and social justice. Here we synthesise theoretical perspectives and provide some illustrations of curriculum statements. The second example draws on a project of a pre-service teacher's teaching practice illustrating the instructional resources developed to teach about social-institutional aspects of NOS.

6.3.1 Curriculum Policy Statements

In developing a set of curriculum policy statements, we focus on selected frameworks on social justice and NOS: the social justice frameworks proposed by Rawls (1985) and Miller (2001), and the NOS framework proposed by Erduran and Dagher (2014). In developing this set of curriculum statements, we sought to determine some common themes that provide an overlap of different categories of social justice and NOS. Essentially, the particular concepts from both social justice and NOS approaches could potentially unite under a broader overarching concept. In the case of "human rights" as a social justice category (i.e. from Rawls 1985) for instance, the relevant NOS concept is "social values" (Erduran and Dagher 2014) that a community of scientists must abide with such as respect for communality. One overarching common theme is "respect" which applies to both categories. In order to address the synthesis of social justice and NOS concepts for science education, we illustrate some applications on the synthesis of the themes for considering some potential curriculum statements (see Table 6.3). With respect to the "human rights" and "social values" categories, we propose the statement "*Students will understand that scientists should have the right to express their research without feeling threatened about potential backlash.*" Here the scientists are positioned to have basic human rights in performing their professional tasks and in being part of a respectful community. In a similar vein, we took all of the social context

Table 6.3 Suggested Curriculum Statements on NOS and Social Justice

Overlapping theme	**NOS category (Erduran and Dagher** 2014)	**Social justice category**	**Potential curriculum statements**
Diversity	Social certification and dissemination	Difference principle (Rawls 1985)	*Students will understand that scientists with diverse social positionings and backgrounds may debate and enrich the scientific enterprise collaboratively.*
Respect	Social values	Human rights (Rawls 1985)	*Students will understand that scientists should have the right to express their research without feeling threatened about potential backlash.*
Identity	Professional activities	Solidaristic community (Miller 2003)	*Students will engage in activities such as writing, presenting and communicating results of investigations to other teams and demonstrate social responsibility in contributing to the school community.*
Equity	Political power structures	Equal liberties (Rawls 1985)	*Students will be respectful of people from different backgrounds such as gender, class, national origin and race, and understand the injustices resulting from discrimination and exclusion.*
Ethos	Scientific ethos	Freedom (Miller 2003)	*Students will understand that scientists and citizens should have freedom of expression of ideas.*
Opportunity	Social organisations and interactions	Instrumental associations (Miller 2003)	*Students will understand that scientists have aims and purposes that can best be realized through collaboration with others.*
Economic fairness	Financial systems	Share of benefits (Rawls 1985)	*Students will understand that scientists and societies rely on economics but that there should be justice in how commodities are distributed and traded among communities.*

categories from Erduran and Dagher's (2014) NOS framework and mapped the social justice categories from Rawls (1985) and Miller (2001). Another example concerns the overlapping theme of "ethos". In this case, Erduran and Dagher's (2014) category of "scientific ethos" is similar to Miller's (2001) category of "freedom". For example, in scientific communities, ideas and evidence are meant to be exchanged freely without being restricted on ideological grounds. Freedom of expression is an important aspect of scientific ethos as well as a socially just society. A potential curriculum statement to capture the overarching theme is "*Students will understand that scientists and citizens should have freedom of expression of ideas.*" Overall, the intersection of social justice and NOS ideas lead to a set of broad themes such as diversity, respect, community, equity, ethos, opportunity and economic fairness that can provide a comprehensive set of ideas for that might be used as input for setting curricular goals.

The question then becomes: in what ways (if any) have these concepts and goals related to diversity, inclusion and social justice found their place in visions for reform across the world and within science curricula? As Kayumova et al. (2018) argued, an analysis of the various policy documents that have been published in various parts of the world (e.g., *New Generation Science Standards* in the U.S. context, *Responsible Science and Innovation* in Europe) there exists a discrepancy between contemporary global challenges and reform efforts, as reform efforts emphasize goals related to economic competition instead of goals related to students social justice. An example is found in the report by the European Commission called "Science Education for Responsible Citizenship" (EC 2015), which offers a twenty-first century vision for science for society within the broader European agenda. The report places emphasis on the process of aligning research and innovation to the values, needs and expectations of society, referred to as "responsible research and innovation". These reform recommendations, however, do not reflect how global challenges (e.g., migration, refugee crisis) have shaped this vision for science for society and therefore lack attention to the need for more inclusive, equitable, and just societies. As Kayumova et al. (2018) argued, "*goals related to reducing inequality, promoting social change and social justice are completely absent*" (p. X).

Likewise, an analysis of science curricula reveals minimal attention on the social-institutional aspects of NOS. For example, Kaya and Erduran (2016) illustrated that there exists a distinct underemphasis on the social categories of curriculum statements in science curricula in the context of the US, Ireland and Turkey. Consider, for example, the following three example statements from the Turkish curricula, as identified by Kaya and Erduran (2016):

- "*To enable students' appreciation of how science is developed collaboratively among scientists from different cultures*" (Social certification and dissemination)
- "*Scientifically literate person is aware of how social values of the culture and societal structures and beliefs influence how knowledge is cognitively processed*" (Social values)

- *"The students investigate and present the studies conducted by public/private institutions and civil society organizations that contribute to the development of chemical industry in our country"* (Social organisations and interactions)

These examples were fairly rare in a sequence of curriculum documents. The first one relates to diversity, the second one relates to equality and the third one relates to opportunity. A potential contribution of our current analysis is that specific curriculum statements can be generated that would be inclusive of social justice and NOS themes concurrently. For example, for the theme "ethos", a curriculum goal could be "*Students will understand that scientists and citizens should have freedom in expression of their ideas.*" Table 6.3 consists of further examples of potential curriculum statements on the various themes.

6.3.2 Teaching and Learning Resources

Alayoglu (2018) developed a series of lesson activities on the inclusion of social-institutional systems of NOS in science education 12 year old students in Istanbul, Turkey. Using a pre-test post-test quasi-experimental research design, the effectiveness of the resources were evaluated following a 4-week intervention. The results showed that there were statistically significant differences between the study groups in favor of students in the experimental group on both study variables. In other words, integration of the social-institutional aspects of science into science lessons enhanced students' understanding of the social dimension of science. In what follows, we describe one of the activities on Moon Mining that incorporated elements of the social-institutional aspects of NOS.

The activity begins with engaging students in a discussion about the moon and asteroids being rich with minerals that are rare on earth. Because of this, some big companies and governments aimed to remove these valuable minerals from space. At the beginning of the lesson, Alayoglu, as a teacher-researcher, offered brief information about the role of politics in science. In the classroom discussion, the point was raised about how science and technology have been historically linked to governments and states. For example, Galileo sharpened his telescope to see distant enemy better. The lesson resources included a range of activities for students. For example, a series of questions were produced to elicit the specific social-institutional category as illustrated in Table 6.4. In our interpretation of the practical questions developed by Alayoglu (2018), there are links to the NOS categories developed by Erduran and Dagher (2014) and the social justice categories discussed by Rawls (1985) and Miller (2003). In other words, these questions which are practically usable at the level of the classroom are also theoretically related to NOS and social justice categories. In this table, the social-justice categories embedded in these activities are presented next to the NOS categories.

What this example illustrates is that even for a topic that is seemingly devoid of social context (i.e. space explorations are not situated in an obvious way to lend

Table 6.4 Example questions for teaching and learning to incorporate NOS and social justice: Synthesis of practical instructional resources and theoretical perspectives

Pedagogical questions (From Alayoglu 2018)	**NOS aspects (From Erduran and Dagher** 2014)	**Social Justice (From Rawls** 1985; **Miller** 2003)
Do astronomers and scientists work alone or within an organization or community? How? Which institutions investigate space and other planets?	Social organizations and interactions	Instrumental associations
Do you know any other scientific institutions like NASA in which many scientists work together?	Social organizations and interactions	Instrumental associations
In February 2012, The Australian Centre for Space Engineering Research (ACSER) in Sydney organized a meeting on "Searching for Mine" and brought together famous companies, scientists, engineers and robotic experts. Why did many people from different disciplines meet? What do you think was discussed at this meeting?	Professional activities & social certification and dissemination	Solidaristic community
Who will benefit from mines being removed from the moon or asteroids?	Financial systems	Share of benefits
Why do China and USA are interested in space mining?	Political power structures	Equal liberties
Could space mining be dangerous? Could it harm the environment?	Scientific ethos & social values	Human rights

themselves to social justice issues), there is potential for their articulation for social justice. For example, the question of *"Who will benefit from mines being removed from the moon or asteroids?"* raises questions about share of benefits. The NOS and social justice categories along with their examples can potentially provide a toolkit for teachers to organise their questioning in lessons to elicit social justice themes through the teaching of NOS. Teacher education at both pre-service and in-service stages needs to support science teachers' learning of strategies that promote students' understanding of NOS and social justice in unison. There is already a book-length account on the design, implementation and evaluation of a teacher education approach that incorporated NOS from an FRA perspective including the social-institutional aspects of science (Erduran and Kaya 2019). Further examples that focus more closely on social justice issues can be designed and tested.

6.4 Discussion and Conclusions

An examination of the research literature of social justice and NOS reveals a set of concepts that cut across the two knowledge bases, such as equality, social responsibility and human rights. By drawing out parallels between social justice and NOS literatures, we forge potential links that can foster both agendas, and provide concrete curriculum statements to correspond to each category of concepts. Given the scarcity of the social context of NOS in many science curricula from around the world, for instance Turkey (Kaya and Erduran 2016) and Taiwan (Yeh et al. 2019), the paper contributes to the elaboration of potential curriculum statements on the subject while merging NOS goals with social justice goals. The instructional approaches including questions and scoring criteria provide some concrete examples of practical approaches to teaching and learning of NOS and social justice. The curriculum statements can be extended further to a set of practical recommendations that help us to respond to the question: How can science educators provide all learners with equitable opportunities for participating in communities of learners in an increasingly globalized world? As Zembylas and Avraamidou (2008) argued:

> Science education practices and curricula emphasizing professional or Western science alienate underrepresented groups. The premises of these practices are based upon: rigid teaching strategies and uncreative methods; a view of science as a very technical field that is practiced by intelligent individuals who manage to leave their subjectivities outside the field; a context of practicing science that is detached from cultural relevance; and, perspectives in science that lack representation from diverse groups (p. 994)

This is precisely where the role of social justice comes into play in science teaching and learning. Historically, social justice has been conceptualized in various ways and has been used as a theoretical/research constructs in various fields, such as education, philosophy, psychology, and sociology. In this chapter, we asked to what degree is science associated with this goal, how science is connected to greater social issues, and how science falls within political discourses. In doing so, we explored a set of overlapping constructs embedded in conceptualizations of the NOS and social justice. Ultimately, social justice is meant to promote a just and democratic society by valuing diversity, which refers to various aspects of human identity such as race, gender, sexual orientation, nationality, and which have been subjects of discrimination (Harding 1986).

John Rawls (1985) and David Miller (2001) are key theorists on social justice. We have used some central concepts from their work in charting out a territory for intersections with NOS literature in science education. Even though these two theories have distinct differences, they share specific commonalities, such as an emphasis on *equality, citizenship,* as well as the *socio-political* forces that shape societies. Equality, citizenship, and socio-political forces are crucial in shaping educational practice as they provide both a goal and a context for conceptualizing scientific literacy, which remains one of the key goals of science education. Scientific literacy is broadly conceptualized as scientific knowledge in order to identify questions, acquire new knowledge, describe scientific phenomena and draw conclusions from evidence, to understand science as a form of human knowledge and research, to understand the role of science to shape our material, intellectual and cultural environment, and to be willing to engage with scientific ideas and topics and to deal with them in a reflective manner (OECD 2006). Taking scientific literature as a point of departure, in what follows we discuss what it would mean for science education to adopt a social-justice lens and we propose a set of key theoretical constructs that are crucial in contemporary conceptualizations of social justice.

Adopting a social-justice lens to framing scientific literacy would mean that all students, regardless of race, sex, class, gender, sexual orientation, or ability, should have equal access to opportunities through school science for becoming scientific literate. In the field of science education, quite a few researchers have raised important questions and engaged in criticism about the role of school science in society through the concepts of equality, equity, power relations and knowledge production in schools, and how Western science has traditionally excluded many groups of students (e.g., Calabrese-Barton et al. 2003; Harding 1986; Rivera-Maulucci and Fann 2016). This is precisely what a social-justice perspective can do for science education.

Essentially, a social justice perspective in NOS instruction provides us with the theoretical constructs to understand social inequalities in school science as well as science more broadly, and to work towards a social or cultural shift where no student or groups of students are excluded. The importance of a social justice perspective in science education is paramount given existing literature that provides evidence that citizens are inadequately prepared to use scientific knowledge to make informed decisions in their everyday lives; the percentages of under-privileged students, such as girls and minorities, following careers in science remain disproportionately low around the world and science is poorly taught in schools (Eisenhart et al. 1996). Echoing Calabrese-Barton's et al. (2003), we argue that re-conceptualizing the NOS and science teaching through a social justice lens requires an understanding of science as a political activity:

> The implications of such a stance are that science (and any education in science) will only be equitable and empowering if students learn—in addition to the standard knowledge base of ideas and skills—to uncloak those assumptions, to draw strength from their exposure, and to expand understandings of the agreed- upon boundaries for where and how scientific ideas are generated. (p. 136)

The contribution of this chapter, then, lies within an argument about conceptualizing NOS being inclusive of a political activity and enacting NOS instruction for the purpose of promoting social justice. In doing so, we offered definitions of a set of contemporary constructs that might be used by researchers interested in social justice. In addition, we offered concrete examples of potential curriculum statements for various themes that relate to social justice issues, and we suggest an example teaching strategy (i.e. questioning) that teachers can potentially use for eliciting social justice themes through teaching NOS. Our argument is consistent with the position presented by Rita Vilanova and Isabel Martins in the next chapter where they explore the relationship between NOS and citizenship education. In Chap. 7, these authors question the limitations of focusing on epistemological perspectives on science for the purposes of citizenship and argue for the broadening of the content of science textbooks. Similarly, we have advocated the broadening of the science curriculum to forge links between the social-institutional aspects of NOS and social justice to serve citizenship goals. In educational systems framed by neoliberal ideologies, surrounded by a rise in inequality in the distribution of income as well as new socio-political realities caused by the massive migration of refugees, there is already an existing imperative to embrace socially just agendas for science curricula. We do acknowledge that these theoretical conceptualizations and curriculum examples are by no means exhaustive or applicable in all contexts. However, we hope that these serve as a springboard for further explorations of how NOS approaches might serve to promote goals related to social justice in science education.

References

Alayoglu, M. (2018). *Fifth-grade students' attitudes towards science and their understanding of its social-institutional aspects*. Unpublished Master's Thesis. Bogazici University, Istanbul, Turkey.

Allchin, D. (2011). Evaluating knowledge of the nature of (whole) science. *Science Education, 95*(3), 518–542.

Bencze, L., & Alsop, S. (Eds.). (2014). *Activist science and technology education*. Dordrecht: Springer.

Berry, B. (2005). *Why social justice matters*. Cambridge, England: Polity Press.

Calabrese-Barton, A., & Upadhyay, B. (2010). Teaching and learning science for social justice: Introduction to the special issue. *Equity & Excellence in Education, 43*(1), 1–5. https://doi.org/10.1080/10665680903484917.

Calabrese-Barton, A., Ermer, J. L., Burkett, T. A., & Osborne, M. D. (2003). *Teaching science for social justice*. New York: Teachers College Press.

Cheung, K. (2020). Exploring the inclusion of nature of science in biology curriculum and high-stakes assessments in Hong Kong. *Science & Education*. https://doi.org/10.1007/s11191-020-00113-x.

Couso, D., & Simarro, C. (2020). STEM education through the epistemological lens: Unveiling the challenge of STEM transdisciplinarity. In C. C. Johnson, M. J. Mohr-Schroeder, T. J. Moore, & L. D. English (Eds.), *Handbook of research on STEM education*. Abingdon, OXON: Routledge.

Eisenhart, M., Finkel, E., & Marion, S. F. (1996). Creating the conditions for scientific literacy: A re-examination. *American Educational Research Journal, 33*(2), 261–295.

Erduran, S., & Dagher, Z. (2014). *Reconceptualizing the nature of science for science education: Scientific knowledge, practices and other family categories*. Dordrecht: Springer.

Erduran, S., & Kaya, E. (2016). Scientific argumentation and deliberative democracy: An incompatible mix in school science? *Theory Into Practice, 55*(4), 302–310.

Erduran, S., & Kaya, E. (2019). *Transforming teacher education through the epistemic Core of chemistry: Empirical evidence and practical strategies*. Dordrecht: Springer.

Erduran, S., Dagher, Z. R., & McDonald, C. V. (2019). Contributions of the family resemblance approach to nature of science in science education. *Science & Education, 28*(3–5), 311–328.

European Commission. (2015). *Science education for responsible citizenship*. Brussels: Directorate General for Research and Innovation, Science with and for Society. http://ec.europa.eu/research/swafs/pdf/pub_science_education/KI-NA-26-893-EN.pdf.

Harding, S. (1986). *The science question in feminism*. Ithaca: Cornell University Press.

Irzik, G., & Nola, R. (2014). New directions for nature of science research. In M. Matthews (Ed.), *International handbook of research in history, philosophy and science teaching* (pp. 999–1021). Dordrecht: Springer.

Kaya, E., & Erduran, S. (2016). From FRA to RFN, or how the family resemblance approach can be transformed for science curriculum analysis on nature of science. *Science & Education, 25*(9), 1115–1133. https://doi.org/10.1007/s11191-016-9861-3.

Kayumova, S., Avraamidou, L., & Adams, J. (2018). Science education now: Diversity, equity, and the big picture. In L. Bryan & K. Tobin (Eds.), *Critical issues and bold visions for science education: The road ahead* (pp. 285–297). Rotterdam: Brill/Sense Publishers.

Lederman, N. G., Abd-El-Khalick, F., Bell, R. L., & Schwartz, R. S. (2002). Views of nature of science questionnaire (VNOS): Toward valid and meaningful assessment of learners conceptions of nature of science. *Journal of Research in Science Teaching, 39*(6), 497–521.

Matthews, M. (2012). Changing the focus: From nature of science (NOS) to feature of science (FOS). In M. S. Khine (Ed.), *Advances in nature of science research: Concepts and methodologies* (pp. 3–26). Dordrecht, The Netherlands: Springer.

Miller, D. (2001). *Principles of social justice*. Cambridge, MA: Harvard University Press.

Miller, D. (2003). *Principles of social justice*. Boston, MA: Harvard University Press.

OECD. (2006). *Assessing scientific, reading and mathematical literacy: A framework for PISA*. Paris: OECD.

Park, W., Yang, S., & Song, J. (2020). Eliciting students' understanding of nature of science with text-based tasks: Insights from new Korean high school textbooks. *International Journal of Science Education, 42*(3), 426–450.

Rawls, J. (1985). Justice as fairness: Political not metaphysical. *Philosophy and Public Affairs, 14*, 223–251.

Reeve, R., & Sharkawy, A. (2014). Science education for social justice: Using the knowledge-building communities model. *LEARNing Landscapes, 7*(2), 283–298.

Rivera-Maulucci, M. S., & Fann, K. S. (2016). Teaching for social justice in science education: Helping a new teacher develop a social justice identity. In L. Avraamidou (Ed.), *Studying science teacher identity: Theoretical, methodological, and empirical explorations* (pp. 111–129). Rotterdam: Sense Publishers.

Scherlen, A., & Robinson, M. (2008). Open access to criminal justice scholarship: A matter of social justice. *Journal of Criminal Justice Education, 19*(1), 54–74.

UNHCR, UNICEF, & IOM. (2017). *Refugee and migrant children in Europe accompanied, unaccompanied and separated.* Mid year Overview of Trends January–June 2017. Available at: https://data2.unhcr.org/en/documents/download/60348. Accessed online 30 July 2018.

United Nations. (2006). *Social justice in an open world: The role of the United Nations*. New York: United Nations.

Yeh, Y., Erduran, S., & Hsu, Y. S. (2019). Investigating coherence on nature of science in the science curriculum documents: Taiwan as a case study. *Science & Education, 28*(3–5), 291–310.

Zembylas, M., & Avraamidou, L. (2008). Postcolonial foldings of space and identity in science education: Limits, transformations, prospects. *Cultural studies in Science Education, 3*(4), 977–998.

Chapter 7
A Discursive Analysis of Relationships Between Nature of Science and Citizenship Education: The Case of Brazilian Science Textbook Evaluation Policies

Rita Vilanova and Isabel Martins

7.1 Introduction

In Brazil, textbooks represent an important element in the organization of school knowledge and can be considered a structuring part of curricular dynamics and pedagogical practice (Martins 2020). The production of textbooks also represents an important part of the Brazilian publishing market. In the history of Brazilian education, several policies have been implemented to regulate publication and distribution of books. Since the year 1985, the evaluation, purchase and distribution of textbooks for public schools has been carried out by the Brazilian Textbook Program (*Programa Nacional do Livro Didático* – PNLD).[1] Throughout the years, Education for Citizenship and an awareness of aspects related to the Nature of Science (NoS) have become important educational goals, which materialize in evaluation criteria for textbooks. In this paper, we seek to understand the ways through which textbooks policy texts address and relate these two goals. In order to do that, we conducted a discursive analysis in the 2008 edition of the Brazilian Textbook Guide, the first textbook policy document to explicitly mention NoS. Using Critical Discourse Analysis as our main theoretical and methodological framework, we analyse both conjunctural and textual dimensions of the 2008 Brazilian Textbook Guide so as to problematize both philosophical and educational bases of articulations between NoS and Education for Citizenship. The text is organized in three parts: first we present a brief summary of NoS related discourses present in the literature and explore possible relationships between NoS and Education for Citizenship. Then,

[1] http://www.fnde.gov.br/component/k2/item/518-histórico

R. Vilanova (✉) · I. Martins
Instituto Nutes de Educação em Ciências e Saúde, Universidade Federal do Rio de Janeiro, Rio de Janeiro, Brazil
e-mail: vilanova.rita@gmail.com; isabelgrmartins@gmail.com

H. A. Yacoubian, L. Hansson (eds.), *Nature of Science for Social Justice*, Science: Philosophy, History and Education,
https://doi.org/10.1007/978-3-030-47260-3_7

the conceptual basis and methodological framing of Critical Discourse Analysis are informed. Finally, results and discussions are presented, in order to explore limits and possibilities of articulating NoS and Education for Citizenship in school science.

7.2 Theoretical Bases and Aims of the Study

7.2.1 Discourses of Nature of Science

According to Rafael Amador-Rodríguez and Agustín Adúriz-Bravo (2017) and to Cristiano Moura (2019), several influential models of NoS can be identified in the Science Education literature. Amongst them, there are the 'consensus' view (McComas 2008), the 'whole science' view (Allchin 2011), the 'family resemblance' approach (Irzik and Nola 2011, 2014) and its expanded version (Erduran and Dagher 2014), and the "features of science" model (Matthews 2012).

The 'consensus' view has been summarised by William McComas and collaborators (1998, 2008) in terms of points, which derive from propositions by Norman Lederman and collaborators, Fouad Abd-El-Khalick and Jonathan Osborne (Lederman et al. 2002; Lederman and Abd-El-Khalick 2007; Osborne et al. 2003). It includes claims about: (i) the tentative character of scientific knowledge and its reliance on experimental evidence and rational argumentation, (ii) the rejection of the idea of a universal step-by-step method, (iii) the importance of openness and peer scrutiny, (iv) the recognition that observations are theory-laden and that scientific ideas are part of culture, (v) an acknowledgement of the role played by subjectivity in scientific practice, (vi) the ways through which scientific development is affected by historical and social aspects. Usually presented in the form of a list of tenets, the "consensus view" seeks to express points of agreement about what science is and how it works, and has been widely used to provide guidelines for dealing with NoS related knowledge in basic education.

Despite its widespread use and influence, the consensus view has been criticised by many scholars for both epistemological and sociological limitations. Derek Hodson and Siu Ling Wong (2017) argue that the "consensus view" may, in some respects, misrepresent aspects of scientific practice and lead to normative views about NoS in school curricula and assessment practices. Alternatively, they call for approaches that emphasise the community-regulated character of practices which are involved in the production of scientific knowledge by scientists, especially those concerning the particular nature of linguistic conventions adopted by the scientific community, the role of inquiry and socio-political interests in scientific activity.

Douglas Allchin (2011) has advocated taking multiple dimensions of scientific practice into accounts of NoS, so as to cover the "totality" of science. His "whole science" approach encompasses experimental, conceptual and social aspects of science and involves students in both reflective inquiry-based practices and analyses of historical case studies. In doing so, the author calls our attention to the ways through which conceptual, sociocultural and methodological aspects of science relate to epistemic features and to the potential of socio-scientific issues to help broaden the scope of NoS (Amador-Rodriguez and Adúriz-Bravo 2017).

Michael Matthews (2012) has criticized the essentialist character behind some approaches connected to the "consensus view" of NoS and proposed both an alternative terminology and a new focus, in order to overcome philosophical and educational pitfalls typically connected with NoS research. According to Matthews, an approach that focused on contextual and heterogeneous "features of science" would allow for a more balanced debate on the issue of methodology as well as help disentangle epistemological, sociological, psychological, ethical, financial and philosophical characteristics of NoS. The "features of science" model problematises the idea that students' learning about NoS could be assessed by their ability to identify a number of declarative statements. The argument is that items present in the "consensual view" lists (e.g. cultural embeddedness, tentativeness, theory dependency etc.) should be expanded to include psychological, economic and technological features, amongst others. Moreover they should be reconceptualised as features to be elaborated and discussed in teaching and not as a set of tenets to be observed or goals to be achieved.

The criticism that the "consensus view" may lead to an oversimplified, even reductionist, view of NoS is also present in Gürol Irzik and Robert Nola's proposition that the Wittgensteinian notion of family resemblance may be more apt a notion to discuss NoS (Irzik and Nola 2011). According to this view, which is also subscribed by Matthews (2012), demarcating necessary and sufficient conditions to inform a definition is not the best way to approach a complex issue such as NoS. Irzik and Nola argue that, instead of searching for individual features that would be common to all scientific disciplines, NoS related research and practice should aim at a more comprehensive and systematic accounts of aspects related to activities, aims and values, methodologies and products of Science. This way, differently to the "consensus view" it would capture "the dynamics and open-ended nature of science" (Irzik and Nola 2011, p. 602) and allow more freedom to teachers in the planning of activities in the classroom. According to Sibel Erduran and collaborators, the Family Resemblance Approach views NoS as "a set of aims and values, practices, methodologies, and social norms that are worthy of inclusion in the science curriculum" and stresses both cognitive-epistemic and social-institutional dimensions of science (Erduran et al. 2019).

7.2.2 *Possible Relationships Between Nature of Science, Education for Citizenship and Social Justice*

The brief outlook in the previous section shows that, despite the polysemy around the expression NoS, this is a fast and steadily expanding field in Science Education research. In addition, there is a wide spectrum of curriculum and pedagogical ramifications involved in making relationships between NoS, Education for Citizenship and Social Justice more explicit.

In their attempt to develop a framework for exploring relationships between Education for Citizenship and Social Justice, Liliana Jacott and Antonio Maldonado (2012) identify three main conceptualisations of social justice in the literature. The first one problematizes the issue of inequality and focuses on fairness in the (re) distribution of resources. The second one emphasises the importance of acknowledging diversity and valuing identities and cultures that face exclusion. Finally, the third conceptualization focuses on a political stance concerning people's participation in decisions that may have an effect on their lives.

Such political dimension has been present in the attempts made by the Science Education community to establish implications of teaching and learning about NoS beyond its epistemological and conceptual dimensions. Connections between NoS and democratic citizenship education go back to Rosalind Driver's arguments for both the inclusion of NoS as a goal for science education and the idea that people must understand NoS to make sense of socio-scientific issues and participate in the decision-making process. Since then, a more holistic approach to NoS, that balances epistemic and social domains of science seem to have gained support by the community (Dagher, Chap. 3 this volume).

Stein Kolstø (2000, 2008) was one of the first to develop an argument that Science Education could play an important role in Education for Citizenship in so far as it involved preparing students for an active, informed, critical and responsible participation in situations in which different aspects of science can improve the quality of student's participation in society. The author claims that one of the main arguments for the relevance of Science Education for promoting Education for Citizenship more generally lies in the recognition of the very nature of democracy itself. He also emphasises relationships between science and society by acknowledging, for instance, personal, social and political dimensions that are related to environmental and human health problems. Furthermore, Kolstø (2008) stresses that, for most people, contacts with science happen in the context of several socio-scientific issues they are confronted with as individuals and as members of society. For this reason, a science curriculum for citizenship would ideally include, amongst other issues, the emphasis on democratic participation, in the sense of providing both psychological and moral bases for taking action. In addition, SE for democratic citizenship presupposes relevant knowledge coupled with skills and attitudes. For the author, rational argumentation about socio-scientific questions would depend on understanding NoS and relationships between science and society. In other words, the awareness of the social aspects of science could help citizens to engage with socio-scientific issues. That would involve not only greater understanding of both conceptual and contextual aspects of science in the consideration of the legitimacy of the scientific debate but also the recognition of uncertainty and argumentation as constituent features of scientific work. Finally, the author argues for the importance of the awareness about the existence of different modes of scientific research and about the role played by conflicting interests in post-academic science (Kolstø 2008).

Despite its widely established importance, there are many questions practical problems involved in articulating Science Education and Education for Citizenship. According to Kolstø (2000), one of them refers to the issue of which content selection would increase students' ability to interpret situations related to science in everyday life. The reason is that issues concerning, for example, values, limits or decision-making strategies in science are too broad to serve as references for classroom work. Thus, it becomes necessary to specify which values and which limits to emphasise, and even which topics would be most relevant. Another worrying issue is the definition of criteria for establishing the relevance of which and how elements of everyday knowledge are to be included in science curricula. According to the author, what seems to be missing is a discussion of how each one of the topics that ends up in the curriculum can contribute to the problems that students are likely to encounter in their adult lives. The third point pointed out by Kolstø (2000) refers to the amount of scientific content needed to deal with a socio-scientific question:

> "It is important to try to identify a basic structure that is within the reach of most students. Decision-making on socio-scientific issues is based on values and to the present, we do not know to what extent knowledge of science can improve the decision-making process. It is important, however, that students do not judge their knowledge insufficient for engagement in matters of their interest (Kolstø 2000, p. 293).

The issue of how the understanding of science itself – as well as of the nature of scientific knowledge – can be transformative of people's lives has been further problematised from a number of other perspectives. For instance, they involve analyses of the role of activism and of radical views on science and technology education in facing contemporary socio-ecological challenges (Alsop and Bencze 2014; Bencze 2017), a critique of capitalist influences on the fields of science and technology and their adverse effects on social justice (Bencze and Carter, Chap. 4 this volume), case studies about ways through which science can be transformative of children living in poverty in urban areas (Barton et al. 2003), considerations about the limitations of functional perspectives of scientific literacy in decision making (Martins 2011) and implications for teachers' education (Zimmermann 2000; Carvalho 2001).

7.2.3 *Aims of the Study*

In our earlier work, we have conduced discursive analyses of ways through which Science Education research results, in general, of NoS related research, in particular, have been recontextualised in the curriculum. Following Bakhtinian assumptions about the nature of language, we have dealt with textbooks as discursive materializations of knowledge, practices, histories, pedagogies as well as both epistemic and contextual values in science (Martins 2006). In addition, we have explored critical discourse perspectives to analyse examples of educational policies in Brazil that have established links between the inclusion of NoS in the curriculum and the achievement goals related to Education for Citizenship (Vilanova 2011).

In this chapter, we problematize some of the philosophical and educational bases of articulations between NoS and Education for Citizenship by means of a critical discourse analysis of Brazilian public policy documents concerning textbook evaluation and distribution. Critical Discourse Analysis is based upon the premise that texts and discourses are moments of social life and, as such, not only do they reflect but also refract social problems, providing important clues to understand and hopefully solve them. This idea justifies (i) our interest in linguistic as well as in historical, cultural and socio-political aspects of texts and (ii) structuring the analysis in terms of the search for links between such aspects.

7.3 Conceptual Bases and Methodological Framing of Critical Discourse Analysis

Having established the origins and the nature of our interest in the relationships between NoS and Education for Citizenship – and of its connections with social justice - we present Critical Discourse Analysis theoretical bases in order to introduce our analysis of both conjunctural and textual dimensions of textbook policy documents.

To a greater or lesser extent, texts are part of all social events. Texts involve both linguistic and non-linguistic elements and aspects of action. According to Norman Fairclough (2003), texts are modelled by, on the one hand, social structures and social practices, and, on the other hand, by social agents, that is, people involved in social events. Social structures are quite abstract entities. We can think of a social structure as defining a potential or a set of possibilities. Examples are the economic structure, the political structure or a language. In addition,

> [...] the relationship between what is structurally possible and what actually happens between structures and events is very complex. Events are not simple and direct effects of abstract social structures. Their relationship is mediated – there are intermediate organizational entities between structures and events. Let's call them 'social practices' (Fairclough 2003, p. 23).

In order to explore linguistic relations that are established in the sphere of social practices, Fairclough (2003) mobilizes the Foucaultian concept of order of discourse. An order of discourse is a network of social practices in its linguistic aspect, that is, a form of organization and social control of linguistic variation (Fairclough 1992, 2003). From this perspective, it is possible to distinguish boundaries and relationships in the totality of a structured set of social practices in specific social domains. However, these systems undergo a series of regulations, which depend on historical and social scenarios. In other words, they are open and subject to re-articulations and modifications in both its discursive and non-discursive (social) aspects. Therefore, textual analysis is limited when employed without observation of the context in which the text is embedded. For instance, ideological effects of texts cannot be grasped only by textual analysis. In order to discuss how meanings are constructed, it is necessary to observe how texts relate to particular areas of social life. Thus Critical Discourse Analysis seeks to understand social issues through an analysis of their semiotic (textual) representations (Fig. 7.1).

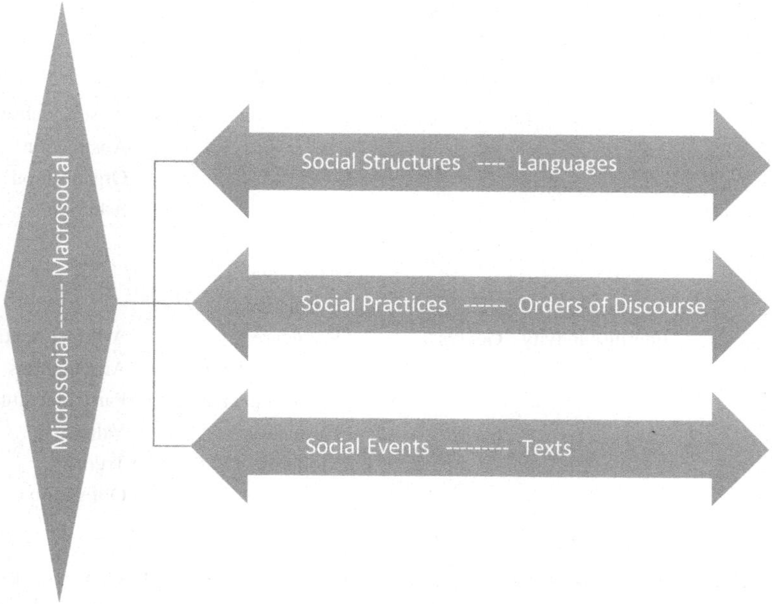

Fig. 7.1 Relationships between social structures and texts

The goal of Critical Discourse Analysis is to explore the nature of the relationships between the language(s) and society. This is done through the articulation of the analysis of conjunctural aspects with the analysis of textual aspects. Conjunctural analysis involves the consideration of social, political, historical, economical aspects of the problem under investigation. Textual analyses problematise both the nature of texts and their inherent heterogeneity. Table 7.1 summarises the subset of categories that were used in the textual analysis performed in this study. Intertextuality and interdiscursivity are the essential elements for understanding processes of textual production. Interdiscursivity allows for an understanding of how discourses dialogue with one another, that is, how different discourses are assimilated in a given discourse. Textual heterogeneity relates directly to the concept of order of discourse and is operationalized in Critical Discourse Analysis through the analysis of genres, styles and discourses. Interdiscursive processes can reinforce hegemonic perspectives, thus neglecting the historical nature of the order of discourse, as well as favour the emergence of new configurations of elements of the order of discourse (Fairclough 1992). Critical Discourse Analysis states that genres are the specifically discursive aspect of actions and interactions in the course of social events. Thus, when we analyse a text in terms of genre, we are asking how it constitutes and contributes to action and interaction in social events (Fairclough 2003). Discourse genres can be grouped into three categories: (i) pre-genres, which encompasses more abstract categories and may be identified as narrative, argumentative,

Table 7.1 Categories used in textual analysis

Textual heterogeneity	Intertextuality	Discursive representations	Themes	
			Perspectives	
			Recontextualisations	Presence/absence
				Abstraction
				Organisation
				Addition
		Presuppositions	Propositional	
			Existential	
			Evaluative	
	Interdiscursivity	Genres	Pre-genres	Abstract relation
			Disembedded genres	Articulations
			Situated genres	Particular situations
		Styles	Evaluation	Values
			Modality	Truth
				Obligation

descriptive etc.; (ii) disembedded genres, which are associated with less abstract categories, such as interviews, scientific articles, lectures; (iii) situated genres, which are specific to particular networks of practices, such as an ethnographic interview or a textbook. Styles, in turn, characterize the extent to which people address each other, which depends on the nature of events, on the relationships between social practices and social structures, and on the capacities of agents. These factors have implications for established dialogues and social differences. Thus, relevant questions are: to what extent is there symmetry among the agents involved in social events? Which communicative strategy results in the reduction of difference or lack of dialogue? In Critical Discourse Analysis, styles can be analysed in terms of evaluations and modalities. Evaluations are the more or less explicit or implicit forms through which authors commit themselves to values. Evaluations are marked in the text, especially in terms of what is desirable or undesirable. Modalities also rely on epistemic (commitment to truth) and deontic (commitment to obligation/necessity) levels.

Finally, discourses, according to Fairclough (1992), configure particular forms of text construction, in the sense that they emphasise certain contents and areas of knowledge according to institutional interests. In this sense, discourses can be understood as the set of statements that, articulated through language, express the values and meanings of different institutions (Meurer 2005). To identify different discourses in a text, we can think of discourse as representing: (a) a particular area of the world (including areas of social life), i.e. the 'main themes', and (b) a particular perspective on the themes (Fairclough 2003). According to Critical Discourse Analysis, we can understand discursive representations as recontextualisations. That is to say, discourse elements are selectively filtered and modified depending on how they are appropriated and incorporated in new contexts. This explains the different ways in which a particular type of discourse can be represented in different fields, networks of social practices, and genres. From this perspective, these selection practices can be analysed through the following categories: (i) presence, that is, the prominence or 'flatness' of discourse elements that are involved in the text; (ii) abstraction, or degrees of generalization of a concrete event; (iii) organization, i.e., how texts are organized; (iv) additions, that is, what is added to the represented speech, for example, explanations, legitimations, evaluations. Presuppositions are represented in the text by propositions that are considered by their producers as something already established. They are a way of incorporating elements – that may or may not have been created by the subjects involved – in the production of the text to be analysed. Systems of values associated with presuppositions can be seen as pertaining to particular discourses. For example, in a neoliberal political and economic discourse, we have the assumption that anything that favours efficiency and adaptability is desirable (Fairclough 2003). In many cases, presuppositions and discourses with which they are associated are considered ideological. According to the author: "The ideological work of texts is connected with hegemony. Search for hegemony is a matter of seeking to universalize particular meanings with the aim of acquiring and maintaining domination, and his work is ideological" (Fairclough 2003, p. 58). Presuppositions are often difficult to challenge and may be sincere or manipulative. For Critical Discourse Analysis, presuppositions can be propositional (assumptions about what something is, can, or could be); existential (presuppositions about what exists) or evaluative (presuppositions about what is good or desirable). Finally, for Critical Discourse Analysis it is essential to be explicit about the perspective through which the problem is to be approached – including the issues in focus, social theory and the discourse theory upon which analyses are based. In other words, the analysis cannot be reduced to applying categories contained in a pre-existing analytic structure.

7.4 Results and Discussion

7.4.1 Conjuncture Analysis

In this paper we present an analysis of conjunctural aspects of Brazilian educational policies and recommendations regarding both NoS and EC, followed by textual analyses of textbook policy texts. While the former set of analyses highlights cultural and socio-political aspects of educational discourse on the theme, the latter explores heterogeneity in terms of discursive representations in texts. Doing so makes it possible to discuss relationships between NoS and Education for Citizenship in two ways. Firstly by identifying which views of citizenship and of Education for Citizenship are mentioned in the policy texts or, in other words, to which discourses textbook policies seem to respond. Secondly by discussing meanings which are constructed by policy texts through connections and interfaces between SE and Education for Citizenship.

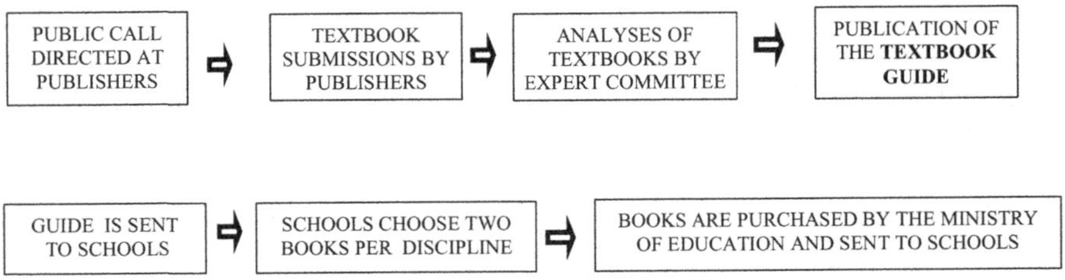

Fig. 7.2 Stages of the Brazilian Textbook Evaluation Program operationalization

The current Brazilian Educational Act (Brasil 1996) was promulgated in 1996. It reinforces ideals of curriculum flexibility and introduces new evaluation practices. Following this Act, the Brazilian Ministry of Education issued curriculum guidelines (Brasil 1997) containing recommendations for schools to elaborate their own pedagogical projects and curricula. Policy texts produced in this period were marked by guidelines that emphasise the affirmation of citizenship[2] and the consolidation of democracy, under the regulation of the 1988 Constitution (Brasil 1988), promulgated after the end of the military dictatorship in 1985 and popularly known as the Citizen Constitution.

New evaluation policies were also put forward, including the National Textbook Evaluation Scheme.[3] This public policy regulated the evaluation, purchase and distribution of books for Brazilian public primary and secondary schools. It is, in fact, a program of huge proportions, which distributes books on the scale of millions (Hofling 2006). The following diagram shows the two stages of the process (Fig. 7.2).

Following a public call, publishing houses submit exemplars of their relevant titles (Student's and Teacher's Books) to be evaluated. The evaluation team is chosen amongst a number of institutional applications through yet another public call. Reviewers have expertise in science education research and school science teaching. It is not uncommon that panels are composed of science education researchers, teacher educators and experienced science teachers.

[2] According to National Curriculum Guidelines, "The aims of Natural Sciences in primary education are designed so that the student develops skills that allows him to understand the world and act as an individual and as a **citizen**, using scientific and technological knowledge" (Brasil 1997, p. 31).

[3] It is important to mention that actions of regulation of the production of textbooks in Brazil are a part of the recent history of public education in this country and started in 1929 when the National Book Institute was created. It stimulated the growth of the production and circulation of national textbooks. Since then, textbook distribution has been carried out by different governmental agencies through various initiatives. In 1938, the National Textbook Commission was created, establishing the first national policy on legislation and control of the production and circulation of textbooks in the country. In 1945, the State consolidated the legislation on the conditions of production, importation and use of the textbook, restricting to the teacher the selection of the books to be used by the students and in the 1970s, the government started purchasing and distributing textbooks to schools, though not all of them were supplied with books (for a full account of Brazilian textbook Evaluation and Distribution policies, please see http://www.fnde.gov.br/component/k2/item/518-histórico)

Evaluation is based on compulsory criteria established by the Ministry of Education, such as: (i) absence of conceptual errors, including in images and proposed activities; (ii) methodological coherence between theoretical-methodological bases and pedagogical proposal; (iii) respect to ethical issues concerning gender, ethnicity, secularism etc. There are also qualifying criteria, which are used to classify the titles as recommended or non recommended. They are: (i) contribution to the promotion of citizenship; (ii) suitability of pedagogical approach and of suggestions presented in the Teacher's Book and; (iii) quality of the editorial project and layout/publishing aspects. Moreover, in the case of Natural Sciences, additional specific criteria include: balanced treatment of concepts from different subject areas (Physics, Chemistry, Biology), use of analogies, interdisciplinary approaches, inclusion of practical activities; development of communicative skills in the languages of science; construction of abilities and scientific values, orientations for critical use of internet sources, amongst others. Furthermore, it is possible to find references to elements that are discussed in SE research. For instance, references to social aspects of science and on critical views of technology connect to the STS (Science Technology Society) agenda. Emphases on the historical character of scientific activity relate to HPS (History and Philosophy of Science) approaches. The idea that science is a collective enterprise and the importance attributed to experimentation can be seen as linked to NoS tenets. Such references are not unlikely to be made in Brazil, as there is quite an active and productive research community, well in tune with international research trends.

Written reports about the main features of recommended textbooks are then published in the form of a guide, which also contains details of the objectives and methodology used in the evaluation. The guide is sent to schools so that teachers have qualified information to select textbooks they wish to work with. Teachers are not obliged to choose amongst recommended textbooks. However, if they do so, there is a commitment by the Ministry of Education to buy them and send them to the schools. Each selection round is valid for 3 years and textbooks should be reused during this period. Thus, the evaluation process has a strong influence on the choice of textbooks that will be actually used in school for a period of 3 years. There is a wide recognition that the continuous and systematic evaluation allowed for an improvement in quality of textbooks over the years. Moreover, such distribution programmes represented an increase in access to pedagogical resources especially by disadvantaged social groups.

7.4.2 Textual Analysis

7.4.2.1 The *corpus*

In this research we analyse the 2008 Brazilian Textbook Guide (Brasil 2007). The reason for choosing this edition is twofold. Firstly, it presents Education for Citizenship as one of the goals for basic education. Secondly, it is the only one, amongst the other six publications of the period that followed the Brazilian Educational Act, to explicitly mention the term Nature of Science.

The 2008 Textbook Guide is organized in a six-piece volume and is available online. The first volume explains the overall objectives and methodology of the program and the remaining five give details about both the assessment and the results of the evaluation processes conducted for books in each one discipline of the four final years of Primary Education[4] (History, Geography, Science, Portuguese Language and Mathematics).

The volume that corresponds to Science is organized in three main sections, which addresses the following issues: (i) the epistemological, curricular and pedagogical assumptions that guided the elaboration of the evaluation criteria; (ii) a comparison between the books in terms of the evaluation criteria and; (iii) results of the evaluation of each one of the textbooks that were recommended.

Our *corpus* consists of excerpts from the 2008 Science Textbook Guide that presented references to and articulations between NoS and Education for Citizenship. The analyses is organised by the categories of genres, styles and discourses.

7.4.2.2 Discourses

The presentation of selected excerpts follows their sequence of appearance in the Guide itself. Excerpts 7.1, 7.2, 7.3, 7.4, 7.5, 7.6, and 7.7, from the first section, refer to underlying assumptions in the elaboration of evaluation criteria whereas excerpts 7.8 and 7.9, from the second section, present evaluation criteria themselves.

Excerpt 7.1

1	The analysis focused on scientific, methodological pedagogical, ethical and aesthetic aspects
2	defined according to the new presuppositions for the teaching of Sciences,
3	configured by the research in the field
4	and current curricular guidelines (Brasil 2007, p. 11).

In lines 3 and 4, we find an example that articulates, on the one hand, research results in the field of science education and, on the other hand, current curricular guidelines. It suggests that such discourses – which are linked to different social practices, namely, academic research and curriculum policy – are relevant and necessary for the proposition of a new set of assumptions for science teaching. In doing so, they establish science teaching as a locus of encounter/production of new discourses, which are legitimised as providing not only the basis but also a focus for the evaluation.

[4] Brazilian Primary Education runs from 5 to 14 years old.

Excerpt 7.2

1	In a society in which **the daily lives of citizens** are increasingly imbued with information
2	and artefacts derived from knowledge produced by science and technology,
3	science education stands out as one of the important components of basic education.
4	As the textbook is one of the resources mostly used by teachers
5	it represents a challenge for the advancement of learning in science,
6	both in terms of its production, of its use and of its adequacy to the contexts of school education.
7	To face this challenge it is necessary to bring to the debate questions such as
8	• What is learning and teaching science?
9	• **How do language and new modes of communication integrate into science teaching and learning**?
10	• **What are the relationships between the nature of science, school science, and its teaching?**
11	• What contents to teach in Science?
12	• **What ethical and citizenship issues pertain to teaching and learning in science?**
13	• What is the role of the teacher in science education?
14	(Brasil 2007, p. 13)

This excerpt reveals the importance of science as an essential component of general education. Also, there is an acknowledgement of the pressure suffered by textbooks, given their widespread presence and use in schools, to respond to current educational challenges. In line 1, the text stresses the role of scientific knowledge in the life of citizens, establishing the need for dealing with science and technology in a daily basis. Facing up to the challenge requires input from science education research, with a particular focus on issues such as (i) the role of language and communication in science teaching and learning (line 9), (ii) the nature of science and the nature of school science (line 10) and (iii) relationships between ethical and citizenship issues with science teaching and learning (line 12).

Other aspects concerning forms through which research can be connected with the facing up of such challenges are analysed in excerpts 7.3, 7.4, 7.5, 7.6, and 7.7.

Excerpt 7.3

1	A scientific education implies understanding learning as a process of construction,
2	with gradual **appropriation of the discourse of science**.
3	This implies **breaking with a culture routinized by some theories and practices**
4	**that are presented uncritically in schools** and introducing the practice of research into teaching actions
5	(Brasil 2007, p. 14)

In excerpt 7.3 we have two presuppositions that establish what is involved in teaching science. The first, of existential nature, highlights what is considered indispensable in a scientific education, namely, the appropriation of the discourse of science (line 2) and the rupture with an uncritical and routinized culture (lines 3 and 4). The second, which has a valorative character, presents research as both a desirable and necessary component of teaching (line 4).

Excerpt 7.4

1	Asking questions and producing answers, communicating them to others
2	is the essence of the construction of a process of argumentation and communication that assists
3	in the **learning of knowledge and procedures of science**.
4	**These processes help the citizen to understand how science works and how to use it in solving**
5	**the problems of their daily life, as well as to give an opinion on the questions that involve**
6	**scientific and technological knowledge.**
7	(Brasil 2007, p. 14)

In excerpt 7.4 we identify intertexts with 'consensus view' related discourses, with an emphasis on the idea that the way science works is grounded on a procedural dimension (lines 3 and 4). In lines 2 and 3 there is a propositional presupposition that distinguishes between a conceptual and an instrumental dimension of science. Nonetheless, both dimensions are articulated with Education for Citizenship (lines 4–6) through the claim that such an approach to science could help citizens in two ways: in coping with problems found in daily life and in forming opinions about scientific issues.

Excerpt 7.5

1	Understanding and accepting that science education is to operate with scientific knowledge rather than simply storing it
2	implies an **understanding of the nature of science and its modes of functioning**
3	(Brasil 2007, p. 16)

Excerpt 7.6

1	A science education based on such assumptions, together with a reconstruction of student knowledge,
2	also **aims at understanding the nature of science, its modes of functioning.**
3	**Learning science is also to know the methods of science, its difficulties.**
4	**and how scientific theories are produced over time.**
5	It is learning how to manipulate their tools and how to use them in social reconstruction.
6	It is to **recognize that every discovery has an author and a social and historical context.**
7	**When science is conceived as human production, the vision of its nature and its limits is enlarged.**

8	Learning sciences thus ceases to be an activity that aims simply to dominate specific knowledge
9	to be understood as the appropriation of another culture
10	(Brasil 2007, p. 16).

Excerpt 7.7

1	Overcoming the understanding of learning as storage of knowledge and *accepting* that
2	**to be involved in the processes of science implies broadening the concept of**
3	**what is to be learned in science, including, in addition to concepts, principles and theories,**
4	**attitudes and values of science.**
5	(Brasil 2007, p. 17).

In excerpts 7.5, 7.6 and 7.7, we highlight the claim for understanding the nature of science as both a component and a goal of science education. In excerpt 7.5, by contrast to more traditional science education approaches – in which learning is understood solely as a kind of knowledge storing process (lines 1 and 2) – there is a demand for introducing NoS related topics in textbooks. It is worth noticing that, in addition to understanding, learners must accept (excerpt 7.7, line 1) that science education involves broadening the scope of the object of science teaching and learning beyond conceptual dimensions in order to include attitudes and values. This configures a propositional presupposition. Finally, excerpt 7.7 represents the discourse of NoS 'consensus view' by assuming that learning science is also to know aspects related to science methods, difficulties, contexts and culture (excerpt 7.5, line 2; excerpt 7.6, lines 2, 3, 4, 6 and 7; excerpt 7.7, lines 2–4).

Excerpt 7.8

1	**Assuming learning as a process of knowing how to with scientific knowledge implies**
2	**valuing the practice of citizenship in science education** and
3	aiming at the formation of social subjects that are participants in society.
4	(Brasil 2007, p. 18)

Excerpt 7.9

1	**By learning sciences from this perspective**, students become part of human beings' efforts to
2	increase their understanding of the environment in which they live and to be able to intervene in it.
3	**It is not enough to acquire knowledge, but you must know how to handle it in order to**
4	**resolve new problems that constantly emerge in your environment.**
5	**This is a true practice of citizenship** (Brasil 2007, p. 18)

Excerpt 7.8 presents a propositional presupposition that establishes a link between the inclusion of NoS in the curricula (line 1) and the practice of citizenship (lines 2 and 3), mobilising elements of discourses of democratic and participative citizenship (Vilanova 2011). Excerpt 7.9 elaborates this assumption, stating that considering learning from this perspective allows people to operate with scientific knowledge in order to solve new problems and to intervene in reality (lines 2 to 4), implying that to be a practice of citizenship proper.

7.4.2.3 Genres and Styles

According to the definitions presented, the 2018 Textbook Guide would be an example of a disembedded genre, that is, a less abstract type of genre as it circulates within a given order of discourse and is tied to specific sets of social events. Guides generally provide orientations, suggestions and descriptions, according to the purposes of the social activities in which they play a part. There is an inherent normative dimension to them as well as an expectation that readers should follow the advices they present. In this particular case, there is yet another relevant aspect, namely the nature of the discursive practices in which this guide is supposed to take part in. They involve particular discursive interactions between members of schools' pedagogical teams, mediated by the text itself, in a context of making decisions about which textbook is to be chosen. Apart from its role in providing relevant input to inform teachers' selections, there is the potential contribution to foster reflection and to contribute to teachers' professional development by means of offering research based insights, besides suggestions for change. Some suggestions are presented as recommendations towards a renewal of practices while others call for the substitution of what are deemed to be inadequate approaches and strategies. Such features can be grasped through the analyses of styles, especially of the forms of modality in excerpts presented at the first section of the Textbook Guide and which are described below:

1. This **implies breaking** with a culture that is routinized by some theories and practices (Brasil 2007, p. 14).
2. A science education **based** on such assumptions, together with a re-construction of student knowledge, also aims at understanding the nature of science, its modes of functioning. **Learning science is** also to know the methods of science, its difficulties and how scientific theories are produced over time. It is learning how to manipulate their tools and how to use them in social reconstruction. **It is to recognize** that every discovery has an author and a social and historical context (Brasil 2007, p. 16).
3. **Overcoming the idea** of learning as storage of knowledge and accepting that it is important to get involved in the processes of science implies broadening the concept of what is to be learned in science (Brasil 2007, p. 16).
4. **Assuming** learning as a process of knowing how to operate with scientific knowledge **implies** valuing the practice of citizenship in science education (Brasil 2007, p. 18).

5. To this end, **it is essential to replace** the conventional curriculum – centred on the transmission of classical scientific knowledge, which is of little relevance to students' lives – by a curriculum centred on the development of knowledge, skills, values and attitudes, aiming at the formation of citizenship (Brasil 2007, p. 18).

The analysis of styles allows the identification of communicative strategies mobilized in the text. For instance, there is a prescriptive tone in statements 1, 3 and 4. Not only do they address the need for changes in science teaching, but they also assert what are indispensable goals to be achieved. Furthermore, the idea of change is presented as necessary and marked through the use of verbs such as *to break*, *to overcome* and *to assume*. These deontic modalities, that is, formulations that denote the normative character of something, express a commitment to what is considered essential in science education. In example 2, we find a series of assertions that define what teaching science is. Modality here is epistemic, that is, the statement expresses a degree of commitment to something that is true. In this example, we notice a high degree of commitment to the insertion of NoS related perspectives in the textbooks. Likewise, Example 3 shows the same degree of commitment by claiming that replacing the curriculum is essential for change to happen.

7.4.3 Discussion

Our analyses raises a discussion about how NoS and Education for Citizenship have been articulated in Brazilian curricula, in general, and in the context of the 2008 Brazilian Textbook Guide, in particular.

A first point to be made is that the Guide both reflects and refracts conjunctural aspects of the Brazilian educational scenario in so far as there are instances of both public policy standards and of Science Education research results being recontextualised in a document that effectively reaches teachers and demands their positioning with respect to decisions about what to teach in their classrooms. Such intertexts may be linked to the presence of science educators and science education researchers in the expert panels.

One interesting point is that many aspects of the relationships between NoS and citizenship are overtly presented and explicitly addressed in the evaluation criteria. Therefore, it is possible to claim that, in the context of the Guide, NoS is seen as key element in Education for Citizenship.

Although relationships between policy texts and science education research discourses can be regarded as positive, we also believe that they must be problematised and qualified when it comes to establishing links between NoS and Education for Citizenship. The reason is that working with NoS in the curriculum may not be sufficient as an enabling factor for students' participation in society. One reason is that NoS related discourses found in the Guide are mostly connected to tenets listed under the 'consensus view' formulation, which reinforce epistemological dimensions of science over socio-political, economic or historical ones (see excerpts 7.4, 7.5, 7.6, and 7.7). We tend to think that it is unlikely that focussing mostly on epistemological aspects could alone educate students to participate in societal matters. Such participation would depend on several other skills such as the development critical thinking and argumentation, as well as on considerations about features that characterise the democratic regime under which students live. The debate about problematising the possibilities of decision making in different (democratic) societies has been expressed in previous studies, where we have stressed the importance of considering cultural, historical and political dimensions in the promotion of scientific literacy related research and practices (Martins 2011).

We also think that an additional focus on cultural and social dimensions of NoS would allow for the consideration of relationships between scientific knowledge and other types of knowledge in a way that did not pose strong asymmetries between them. Scientific knowledge has enjoyed a higher degree of social legitimation and is seen as privileged in terms of objectivity and reliability when compared to other types of knowledge, for example those labelled as indigenous knowledge, popular knowledge or even though those connected to the Humanities. This point should be reinforced as there are cases in which scientific knowledge may not be the most important to inform decisions, for instance, when economic, social and cultural aspects tend to play an important part. In other words, contextual factors sometimes play a more important role in decision-making than considerations about issues of neutrality and objectivity or any specific knowledge held. Understanding the characteristics and limitations of science and technology, however, can serve as a tool for citizens' engagement, both in the production of more reasoned arguments and in the identification of technocratic arguments in debates on controversial issues (Aikenhead 1985, p. 451). Moreover, the consideration of situations where scientists are dealing with real-world problems, that is, where scientific models are articulated to a given context, could contribute to student participation in two ways: as a tool for interpretation of situations when scientific predictions fail and as the perception of the pertinence and legitimacy of applying scientific models to new and/or complex situations.

Finally, the analyses warn us that some articulations between NoS and Education for Citizenship would run the risk of remaining under the logic of establishing which is "the best possible knowledge" for decision-making. This is to say, teaching about NoS would not necessarily lead to critical positioning of citizens on scientific topics, if what is being done is to replace conceptual content with epistemological content, allowing once again the critical capacity of students to be eclipsed by contents that "must be learned".

7.5 Limits and Possibilities for Articulating the Nature of Science and Education for Citizenship

The analysis of the styles highlights a series of prescriptions and assertions about science education. For Critical Discourse Analysis, styles characterize ways through which people address each other, which depends on the nature of the events and on the capabilities of the agents. These factors may have implications for the discursive dynamics between reader and text author, which is mediated by the text. In this respect, we must consider the actual scenario in which the Guide is read and textbook choices are made in schools.

The analysis of styles show that the text leaves no room for views that may be contrary to the belief that an articulation between NoS tenets and Education for Citizenship principles would favour educational practices related to decision-making and students' participation in society. However, the use of imperatives and the disqualification implied by the verbs 'overcome' and 'replace' when employed in connection with teachers' practices suggest that the text in the Guide should be read as a prescription. One undesirable consequence could be to make teachers feel destitute of their prerogative of choosing the textbook according to their own readings of the experts' evaluations. In other words, the analysis of the discursive representations, of the presuppositions and relations between genres, styles and discourses in the texts suggest that there could be universal parameters based upon which to teach about science in textbooks. They would be granted by an external source of knowledge, namely, the consensus of experts (Alters 1997, cited by Moura 2019).

It is also possible to argue that tensions that are present in current Science Education debates about curricular tradition and teacher education are silenced in the text. This is textually realised, in terms of style, by means of a communicative strategy, typically associated to curriculum reform texts, which, by foregrounding ruptures and substitutions, end up creating quite asymmetrical positions between the author of text and the reader. As a consequence, the text lacks dialogicity and goes against the recommendations to foster critical thinking involved in both NoS and Education for Citizenship discourses. Because policy texts are continuously recontextualised as they move across the various institutions of the school system, this lack of dialogicity can lead to a low degree of engagement with such discourses in classroom work.

In summary, the analyses suggest that 2008 Textbook Evaluation Guide values articulations between NoS with Education for Citizenship. Furthermore, it reflects a dialogue between public policy and educational research results. As decisions about content, methodologies, and innovations to be implemented in science classes are embedded in socio-political contexts (Moura et al., Chap. 8 this volume), further research could explore the extent to which conjunctural aspects might have been influential in encouraging different types of articulations between NoS and Education for Citizenship. A discussion of curricular and pedagogical implications of such decisions seems of utmost importance in the context of recent changes in the Brazilian contemporary educational scenario (Brasil 2018).

References

Aikenhead, G. (1985). Collective decision making in the social context of science. *Science Education, 69*(4), 453–475.

Allchin, D. (2011). Evaluating knowledge of the nature of (whole) science. *Science Education, 95*(3), 518–542.

Alsop, S., & Bencze, L. (2014). Activism! Toward a more radical science and technology education. In J. Bencze & S. Alsop (Eds.), *Activist science and technology education. Cultural studies of science education* (Vol. 9). Dordrecht: Springer.

Alters, B. J. (1997). Nature of science: a diversity or uniformity of ideas? *Journal of Research in Science Teaching, 34*(10), 1105–1108.

Amador-Rodríguez, R Y., Adúriz-Bravo, A. (2017). Concepciones emergentes de naturaleza de la ciencia (NOS) para la didáctica de las ciencias. *Enseñanza De Las Ciencias: Revista De investigación Y Experiencias didácticas*, [en línea], 1, n.° Extra, pp. 3499–3504.

Barton, A. C., Ermer, J. L., Burkett, T. A., & Osborne, M. D. (2003). *Teaching science for social justice*. New York: Teachers College Press.

Bencze, L. (2017). *Science and Technology Education Promoting Wellbeing for Individuals, Societies and Environments: STEPWISE*. Netherlands: Springer.

Brasil. (1988). http://www.planalto.gov.br/ccivil_03/constituicao/constituicaocompilado.htm. Accessed 12 June 2018.

Brasil. (1996). http://www.planalto.gov.br/ccivil_03/Leis/L9394.htm. Accessed 12 June 2018.

Brasil. (1997). http://portal.mec.gov.br/seb/arquivos/pdf/livro04.pdf. Accessed 12 June 2018.

Brasil. (2007). http://www.fnde.gov.br/home/index.jsp?arquivo=livro_didatico.html#pnld. Accessed 12 Dec 2018.

Brasil. (2018). http://basenacionalcomum.mec.gov.br/wp-content/uploads/2018/12/BNCC_19dez2018_site.pdf. Accessed 21 Jan 2018.

Carvalho, L. M. de (2001) A natureza da ciência e o ensino de ciências naturais: tendências e perspectivas na formação de professores. *Pró-posições, 12*(1) (34), 139–150.

Erduran, S., & Dagher, Z. (2014). *Reconceptualizing the nature of science for science education: Scientific knowledge, practices and other family categories*. Dordrecht: Springer.

Erduran, S., Dagher, Z. R., & McDonald, C. V. (2019). Contributions of the family resemblance approach to nature of science in science education – A review of emergent research and development. *Science & Education, 28*, 311–328. https://doi.org/10.1007/s11191-019-00052-2.

Fairclough, N. (1992). *Discourse and social change*. London: Polity Press.

Fairclough, N. (2003). *Analysing discourse: Textual analysis for social research*. London: Routledge.

Hodson, D., & Wong, S. L. (2017). Going beyond the consensus view: Broadening and enriching the scope of NOS-oriented curricula. *Canadian Journal of Science, Mathematics, and Technology Education, 17*(1), 3–17. https://doi.org/10.1080/14926156.1271919.

Hofling, E. (2006). A trajetória do ProgramaNacional do LivroDidático do Ministério da Educação no Brasil. In H. Fracalanza & J. Megid-Neto (Eds.), *O livro didático de ciências no Brasil* (pp. 20–31). São Paulo: Komedi.

Irzik, G., & Nola, R. (2011). A family resemblance approach to the nature of science for science education. *Science & Education, 20*(7–8), 591–607.

Irzik, G., & Nola, R. (2014). New directions for nature of science research. In M. Matthews (Ed.), *International handbook of research in history, philosophy and science teaching* (pp. 999–1021). Dordrecht: Springer.

Jacott, L., & Maldonado, A. (2012). Social justice and citizenship education. In P. Cunningham & N. Fretwell (Eds.), *Creating communities: Local, national and global* (pp. 511–517). London: CiCe.

Kolstø, S. (2000). Consensus projects: Teaching science for citizenship. *International Journal of Science Education, 22*(6), 645–664.

Kolstø, S. (2008). Science education for democratic citizenship through the use of the history of science. *Science & Education, 17*(8–9), 977–997.

Lederman, N. G., & Abd-El-Khalick, F. (2007). Nature of Science: Past, Present and Future. In S. K. Abell & N. G. Lederman (Eds.), *Handbook of Research on Science Education* (pp. 831–880). Mahwah, NJ: Lawrence Erlbaum Associate Publishers.

Lederman, N. G., Abd-El-Khalick, F., Bell, & Schwartz, R. (2002). Views of nature of science questionnaire: Toward valid and meaningful assessment of learners' conceptions of nature of science. *Journal of research in science teaching, 39*(6), 497–521.

Martins, I. (2006). Analisando livros didáticos na perspectiva dos Estudos do Discurso: compartilhando reflexões e sugerindo uma agenda para a pesquisa. *Pro-Posições, 17* n. 1 (49), 117–137.

Martins, I. (2011). Literacy as metaphor and perspective in science education. In C. Linder, L. Östman, D. A. Roberts, P.-O. Wickman, G. Ericksson, & A. McKinnon (Eds.), *Exploring the landscape of scientific literacy* (pp. 90–105). New York: Taylor and Francis.

Martins, I. (2020). Science textbooks: A discursive perspective. In C. N. El-Hani, M. Pietrocola, E. F. Mortimer, & M. R. Otero (Eds.), *Science education research in Latin America* (pp. 325–342). Leiden: Brill/Sense.

Matthews, M. (2012). Changing the focus: From Nature of Science (NOS) to Features of Science (FOS). In K. S. Taber (Ed.), *Advances in nature of science research* (pp. 3–26). https://doi.org/10.1007/978-94-007-2457-0_1.

McComas, W. (2008). Seeking historical examples to illustrate key aspects of the nature of science. *Science & Education, 17*(2–3), 249–263.

McComas, W., Clough, M., & Almazroa, H. (1998). The role and character of the nature of science in science education. In W. McComas (Ed.), *The nature of science in science education* (pp. 3–40). Dordrecht: Kluwer.

Meurer, J. (2005). *Gêneros textuais na análise crítica de Fairclough.* In J. Meurer, A. Bonini, & Motta-Roth (Eds.), *Gêneros: teorias, métodos, debates* (pp. 56–69). São Paulo: Parábola Editorial.

Moura, C. (2019). Educação científica, história cultural da ciência e currículo: articulações possíveis. Unpublished PhD thesis, Centro Federal de Educação Tecnológica Celso Suckow da Fonseca.

Osborne, J., Collins, S., Ratcliffe, M., Millar, R., & Duschl, R. (2003). What "ideas- about-science" should be taught in school science? A Delphi study of the expert community. *Journal of research in science teaching, 40*(7), 692–720.

Vilanova, R. (2011). *A cidadania nos livros didáticos de ciências: mudança discursiva, mediações e tensões na dinâmica de produção de coleções didáticas para a educação pública.* Unpublished PhD thesis, Pontifícia Universidade Católica do Rio de Janeiro.

Zimmermann, E. (2000). Modelos de Pedagogia de Professores de Física. Caderno Catarinense de Ensino de Física. *Florianópolis, 17*(2), 150–173.

Chapter 8
Teaching About Sciences in/for the Global South: Lessons from a Case Study in a Brazilian Classroom

Cristiano B. Moura, Iamni Torres Jager, and Andreia Guerra

8.1 Introduction

(…) it's very difficult to be creative having been in prison for so long.

On Tuesdays, as usual, at 7:00 a.m., I pass by guards upon entering the school where I work. This school, located in Rio de Janeiro, Brazil, is a public institution located inside a female prison. This prison unit was built in the 1980s and did not anticipate a school space, therefore one of the cells had to be turned into a school after its construction in order to conform to legislation. For me to arrive in this cell-school, I leave all my personal belongings, in the entrance porch of the penitentiary unit, then I pass through a metal detector and several gates, which are opened by the security agents and locked up again after my passing. On several occasions, during work hours, I forget that I am inside a maximum-security prison.

The story of this article begins with the narrative of a teacher entering a classroom located in one of the largest countries in the world in terms of territorial extent, biodiversity and population. This is also one of the countries of greatest social inequality in the world. According to data from the World Inequality Report, coordinated by Thomas Piketty among others,[1] 30% of Brazilian income is concentrated in the hands of 1% of the country's inhabitants. This concentration index is charac-

[1] Piketty, T., Alvaredo, F., Chancel, L., Zucman, G. World Inequality Report (2019, February, 2nd) Retrieved from https://wir2018.wid.world/

C. B. Moura (✉) · A. Guerra
Centro Federal de Educação Tecnológica Celso Suckow da Fonseca (CEFET/RJ), Graduate Program in Science, Technology and Education, Rio de Janeiro, Brazil
e-mail: cristiano.moura@cefet-rj.br; andreia.moraes@cefet-rj.br

I. T. Jager
Colégio Estadual Primeiro Tenente PM Hailton dos Santos, Rio de Janeiro, Brazil
e-mail: iamni.jager@gmail.com

H. A. Yacoubian, L. Hansson (eds.), *Nature of Science for Social Justice*, Science: Philosophy, History and Education,
https://doi.org/10.1007/978-3-030-47260-3_8

terized by the greatest kind of concentration in the world. Considering wages as an indicator of poverty, we realize that poverty in Brazil is not distributed equally among the different gender and ethnic groups, even between the people with the same academic qualifications: 42% of the workers are female and 33% are male, and in the range of 10 minimum wages, 3% of the workers are men and 1.5% are women. In the case of the dissimilarity between blacks and whites, the inequality is not different. In the range between 1 and 1.5 times the national minimum wage, 40% of workers are black and 31% white, but in the range of more than 10 minimum wages, 1% are black and 4% are white. Brazilian taxation reflects the country's economic inequality as well. Reports on Brazil's socioeconomic profile, such as that made by Oxfam,[2] show that the country imposes a higher tax burden on the poor than on the rich. In discussions about the country's economic situation, inequality is often not the priority issue. Since 2016, the government has put an austerity program in place that has imposed a budget ceiling, interfering directly in social programs that aimed at actions to contain Brazilian social inequality. As a consequence, they have reduced corporate oversight of environmental issues, aggravating national problems. Implementing an austerity program, the government has set since 2016, a debt ceiling that has imposed budgetary conditions.

The Brazilian environmental issue has a great impact to the world. Brazil, alongside Canada, Russia, USA and Australia, is part of a select group of countries that are strategically sensitive to the planet in terms of environment. Studies show that these countries have a proportion of old-growth forests that are fundamental to actions to combat global warming (Watson et al. 2019). In the case of Brazil, there is a large part of the Amazon forest that is under these conditions, and it is fundamental that this area continues with the presence of indigenous people, but without industrial or mineral exploitation. However, the laws mentioned above end up threatening this Brazilian territory, by weakening the guarantee of the right to land for traditional people. These considerations show that Brazil, an economically and socially unequal country where 16 million people live below the poverty line, plays a strategic role for rich nations.

The Brazilian situation frames the science classroom that supports this work and motivates us to choose social justice as one of the main reasons why we should teach science. To develop a science education with a focus on social justice, we understand that the data on inequality presented previously are the product of historical conditions in which scientific knowledge was also involved. Many economic and environmental policy options have been backed by scientific or pseudo-scientific discourses, and scientific and economic development in the world has been connected to the exploitation of diverse resources from elsewhere in the world. In this way, science education should discuss how science was performed throughout history. It is then that we find a field of research in science education that is

[2] OXFAM – Report "País estagnado: retrato das desigualdades brasileiras 2018" (2019, February 2nd). Retrieved from https://www.oxfam.org.br/pais-estagnado

fundamental to carry out the discussion on science. We refer to the field that deals with the well-known and oft-quoted jargon "Nature of Science" (NOS, hereafter).

The discussions about NOS are linked to the so-called social and cultural studies about science. These studies have been considering the situated and partial nature of scientific knowledge for some time (Longino 1991), that is, unveiling questions related to objectivity/subjectivity, universalism, power, and knowledge, as well as the connection between these concepts in the development of science. In the same sense, post-colonial studies have promoted discussions about the asymmetries of power throughout history, regarding the production and circulation of what is now known as modern science (Santos and Meneses 2010). These discussions refer to the politics of knowledge, which have only recently been more explicitly and prominently included in the discussion of NOS (Dagher and Erduran 2016; Yacoubian 2015).

Considering the importance of improving discussions about the politics of knowledge, we propose studying NOS from a postcolonial perspective that, in our defense, helps to create connections of knowledge with contexts in the Global South, historically exploited by the Global North. In addition, we propose addressing NOS as a look at the context of students that is open enough to allow a radical contextualization, where students' life histories are present and where they can raise questions relevant to their training as human beings in an unequal society. In this way, we describe in this paper a historical study on Botany during eighteenth and nineteenth centuries, in which we tried to address the asymmetries of power inside science construction emphasizing, primarily, the role of gender and race. In an ethnographically-inspired study in a school inside a prison, we describe a course in which the historical-cultural approach to Biology allowed the linkage of the gender and race issues in science with the own issues of students' lives in an unequal country like Brazil.

As a result, we intend with this work to provide resources for constructing answers to the question: *how can the discussion of science from a historical decolonial perspective promote a teaching about science that is more connected with issues of social justice linked to realities experienced by the global south?* For this, we will begin with a discussion about the historical construction of the sciences from a postcolonial point of view. Then, we will discuss how the production of discourses about what one should teach about sciences ends up neglecting many contexts for which these discourses are produced. This can lead to fundamental biases being introduced that interfere with the communication of what is being taught. From the theoretical discussion, follows a narrative about science classes held in a classroom within a Brazilian prison context then we conclude with a discussion about how such pedagogical experience relates to the theoretical considerations presented and allows us to come up with answers to the central question of this work.

8.2 The History of Western Modern Science Through Decolonial Lenses

Hodson (2014) points out that much of what is being researched and what is being done today in the field known as History, Philosophy and Sociology of Science (HPSS) in science teaching comes from a critical look at the Modern Science, its possibilities and limits. This was one of the great lessons of both World Wars, and had as its reaction the origin of the so-called Science Studies. If in the past it was believed that the approach of the History of Science in teaching could foster doubts or misrepresentations about the scientific endeavor (Kuhn 1962), today its importance is already a matter of concern as a way for the critical formation of students (Allchin 2011). Because we consider the critical formation of students a crucial point in science teaching, we have been working with the historical approach in NIEHCC[3] for some time (Schiffer and Guerra 2015). However, when we opt for the historical approach in science teaching, we are faced with the fact that the History of Science is a field of research by itself and that, therefore, it has changed and has been changing over time. As Nyhart (2016) points, there are research strands in history of science now that tease out "who has benefitted and who has suffered in its formation" (p. 7), showing new (critical) perspectives on the History of Science.

As examples of new concerns in the History of Science (HS), several studies deal with understanding the role of women in the construction of science (Terrall 2011), others studies historicize certain concepts previously considered almost as static in time, such as the concept of objectivity (Daston and Galison 2007). Other studies discuss material culture not merely as a set of objects but as something that articulates with the social and cultural media of a given time, sustaining scientific practices that end up perpetuating themselves (Klein 2003). These new approaches, many of them located within what is classified as Cultural History of Science (CHS) (Moura and Guerra 2016; Pimentel 2010), are part of the movement that Nyhart (2016) classifies as the task of understanding how science came to be understood and established as science. Also, these studies discuss who are the invisible people in the construction of science (Milne 2015) and why they were hidden, besides highlighting how the material conditions were implicated in these stories.

This important movement in the HS reveals new ways of writing history, based on the concept of scientific practice. According to Gavroglu (2007), the historiographical category of scientific practices focuses the historian's attention on the daily practices of scientists. Thus, instead of considering science as an exclusively or predominantly intellectual activity, the approaches that analyze science as a set of human activities, of which intellectual activity is part, have begun to gain ground. The formation of the identity of the scientists also begins to be analyzed from the dynamics of the relations within the scientific community that support them. Thus, it becomes fundamental to understand what types of values, communicative codes

[3] Acronym in Portuguese for: Research Group in Science, History of Science and Culture (http://niehcc.wordpress.com/).

and practices sustained over time differentiate each academic community (Gavroglu 2007). Perceiving these nodal points can help in identifying the characteristics involved in the formation of new scientists to act in each area, as well as the validation of the research developed in the various branches of knowledge. In addition, understanding this dynamic involves shedding light on the power relations that permeate the scientific enterprise, such as institutional relations, training processes, hierarchy in universities and laboratories, processes of evaluation of scientific production and evaluation for financing applications, among others (Pimentel 2010).

Parallel to this movement on HS, the so-called postcolonial studies have drawn attention to the relationship between scientific development and the historically constituted model of society, with subordination of the Global South (Santos 2002). In this sense, Nyhart (2016) points out that when historians began to look beyond European laboratories and the social structures surrounding them, they were able to find a number of women and men who helped to construct science as collectors, servants, translators, and several other "go-betweens". Even looking at European science with this new perspective on the scientific practices makes it possible to perceive that many women and other historically subalternized social actors participated in the construction of science, such as illustrators, scientists' assistants, among others (Nyhart 2016). This implies realizing that the very image of the scientist or those who do science is rethought in this approach.

Specifically, regarding postcolonial readings on the history of modern science, there are important accounts in the literature that point to the historical conditions that gave rise to this way of knowing the world. According to Santos (2010), Europe experienced modernity only when it came into contact with the "new world", i.e., the Americas, because it was through this contact that the lines of difference that put the colonized peoples under subaltern conditions were established. Santos (2010) brings examples of the law, showing that the American autochthonous ways of life were pointed out as the "Natural State", that is, understood as a pre-civilization stage in which basic notions of social organization were followed. Thus, two paradigms are established: that of universalism, since the European civilization and cognitive model is taken as the standard in relation to other knowledge, and also the dichotomous thinking, which narrows the possibilities of modes of being and knowing to a binary analysis, as in civilized/barbarian, European/non-European, among others (Santos 2002).

From the point of view of Modern Science, Pimentel (2007) outlines a particular profile for the so-called Scientific Revolution. Based on the CHS, the author points out the importance of the process of colonization and exploration of the Americas in European scientific development around the sixteenth century. This process allowed for contact not only with several animals and plants not known by Europeans, but also with new people, traditions and knowledge. More than that, it has also enabled the Great Navigation to promote the development of instruments, astronomy and technical knowledge related to everything that involved navigating previously unexplored seas. New spaces such as botanical gardens, which combine scientific development, recreational and cultural spaces, arise in this context with much material coming from the colonies; the curiosity offices start a collecting practice with materials not only from the colonies, but also with the European innovations in terms of machines and other artifacts. (Pimentel 2007). In addition to a passive position, where colonizers collected plant and animal, as well as records of *in situ* observations, there are studies that show that colonies have contributed more actively to the process of building modern science. Bleichmar (2011), for example, cites that natives of Spanish colonies in America were responsible for the collection and dissection of many unknown specimens in Europe, which were later sent to renowned botanists to be classified and published as new species.

The deletions that occur in the context of colonization are analogous to those occurring in other spheres in the HS. Foundational activities that were performed in history that make up parts of modern science can ultimately be left out of science teaching and therefore the concept of modern science that is taught. This can happen when the role of women in the construction of science is not documented in the official report of a survey (Terrall 2011), or the local colonies knowledge is erased from official records (Bleichmar 2011) This observation highlights the need to extend the historical analysis of science through the concept of scientific practices from a historic point of view. With this extension, a potential for a broader understanding of NOS opens up, highlighting science beyond an "activity performed by scientists". Discussing these deletions is to address the social structures that have been maintained throughout history, perpetuating inequalities and creating a democratic deficit in the power spaces of society. We understand that this movement of discussion about science is fundamental to achieve social justice, since, according to Santos (2010), the abyssal lines that produce the differences and subaltern places in the world continue to be produced all the time, for example, between white and black people, between man and woman, between European and immigrant. Even after decolonization (which has not yet occurred fully), global social injustice continues to be produced from a global cognitive injustice. Therefore, there is no way to pursue global social justice as a goal without doing so in relation to a global cognitive justice. According to Santos (2010: p. 48) "the scientific knowledge has intrinsic limits in relation to the types of real world intervention it makes possible". Santos (2010) also argues that exploring the internal plurality of science, the alternative scientific practices that have been made visible by feminist and postcolonial epistemology is not sufficient to a global cognitive justice. To achieve global cognitive justice, an external plurality of science emphasizing the alternative ways of knowing is necessary, one that includes the knowledge produced by people from the countryside, the forest, among others.

Santos (2010) dissociates the inner plurality of science to the external plurality of it. In an opposite perspective, we consider them as complementary. This is because an understanding of science as a human activity that includes daily practices, that has specific purposes, allowing only certain interventions in the reality, can help to realize that other forms of knowledge, with other assumptions and with other interventions in the reality are equally possible, with diverse consequences for the life of Humankind on Earth. Thus, these two complementary movements are part, in our defense (Moura 2019), of the same movement about uncovering what was concealed in the HS, in order to seek sustainable and socially fair alternatives for humanity.

The proposal of the approach to science in a historical-cultural perspective, as we understand in this text, attempts to reduce this distance through the discussion of scientific practices, seeking to circumvent the various obstacles discussed here. From this approach, therefore, it is possible to discuss the role of women, Blacks, Amerindians in science and also the practices that have also been subalternized, notably practices of dissemination, collection, illustration, among others, highlighting "there is no knowledge without practices and social actors" (Santos and Meneses 2010: p. 9).

8.3 Local Contexts and Ways of Approaching Science Contextually: Mapping Tensions

From the beginning, the approach between the fields of HPSS and Science Education has fulfilled a demand to promote a less abstract and more contextualized science education. In this way, the attention should be paid not only to the products of science, but also to its modes of production, its historicity, and the epistemological and social foundations that underpin it (Matthews 1992). This movement supports the idea that the inclusion of the History of Science in curricula, beyond the Philosophy and Sociology of Science, promotes the humanization of science education. Following Matthews (1992), many have come to refer to this type of approach, blending the History, Philosophy and Sociology of Science into what has been called the "contextual approach".

As a polysemic word, the term "context" refers to senses other than the historical one that Matthews (1992) sought to bring in their work. The discourse on the need for contextualization has been continuously reverberated in the field of science education, as a necessary effort to promote students' understanding on the relations between science and society, be it from a historical or a contemporary point of view. We can therefore circumscribe the so-called historical contextualization to the broader context of education in science.

However, Barton (1998) points out the need to think about contextualization in a broader sense. According to her, in order to approach science for all, an effort is commonly made to contextualize science, which does not take into account the relation of different groups of students to knowledge. That is, the relation of knowledge to students is thought of as a one-way relationship, where something is constructed *a priori* to be taught to students, failing to consider the students' own relationship with the knowledge and the context in which the knower and the known (the students and the knowledge) relate. We understand that the crucial point of criticism is that this type of relationship cannot be determined in advance.

Barton's (1998) criticism is also made by Costa and Lopes (2018) when the authors assert that contextualization is generally treated as an attempt to control the other. That is, in terms of curricular policies, it is the attempt to expel the difference from within the curriculum, from what can escape control, seeking to predict to the maximum what should be addressed, and defining what should be the contextual elements for every situation. With this, it is added to the curriculum these elements of contextualization and, afterward, it is delivered a ready proposal, narrowing the possibility of curricular construction. Barton (1998) also suggests that science education should move towards promoting the joint action of students and teachers in the curricular building, with the aim of situating science in the social, political, physical and historical spheres in each context.

Regarding the HPSS approach in science teaching, it is well known that the proposals of this approach have long been related to the debate about NOS in scientific education. Considering this relationship, we argue that it is essential to recognize that the NOS proposals in the literature run the risk of narrowing the educational experience in different educational contexts, in proposing a way of working with NOS in science teaching that can be adopted uncritically in different educational contexts. These proposals, which end up acting as guides on how to approach NOS in education, do not constitute straitjackets. However, by proposing a summary of the debate and including, in some cases, evaluation proposals, they become easy preys for the accountability, what Miller (2014) calls "test culture". In the "test culture", what is important into the educational phenomenon is the attainment of predetermined performance parameters.

This debate, however, is a two-way street. In the first sense, we can say that it is clear that the conceptual schemas on NOS that we are discussing here cannot predict all the contexts of its application. These general ideas and principles need to be adapted to the realities. On the other hand, in understanding them as utterances, it is essential to emphasize that there is no utterance produced that does not have a supposed addressee (Voloshinov 1973). We can infer that the NOS proposals such as speeches that have arisen in North America or dialoguing with North America have as their intended addressee (although not necessarily explicit) a reality that is experienced by their authors. And as Barton (1998) and Costa and Lopes (2018) highlight, it is natural that part of what is pointed out as essential for one context does not apply to another.

In producing a proposal of NOS that contains this supposed addressee, even if implicit, this proposition ends up in the problem of the "invention of the other", as discussed by Castro-Gomez (2005) in determining that the experience of a particular group (i.e. students in the U.S. urban context) would represent the "standard experience" of every student. For Castro-Gomez (2005), the "invention of the other" is an expression that means the mechanisms through which a particular group of people represents the others, establishing a norm about what the other is, generally in opposition to another representation; for example, the barbarian in opposition to the civilized. This is clear, for example, when Allchin (2011) points out that selective news from the media would be found by the "typical citizen" (p. 519) in their common experience. We may question whether this same experience is valid for Canadian indigenous students, Bolivian peasants or students living in an urban favela in Rio de Janeiro. In this case, the invention of the other is by a process of omission or oversimplification of what might mean an expression like "typical citizen".

The considerations outlined here about the historically established theoretical frameworks for NOS research field are related to two fundamental points: the first is the impossibility of full prediction of the contexts for which NOS models have been and are thought. This results in the need for caution when using frameworks, for supporting the making of lesson plans and assessment tools and/or to enlighten the construction of curricular policies for education. We argue that these heuristic synthesis tools of the NOS field are fundamental as epistemological guides in teacher training, but it is necessary to give room to teacher decisions, celebrating their inventiveness and their perception of the needs of each context. It means that in some contexts it will be necessary even to give up with any NOS framework to make possible NOS discussions in classroom. This leads us to the second point, which is directly connected with the first: it does not make sense, therefore, to propose a theoretical model or framework to address metascientific questions in contexts of the Global South, because even the Global South has a heterogeneity that prevents a general proposition for all its contexts.

We understand that if the field of HPSS in science teaching arose with the objective of "overcoming the 'sea of meaninglessness'" (Matthews 1992) of it, this objective will be better fulfilled if we are able to bring this knowledge closer to the realities of the students. Contextualizing historically only makes sense, therefore, if the stories we tell about knowledge can potentially touch the stories of our students. Hence, the importance of the post-colonial perspectives for the Global South, as they touch the issue of historical erasures, the asymmetries of power and the symbolic violence that often permeates the daily life of these students.

The vision we provide here is not directly related to any of those NOS frameworks in the literature, but tends to be closer to the approaches which emphasize the socio-cultural aspects of NOS for Science Education. This is because we understand that addressing NOS in science education aiming at social justice cannot go without discussing the questions regarding knowledge and power, especially in parts of the world which suffered tremendously with the colonization process. The main message of this theoretical development is that one should be attentive to the contexts of implementation of NOS discussion, avoiding the adoption of one or another NOS framework acritically. If we accept that "contextualizing historically only makes sense, if the stories we tell about knowledge can potentially touch the stories of our students", each context will provide different paths and opportunities to discuss knowledge development and, so, the adoption of frameworks could narrow these possibilities, recalling the "addressee" problem of these frameworks we raised in the previous discussion.

Attentive to this and without seeking the formulation of recipes valid for multiple contexts, we seek in this article to constitute an "exemplar" of how such issues were discussed in a particular context than to constitute a certain framework or norm on how these discussions can be carried out in other contexts. This does not mean that we give up the reproducibility of our "experiment". We assume, as research in education postulates (Erickson 2012), that each educational experience is unique and irreproducible, so what we can do is to describe the pedagogical approaches that have succeeded in their objectives to constitute a collection of inspiration for future propositions in other contexts.

We invite the reader, in the next section, to immerse themselves in the mentioned educational context and to get to know a little of the science classes that we will use in this work to justify some points of view that we adopted. The narrative constructed from the science classes was based on an ethnographic study.

8.4 A Look at a Brazilian Classroom

In this section, we return to the classroom that inaugurated this article. The pedagogical action narrated here took place in a high school Biology class, where five students attended the course. Because they are adult women deprived of their liberty, attendance in the classroom is not mandatory, and many opt to perform other activities that occur at the same school time. The participants of the course narrated here were of different ages and were all black and poor women who were arrested for crimes related to drug trafficking. The majority of them declared that this illicit involvement with the drug trade began at the insistence of their boyfriends or husbands, who already practiced these activities previously. The teacher who conducted the work has a degree in Biological Sciences, and she is engaged in social movements since high school (Jager 2018).

The course lasted 20 h, whose main topic was the study of Botany, a subject of the national curriculum for compulsory Biology in the public schools of Rio de Janeiro. In order to deal with the theme, the teacher organized the classes with a historical approach. The CHS approach was chosen because, according to her, it was possible to discuss with these students that the presence and absence of different social actors in the construction of science is also a cultural issue. Considering the educational space, the teacher understood that it would be fundamental that those women could problematize, during the lessons, the participation of women in the construction of science, and reflect on their conditions as Brazilian women without freedom.

The historical case was related to the evolution of Botany in the eighteenth and nineteenth centuries, which made it possible to teach on scientific contents of the school mandatory curriculum. To begin the course, the teacher presented questions about the Enlightenment in England and in France, discussing practices of production, diffusion and dissemination of the scientific knowledge of that time. All discussions in the classroom were performed from paintings, scientific illustrations and excerpts of texts produced in that historical context that portrayed the daily life of the addressed places, the women involved in these contexts and the studied plants.

The teacher presented the views of philosophers Jean-Jacques Rosseau (1712–1778), Immanuel Kant (1724–1804), and Marquis de Condorcet (1743–1794) on women in that cultural context. The teacher chose extracts from the texts of these philosophers to read and discuss with the students, after a brief explanation of where they lived and how they lived. In the case of Rosseau, the outstanding work was *Emilie or de l'Education* (1762), in which the philosopher deals with the education to be given to the man and his future wife, arguing that women should not learn science or philosophy, as their nature would be in feeling and not in reasoning (Gaspar 2009). The students expressed indignation at the author's thought, labeling him as chauvinist. The teacher pointed out that this was not an individual view of that philosopher and that he was immersed in time and space, which somehow endorsed that view towards women. Thus, the teacher read an excerpt from Kant's *Essay on the Feeling of the Beautiful and the Sublime* in which the philosopher explicitly quotes two women: Madame Dacier (1647–1720), who translated into French the works of the Greeks Homer, Plautus and Terence; and Madame du Chatelet (1706–1749), who published books on Physics, stating that women who engage in Greek or mechanical controversies should have beards, for they would express a semblance of depth that they are so engaged.

The teacher interrupted the reading to explain to the students who the women mentioned by Kant were and realized that the students were very angry and agitated in this class, despite the teacher having pointed out that these were philosophers of another time and that their vision was not hegemonic, even in that context. To reinforce her argument, she then read an excerpt from Condorcet's *Sur l'admission des femmes aux droits de cite*, 1790. Then, the teacher emphasized that the French philosopher denied the existence of natural differences between men and women, having fought to guarantee equal rights and education to both. The teacher also highlighted the philosopher's life with two women: Olympe de Gouges and Mary Wollstonecraft. This moment was important so that the teacher could situate the works of these women and present the painting entitled *Mary Wollstonecraft* by John Opie, 1790. One of the students observing the painting exclaimed: "she looks like an old man!" The others laughed and the teacher took advantage of the comment to reinforce that the image portrayed a simple and cultured woman, emphasizing that this woman was the mother of Mary Shelley, author of the book *Frankenstein, Or the Modern Prometheus* (1820), of which the students were aware. The girls found it interesting the fact that a book that was so prestigious was written by a woman whose mother had fought for women's rights.

This discussion allowed the teacher to highlight some women who have achieved prominence in areas such as literature and science in that sociocultural context. In introducing these women, the students pointed out that they appeared to have a good social standing and were very different from them. To allow the students to reflect more on this question, the teacher proposed to them writing a text, in which they should explain how they understood the issue of the feminine nature pointed out in the discussion of the philosophers' texts. One of the students, even before the activity began, said that women and men have the same abilities, but that "women are more warlike". Another student disagreed with the first, stating that men and women are equal but with different opportunities. In the texts, they presented questions discussed by the teacher in the classroom, relating them to contemporary issues regarding gender equity, for instance.

The texts written by the students inspired the teacher to resume the discussion about the look at the nature of the woman in the historical context studied, from the presentation of the image of the index of the book of Eramus Darwin (Charles Darwin's grandfather), 1797, entitled *A plan for the conduct of female education, in boarding schools, private, and public seminaries*. In this book, Eramus Darwin presents ways to a feminine education and whose first topic is entitled *The female character*. The students made fun of the purpose of the book and questioned the teacher about chastity and *The female character*. At that moment, the teacher spoke a little about female education in that socio-cultural context and asked the students about how they were educated. They then narrated a little about their lives and their daily activities before their arrest. One of them said that she did not know any scientists, which was supported by the others. The teacher participated in the debate by speaking a little about her training, highlighting the opportunities she had to graduate in Biology, pointing out the importance of school and university training, so that a person pursues a scientific career. She also highlighted the barriers faced by Brazilian women to pursue a scientific career, due to the difficulty of reconciling their children and their profession, in a country that does not invest much in science and gender equality. In this way, the teacher pointed out that it is not enough the wish to be a scientist to become one, although it is not needed to be a genius to be one.

The teacher resorted again to the presentation of paintings. The paintings were then presented: Jean-Baptiste Chardin's *Young Schoolmistress*, 1736, *Sleep my child*, 1783, and *Lady Reading in the Interior* by Marguerite Gérard, 1795. The students emphasized that now they understood why Eramus Darwin argued that women should study music, "it was to lull the children to sleep", they commented. They also scored on the women's dresses and one of the students recalled the painting depicting Mary Wollstonecraft and compared it to the one produced by Marguerite Gérard, noting that the two paintings portrayed women reading, but one of the women was portrayed all decorated and the other was not. When the teacher questioned how she judged this difference, she argued that Wollstonecraft was more concerned with looking smart than pretty. At that moment, the girls began to discuss a little of their past, when they left school, when they got married or had children. All of them had become pregnant very young, still in their teens, and, according to their accounts, the difficulty in reconciling study, children and housework made them drop out of school.

The second part of the course was focused on Botany and scientific practices in the eighteenth and the nineteenth centuries, highlighting the practices of dissecting, collecting plants in exploratory journeys, scientific illustrations, reading and writing letters and publishing of scientific books. To discuss the practice of scientific illustration, the teacher used two scientific illustrations of the period. The one depicting flowers[4] was drawn by Jane Loudon (1807–1858) for her book, *Ladies*

[4] Loudon, J. W. (1841) Ladies' Flower-Garden of Ornamental Greenhouse Plants. London: William Smith, p. 59. Disponível em: <https://archive.org/details/ladiesflowerga00loud/page/22> Acesso em: 02/01/2020.

Flower-Garden of Ornamental Greenhouse Plants, written in 1848. The other, depicting a bird,[5] was drawn by Elizabeth Gould (1801–1841) for her husband's book, *A Century of Birds from Himalaya Mountain,* written in 1831.

From the students' impressions of the illustrations, the teacher argued that these women learned drawing techniques and that many women at that time worked by assisting their relatives in scientific activity, illustrating works and recording observations. The discussion of the images also made it possible to highlight the exploratory trips of the time, emphasizing their financing and commercial importance. The teacher then emphasized the scientific practices of identifying and acclimatizing new species, food or medicinal, in the lands of the metropolis and colonies. As an example, she highlighted two very used fruits in Brazil, the coconut and the banana. The students thought that these fruits of Indian origin were native to Brazil. This point attracted a lot of attention from the group, and, drawing on this attention, the teacher exemplified another species commonly found in Rio de Janeiro: the jackfruit. The teacher discussed with the students that the jack tree, with African origin, is now considered by ecologists a threat to the native fauna. The students asked many questions and the teacher discussed how the practice of exploratory journeys brought species from one place to another and why the activity of species transposition can interfere with ecosystems.

After this discussion, the teacher highlighted the different social actors involved in the collection of plants in the exploratory trips, the different activities carried out in those trips, noting that, in general, women and the natives were involved in auxiliary activities, as collection, dissection and illustration of the species to be sent to the metropolis. The teacher also commented on the difficulties of the expeditions and the necessary knowledge for the collection and registration of the species. At that moment, Linnaeus' classification system was discussed, detailing the proposed classification system and that, in Linnaeus' work, there are a large number of analogies between human and plant sexualities. To better discuss this system, the teacher took two flowers from different species collected in the prison and showed the students parts of their structure. The students were able to manipulate the flowers and put questions to the teacher. The teacher reported that the students were more and more participatory: if in the beginning they were restricted to answer questions of the teacher, now they raised questions, many of them not directly related to the historical and scientific content taught.

In order to continue the discussion, the teacher asked the students to represent the manipulated flowers by means of an illustration. The first reaction of the group was to reject the activity. However, at the insistence of the teacher, the students gradually began to make the records. With the drawings in hand, the teacher asked them to tell what difficulties were encountered to carry out the activity. The students mentioned: lack of ability to draw, lack of pencils that matched the colors of the plants, lack of an instrument that allowed them to enlarge the structures of the plants and difficulty

[5] Gould, J. (1831) A century of birds from the Himalaya Mountains. London, p. 20. Disponível em: <https://archive.org/details/centurybirdsfro00Goul> Acesso em: 02/01/2020.

in knowing the important details to be recorded. The teacher then narrated her intention with the activity: to discuss that women and men engaged in the work of scientific illustration, who had specific instruction, knew botany and had trained eyes and they could determine what was considered important to be registered. One student highlighted that "once again white women and privileged classes, because they were women who received education in drawing and sciences".

In order to continue the debate, the teacher presented to the students Jane Colden (1724–1760), emphasizing that the American botanist studied the classification system of Linnaeus with her father, who was a doctor and had contact with botanists of the time. She also emphasized that Jane was responsible for the collection and classification of hundreds of species in the English colony, however, she had her share in Botany always associated with the name of her father, who served as her workmate for years. Jane did not have her name recognized in any species; however, her father was honored at the time with the genus Cadalawer (Gronin 2007). A student said: "once again women doing the dirty work for men" and added "I am not a feminist because I understand that men have the ability to do things that women they cannot, but I believe women are smarter and more observant than men." When asked if she believed that women and men differ in nature, and as a result, they were born with different abilities from men, she agreed.

Later, a student asked if everything had a scientific name. The teacher took the opportunity to discuss the role of scientific nomenclature and the controversy over Linnaeus' proposed sexual classification system. At that moment, the teacher highlighted excerpts from Rousseau's book about a compilation of letters exchanged between Rousseau and a duchess to give guidance on Botany to her daughter. This book, directed to the female audience, had many editions. The excerpts read for the students were taken from a translation of the work, made in 1801 in Portugal and signed by "Huma Senhora da Corte" (a lady of this Court). Before even reading the passage, a student intervened, recalling Rousseau's considerations about the nature of the woman seen in previous classes and said that "his book might not talk about sexual system". The teacher corrected the student, emphasizing that he spoke of this system, yet he declared in his writings that it was a forbidden knowledge, a mystery to be unveiled. When asked why the translator had signed "By a Lady of this Court," one student replied, "it was a way for the woman not to look disgusting", and another complemented, "not to appear overconfident".

The teacher resumed the discussion of female participation in Botany and highlighted the names of women who wrote botanical books at the time, but who did not use the sexual classification system proposed by Linnaeus. She also highlighted the book of Maria Jacson, 1797, which used the Linnaeus system. About this book, the teacher reported that Eramus Darwin wrote the preface to the book, seizing the opportunity to read three small parts of the book, in which Eramus Darwin presents an analogy of the human imaginary with the vegetable kingdom, in the form of poetry. One student pointed out, "so it was very easy to record the characteristics of that group of plants." The teacher took advantage of the intervention of this student to ask them to write a poem in the style of the one that they had just read. After some resistance, they asked if they could build the poem together, and, with the teacher's agreement, they fashioned the following poem:

> *Malvaviscus arborius*
>
> *There were several women with red hair and velvety skin that were catching the attention of the various ladies who were around, the ladies were sweet as a goblet of clear water.*
>
> *Ipomoea cairica*
>
> *There were many purple and lilac colors in that multitude of colors in which 5 boys were worshiping a beautiful girl who was there alone and unprotected with her fragile beauty at the mercy of time on a cold night with much wind and that brought the snow that was falling insistently.*

The teacher read the poem aloud. In order to conclude the lesson and in the sense of problematizing the participation of women and social minorities in science, the teacher presented the photo of three contemporary scientists. At first, the students were astonished that there were two black women and a young black man. The teacher then gave a brief description of what they have done and their life trajectories until they came to the profession and have been recognized as scientists. To complement the discussion, the teacher read to them a story from a Brazilian digital newspaper (Odara 2017), in which the careers of black Brazilian women scientists are described, asking the students to point out what those women had in common in terms of life history and scientific career. The teacher highlighted the kind of research these women developed, many of them being also political actors. They answered that they all came from poor families. Given that one of the research proposals was entitled "decolonization of Science" the teacher asked what that research would be. The students failed to give an explanation. The teacher explained what the research meant, referring to a topic discussed during the study of the Enlightenment. She emphasized that this work was intended to gather all the knowledge produced in that context, which understood to be universal, and asked: how can this knowledge be universal if part of humanity was not represented there? They were thoughtful but made no comments.

8.5 What Could We Learn with This Example?

The educational experience presented in the previous section was constructed by a science teacher who took science classes as an opportunity to discuss several processes of exclusion inside science that her students also suffered – in a way or another – in their daily lives. Then, the teacher proposed a course, which aimed to problematize the female condition in society and, particularly, in scientific endeavour. This decision was not taken beforehand but after the first contact with the students, during an activity in which the teacher got to know some details of students' lives, as an example, the fact that all students were arrested because they followed their male relatives, reproducing a cultural behaviour.

From the CHS approach, the teacher organized the science classes giving centrality to scientific practices, showing that a set of human activities constituted science. Instead of individuals or groups making science, attention was given to certain practices that characterize science in that context, what were these practices and also who performed them and who did not. It seems a slight difference but, by doing this, it was possible to tease out, as suggested by Nyhart (2016), "how certain forms of knowledge and practice (…) came to be understood as 'science', what has sustained science socially, culturally, and materially and who has benefited and who has suffered in its formation" (p. 7). This perspective allowed her to discuss, for example, the role of scientific academies in the scientific enterprise of the historical period studied, highlighting who their members were, and their financial support was. In a broad perspective, the teacher discussed other places of production and divulgation of science, such as private science classes, coffee and beer houses. From these debates, the students could recognize the social actors who participated in these activities and the absences of other social actors, such as women, and Black people.

These discussions were performed highlighting that these absences were cultural traces of an epoch, that they were not accidental. In this way, the teacher addressed issues concerning the gender (in)equality in that historical context. From this debate, the students were also introduced to the names of some women who participated in the development of Botany. The students' interventions indicated that they recognized that the sociocultural and economic conditions of those women were essential for them to participate in science. During the activity in which the students drew flowers, they recognized that to perform any scientific activity, even those they judged as simple at first, specific techniques were necessary. Then, they could problematize the fact that not only the scientists should have specific abilities to make science but also those illustrators who performed specific functions essential to the development of science. Besides discussing the situation of women who participated in the development of Botany in the eighteenth and nineteenth centuries, the students brought to science classes issues related to their own conditions as Brazilian Black women. They did not have relatives or friends working as scientists, and most of those who lived close to them had been arrested at least once in their lives. This suggests that the students recognized, during the class, that the absence of some social actors in science is a cultural problem, and the struggle for gender equity is still a contemporary question, both in science and in society.

From the historical study of scientific practices, the students could recognize some reasons why preferably male scientists were honored in the denomination of new plant species. They could also realize that the women and the settlers were important social actors for the Botany development since they participated in collecting, illustrating, and dissecting plants - key activities for the construction of this science. In this way, it was possible to discuss the erasure of some social actors in the diffusion of Botany, and, as Santos (2002, 2010) points out, that this kind of erasure continues in the development of science throughout history.

During the historical study of scientific practices, the students put questions not previously planned by the teacher. One of these questions, the necessity to prevent the extraction of native species, allowed the teacher to bring to science class some

Brazilian contemporary environmental issues, such as the importance of the biodiversity for the world.

We recognize that this is a very specific educational experience, and it is not straightforward to draw the consequences for science education at large. Even though, we argue that this experience allows us to reinforce our argument that it is impossible to anticipate the relevant conditions from each context to teach NOS. Considering, for instance, the consensus view approach for teaching NOS, which tenets would be adequate to address in a context such as the one we described in this paper? Which aspects are off the list but are of great relevance to these students' life histories? We are not only talking about experiences that could be connected to students' lives and "that is all." We propose to construct pedagogical experiences that, through a relevant connection with students' lives, allow the students discuss *about science*, reflecting about the structures of power inside science and society, and what have been their place in these structures. So, we defend that it seems to be hard to address meaningful questions for students' life and, thus, make them really problematize science and society by adopting NOS frameworks as a guidance for planning lessons at schools. In this way, we agree with Bazzul (2019), Chap. 5 of this book, that recognizes pluralistic NOS is indispensable for bringing Social Justice for science teaching. Of course, this discussion is linked to the objectives of education. In our case, we defend an education that through science could be an important way to construct a less unequal society. In this view, science knowledge is not an aim *per se* but a way to achieve a better society, even understanding that what means "a better society" must be a perennial subject of democratic deliberation.

As a summary of the main contributions of this chapter, we understand that the "exemplar" we brought to the chapter allows us, first, to think about the realities to which we teach as holding particularities that invite us to reframe our teaching on NOS as something "alive", which have to meet the relationship of students and knowledge. The stories we tell about the knowledge should have a component that ties itself to the stories of the students we aim to teach, which means that there is no HS in a vacuum of intentionality. Second, that if one aim to teach in a context as the Brazilians ones (an unequal peripheral country with a history of colonization), scientific knowledge should be thought from another point of view that considers the asymmetries of power inside that knowledge and throughout the history of its construction. These two general principles could serve as a starting point to construct a science education from an historical-cultural perspective aiming to nurture social justice.

However, we are not arguing that NOS frameworks are useless. We argue that it is essential to use them with caution. Moreover, the teacher who planned and implemented these lessons was informed by all these discussions in academic literature about NOS Knowledge and Inquiry, including the main frameworks and research results, which indicates that maybe knowing these frameworks and academic discussions can be fundamental to make possible these experiences. Besides, the educational experience indicated that the debates about scientific knowledge performed an important role in that science class making room for those women to recognize their social position in Brazilian society and that this role is historically constructed, so, it can change.

The speeches of students, who participated in this study, indicate that they put questions in science classes related to their perception as Black women and they studied different scientific contents from their understanding of the world and their experience as women. Therefore, we argue that the historical contextualization developed in those science classes have meaning for these students, since the historical narrative developed was linked to their lives' histories. We argue that the educational experience discussed in this paper indicates possibilities to consider the historical erasures, the asymmetries of power and the symbolic violence that often permeates the daily life of those students. Therefore, this experience analyzed from the theoretical framework discussed here suggests that the discussion of science from a historical decolonial perspective could promote a science teaching more connected to social justice since this approach allows discussions about science and society that problematize the abyssal lines which produce the differences between Global south and Global north and subaltern places in the world.

References

Allchin, D. (2011). Evaluating knowledge of the nature of (whole) science. *Science Education, 95*(3), 518–542. https://doi.org/10.1002/sce.20432.

Barton, A. C. (1998). Reframing "science for all" through the politics of poverty. *Educational Policy, 12*(5), 525–541. https://doi.org/10.1177/0895904898012005004.

Bazzul, J. (2019). Political entanglement and the changing nature of science. In L. Hansson & H. Yacoubian (Eds.), *NOS and social justice*. Cham: Springer.

Bleichmar, D. (2011). The geography of observation: Distance and visibility in eighteenth-century botanical travel. In L. Daston & E. Lunbeck (Eds.), *Histories of scientific observation* (pp. 373–395). Chicago: University of Chicago Press.

Castro-Gómez, S. (2005). Ciências sociais, violência epistêmica e o problema da "invenção do outro". In E. Lander (Org.), *A colonialidade do saber: Eurocentrismo e Ciências Sociais. Perspectivas Latinoamericanas* (pp. 80–87). Buenos Aires: CLACSO.

Costa, H. H. C., & Lopes, A. C. (2018). A contextualização do conhecimento no ensino médio: tentativas de controle do outro. *Educação & Sociedade, 39*(143), 301–320. https://doi.org/10.1590/es0101-73302018184558.

Dagher, Z. R., & Erduran, S. (2016). Reconceptualizing the nature of science for science education. *Science & Education, 25*, 147–164. https://doi.org/10.1007/978-94-017-9057-4.

Daston, L., & Galison, P. (2007). *Objectivity*. Brooklyn: Zone books.

Erickson, F. (2012). Qualitative research methods for science education. In B. Fraser, K. Tobin, & C. J. McRobbie (Eds.), *Second international handbook of science education* (pp. 1451–1469). Dordtecht: Springer. https://doi.org/10.1007/978-1-4020-9041-7_93.

Gaspar, M. A. (2009). *A Representação das mulheres nos discursos dos filósofos: Hume, Rpsseau, Kant e Condorcet*. Rio de Janeiro: Uapê SEAF.

Gavroglu, K. (2007). *O passado das ciências como história*. Lisboa: Porto Editora.

Gronin, S. S. (2007). What Jane knew: A woman botanist in the eighteenth century. *Journal of Women's History, 19*(3), 33–59. https://doi.org/10.1353/jowh.2007.0058.

Hodson, D. (2014). Nature of science in the science curriculum: Origin, development, implications and shifting emphases. In M. R. Matthews (Ed.), *International handbook of research in history, philosophy and science teaching* (pp. 911–970). Dordrecht: Springer. https://doi.org/10.1007/978-94-007-7654-8_28.

Jager, I. (2018). Discussing gender with women deprived of liberty: A look from the cultural history of science to the development of botany in the 18th and 19th centuries. Master dissertation, Centro Federal de Educação Tecnológica Celso Suckow da Fonseca, Rio de Janeiro, Brazil.

Klein, U. (2003). *Experiments, models, paper tools: Cultures of organic chemistry in the nineteenth century*. Stanford: Stanford University Press.

Kuhn, T. (1962). *The structure of scientific revolutions*. Chicago: The University Chicago Press.

Longino, H. E. (1991). Multiplying subjects and the diffusion of power. *The Journal of Philosophy, 88*(11), 666–674. https://doi.org/10.5840/jphil1991881114.

Matthews, M. R. (1992). History, philosophy, and science teaching: The present rapprochement. *Science & Education, 1*(1), 11–47. https://doi.org/10.1007/BF00430208.

Miller, J. L. (2014). Curriculum theorization as an antidote in/to test culture. *Revista e-Curriculum, 12*(3), 2043–2063. https://doi.org/10.1080/00220272.2018.1537376.

Milne, C. (2015). Cultural influences on science education. In R. Gunstone (Ed.), *Encyclopedia of science education* (pp. 245–248). Dordrecht: Springer. https://doi.org/10.1007/978-94-007-6165-0_307-2.

Moura, C. B. (2019). *Educação Científica, História Cultural da Ciência e Currículo: Articulações Possíveis* [Science education, cultural history of science and curriculum: Possible entanglements]. Doctoral Thesis, Centro Federal de Educação Tecnológica Celso Suckow da Fonseca, Rio de Janeiro, Brazil.

Moura, C. B., & Guerra, A. (2016). Cultural history of science: A possible path for discussing scientific practices in science teaching? *Revista Brasileira de Pesquisa em Educação em Ciências, 16*(3), 749–771.

Nyhart, L. K. (2016). Historiography of the history of science. In B. Lightman (Ed.), *A companion to the history of science* (pp. 7–22). Oxford: Wiley Blackwell.

Odara, N. (2017). 8 mulheres negras cientistas que você precisa conhecer. *Brasil de Fato*. Retrieved from: https://www.brasildefato.com.br/2017/07/25/8-mulheres-negras-cientistas-brasileiras-que-voce-precisa-conhecer/. Accessed 26 Apr 2018.

Pimentel, J. (2007). La Revolucíon Científica. In M. Artola (dir.), *Historia de Europa: Tomo II* (pp. 163–238). Madrid: Espasa Calpe.

Pimentel, J. (2010). ¿Qué es la historia cultural de la ciencia? *Arbor, 186*(743), 417–424. https://doi.org/10.3989/arbor.2010.743n1206.

Santos, B. S. (2002). Towards a sociology of absences and a sociology of emergences. *Revista Crítica de Ciências Sociais, 63*, 237–280. https://doi.org/10.4000/rccs.1285.

Santos, B. S. (2010). Para além do pensamento abissal: das linhas globais a uma ecologia de saberes. In B. S. Santos & M. Meneses (Orgs.), *Epistemologias do Sul* (pp. 23–72). Coimbra: Almedina.

Santos, B. S., & Meneses, M. (2010). Introdução. In B. S. Santos & M. Meneses (Orgs.), *Epistemologias do Sul* (pp. 9–20). Coimbra: Almedina.

Schiffer, H., & Guerra, A. (2015). Electricity and vital force: Discussing the nature of science through a historical narrative. *Science & Education, 24*(4), 409–434. https://doi.org/10.1007/s11191-014-9718-6.

Terrall, M. (2011). Frogs in the mantelpiece: The practice of observation in daily life. In L. Daston & E. Lunbeck (Eds.), *Histories of scientific observation* (pp. 185–205). Chicago: University of Chicago Press.

Voloshinov, V. N. (1973). *Marxism and the philosophy of the language*. New York/London: Seminar Press.

Watson, J. E. M., Venter, O., Lee, J., Jone, K. R., Robinson, J. G., Possingham, H. P., & Allan, J. R. (2019). Protect the last of the wild. *Nature, 563*(7729), 27–30. https://doi.org/10.1038/d41586-018-07183-6.

Yacoubian, H. A. (2015). A framework for guiding future citizens to think critically about nature of science and socioscientific issues. *Canadian Journal of Science, Mathematics, and Technology Education, 15*(3), 248–260. https://doi.org/10.1080/14926156.2015.1051671.

Chapter 9
Tapping the Potential of Ubuntu for a Science that Promotes Social Justice and Moral Responsibility

Meshach B. Ogunniyi

9.1 Introduction

One of the aims of science education is the development of scientific literacy which promotes equity and social justice. To achieve this goal many developing countries have made spirited efforts to replace the colonial science bequeathed on them with one that is relevant and compatible with their aspirations and the postulates of their hard-won political independence. In light of this, the Department of Education in South Africa explicitly states that the old colonial/apartheid Eurocentric curriculum would be replaced with an inclusive curriculum that exemplifies the spirit of ubuntu and associated values such as humanness, communalism, interdependence, equity, social justice, and moral responsibility-all entrenched in the new Constitution (Act 108 of 1996). According to Motshekga, the Minister of Basic Education, the new Constitution's aim is to:

> heal the divisions of the past and to establish a society based on democratic values, social justice and fundamental human rights; improve the quality of life of all citizens and free the potential of each person; lay the foundations for a democratic and open society in which government is based on the will of the people and every citizen is equally protected by law (Motshekga 2011, p. iii).

Before the demise of the apartheid government in South Africa, there were 19 departments of education with different aims and objectives. The education system then was based on racial lines which favoured the White students over the students of colour. The Black students who were the majority lived in the poorer communities and received the poorest quality of education at the time. However, after 1994, the newly elected government collapsed the 19 Departments of Education into one

---

M. B. Ogunniyi (✉)
School of Science & Mathematics Education, University of the Western Cape, Cape Town, South Africa
e-mail: mogunniyi@uwc.ac.za

H. A. Yacoubian, L. Hansson (eds.), *Nature of Science for Social Justice*, Science: Philosophy, History and Education,
https://doi.org/10.1007/978-3-030-47260-3_9

Department of Education (DOE). In 2011 DOE was split into two, namely, the Department of Basic Education (DBE) and the Department of Higher Education (DHE). The focus of this chapter is on the former.

The new South African government considered quality education as its priority and to achieve this aim, it undertook a series of curriculum reforms. The ultimate goal of these reform efforts was to:

- Transform the old apartheid system of education based on Standards 1–10 to Grades R-12 with all students exposed to the same curriculum materials.
- Replace the exclusivist apartheid curriculum based on racial lines with an inclusivist one based on the principle of inclusiveness so as to meet the demands of a multicultural classroom setting.
- Motivate students to apply their acquired scientific knowledge and skills in ways that are relevant to their life worlds as well increase their awareness of global imperatives.
- Encourage students to work together, in the spirit of ubuntu, to achieve the common good of all.
- Equip all students "regardless of their socioeconomic background, race, gender, physical ability intellectual ability … with knowledge, skills, and values necessary for self-fulfillment, and meaningful participation in society as citizens of a free country" (DBE 2011).

To achieve the aim above, DBE undertook massive curriculum reforms and organized a number of professional development workshops (though inadequate) for teachers to implement the new science curriculum underpinned by the philosophy of ubuntu in their classrooms. In addition, DBE added more language labs to those already built by DOE whereby scientific terms in English or Afrikaans were translated into the nine officially recognized indigenous African languages. However, this effort is still at the preliminary stages.

For ease of reference, this chapter will clarify some issues pertinent to the transformation of the education system intended by the Department of Basic Education (DBE). Some of these include: making ubuntu as the underlying theme for all educational activities; the indigenization of school science to make it relevant to the sociocultural backgrounds of all students regardless of the socio-economic backgrounds; decreasing the influence of colonial/apartheid education system on the current education system; ridding the science curriculum and instruction of their colonial/apartheid legacies especially in terms of the prevailing divisions, inequity, racism and social injustice in the country; intensifying curriculum reforms in all school subjects; the adoption of innovative instructional strategies to meet the demands of multicultural classrooms; and changing the negative image of scientific practice that had prevailed since the Enlightenment period to a more humane one. Other related issues highlighted in the chapter include: the nature of science and science education; racism and racist science education; and knowledge-particularly scientific knowledge and concomitant social responsibility. I shall elaborate these issues further in the sections that follow.

9.2 What Is Ubuntu?

Ubuntu, an Nguni term and the central core of African way of life construes humans as being endowed with virtues such as kindness, respect, sympathy, empathy, cooperative spirit, concern for each other's welfare, and having a sense of moral responsibility. Ubuntu stresses mutual assistance and working together to solve a given problem. Nussbaum (2003) describes ubuntu as the capacity in African culture to express compassion, reciprocity, dignity and humanity in the interest of building a harmonious community where equity, social justice and moral responsibility are cherished and celebrated.

According to Chaplin (2006) ubuntu or being fully human is an African concept with a universal application. The term ubuntu has recently gained prominence in Africa because of people's increased awareness of the need to re-discover their African identity. Venter (2004) has also argued that although ubuntu emphasizes communality rather than individualism as is the emphasis in the Western world, it is equally cognizant of the important role that individuals play in society. In sum, ubuntu is the underlying philosophy that shapes and drives the African traditional educational system and way of life in general. It is the basis for peaceful co-existence, resolution of conflicts and for undertaking daily activities (e.g. Beets and le Grange 2005; Ogunniyi 2004).

9.3 Ubuntu and Science Education

A cursory review of the history of science and science education would easily reveal that science and science education have been used on the one hand to promote human welfare while on the other it has been used to suppress the indigenous communities around the world. This paradox has tended to cause mixed feelings among the indigenous communities about the overall benefits of science. Nevertheless, a ray of hope is currently emerging among the down-trodden indigenous communities around the world about the need to harness the benefits of science and to integrate it with their own indigenous knowledge for their political and economic emancipation from the dominant western hegemonic structures that have kept them bound for so long.

The indigenous scholars, particularly in Africa, decry the general perception in the literature that indigenous knowledge (IK) is no more than a bundle of superstitious beliefs or that all aspects of science are necessarily good or flawless. Rather, their view is that both science and IK have their merits and demerits. Further, they are of the view that the good elements of both knowledge systems should be combined and nuanced in such a way that the outcomes enhance sustainable development of a given community. Development agencies that have ignored this important fact, and have gone ahead to implement programmes (including science education programmes), based solely on scientific models, have had to learn the painful lessons of disastrous consequences of failed projects. All over the African continent and other regions of the world are countless failed white elephant projects as will be shown later. Such failed projects and other adverse environmental disasters caused by scientific and industrial activities have tended to increase people's negative image of science and this is where a teacher has a critical role to play e.g. by engaging his/her students in discussing socio-scientific issues affecting their communities or by asking them to investigate how such disasters could have been prevented.

A teacher's image of science to a large extent influences his/her students' image of science even though such an image may be historically, sociologically or philosophically invalid. For instance, when science is taught as the epitome of truth it presents a faulty image of science. There are copious examples in the literature about how students, in their attempts to get the right answers, have gone to great lengths to forge the results of experiments and to draw figures that resemble what is presented in their text books not realizing that such figures were merely typical examples. Failing to fulfil their distorted image of science, many students, who showed interest in science in the primary school, ultimately become so disillusioned that they do not want to study science beyond the high school level. It is perhaps at this point that teachers could direct their students to the way their communities in the spirit of ubuntu work together to achieve consensus or solve a local problem. A common practice among the indigenous African communities is apprenticeship whereby both experts and novices work collaboratively to perform a given task without making the latter feeling inferior.

First, the new South African science curriculum demands that students be exposed to an ubuntu-based and culturally relevant science education. Second, it recommends that science teachers should present science as a dynamic, tentative and revisionary inquiry that thrives on argumentation, debates, dialogues and collaborative consensus-building. Third, it recommends that students should be made aware of the nature of science (NOS) in terms of its strengths and limitations and the fact that science arises within a whole network of complex social interactions and epistemic practices within the scientific community of practice (Ziman 2000). Fourth, it expects that the new curriculum would help to mitigate the negative and distorted image of science perpetuated through the use of inappropriate pedagogies. Fifth, it encourages teachers to tap the innate propensity of students to be willing to work together in solving a given problem.

9.4 Colonial Education

Before exploring the nature of science (NOS) it is important to provide a brief historical account of Western education in Africa during the colonial period and immediately afterwards. Science was one of the subjects taught in schools established by missionaries and later on by the colonial governments. However, the way science was taught at the time was received with mixed feelings among students because it was in conflict with some of their religious beliefs and cultural practices. Despite this students and the general public were impressed with the stupendous development brought about by science. Of course, this does reduce their concern about the negative impact of scientific and technological activities on their communities and their biophysical environment in general.

The indigenous communities observed the mass deaths resulting from the use of guns and the seizure of their arable lands, forests, sea coasts and water ways by the invaders. They noticed the excavation of their soils by the European companies in search of minerals which resulted in diverse environmental catastrophes such as atmospheric pollution, gully erosion, the destruction of their bio-diversities to mention a few. While these activities enriched the invaders, it steadily increased the poverty of the natives. To make matters worse, the colonial governments imposed heavy taxes on the natives which they could not pay, and as a result the natives had to leave their homes to work in the mines and to depend on the hand-outs of their White masters for their daily survival. The new White landowners soon replaced subsistent farming practiced by the natives with mechanized farming, using cheap Black labour for the cultivation of cash-crops meant for the European markets. These unjust practices ultimately resulted in the inequalities still much abroad in South Africa even today.

Historically, the Western powers using the products of science such as guns and machines of mass destruction have always enslaved the people of colour, impoverished them and dislodged their sense of communality and cultural identity worldwide (Bishop 1990; Diamond 1999). In some countries where the local resistance was difficult to overcome through brute force, the colonial government installed a system of indirect rule whereby the loyal natives, though less able or less educated, were placed over the more literate and restless groups; and consequently setting the whole society in a perpetual state of conflict. The instability created by the colonial governments, to a great extent, has not yet been overcome in many African countries. It is noteworthy that two decades after the demise of apartheid, about 80% of arable lands in South Africa are still in the hands of less than 1% White farmers.

Talking about the ills of British colonial education, Gill and Levidow (1987) assert that science teaching in Britain embraces a subtle form of racist propaganda not easy to detect because science is assumed to be culturally and politically neutral. Gill and Levidow (1987) contend that the hidden political agenda of school science is primarily to: promote the economic interests of the West; justify the subjugation of indigenous sciences on account of being inferior, underdeveloped, less efficient, imprecise and largely superstitious; and to perpetuate the ideology of the superiority of the West over others.

The invasion of Africa by Western powers has led to the atomization of the once cohesive indigenous communities. The Berlin Conference of 1885 created boundaries which split the same cultural groups into different regions or countries which now speak different European languages. Today, Africa has over 50 independent countries using different European languages as their official way of communication, and consequently many indigenous languages have either been destroyed completely or relegated to the background of the official European languages. When people lose their language they not only lose their knowledge, education, values and cultural practices associated with that language but they also lose their overall sense of cultural identity (Ogunniyi 2016). It is in light of this that traditional or indigenous education could prove handy.

Indigenous education is concerned among others, with such things as: the development of a strong moral character; self-development; acquisition of social skills; life adjustment or ability to adapt to changes in one's environment; and social harmony even if it entails self-abnegation. However, these educational practices have been rudely interrupted and largely replaced with the Eurocentric form of education whose focus is mainly the individual ownership of property, knowledge and skills-all for personal benefits. Chinua Achebe describes this whole scenario in his well-known book titled, "Things fall apart" (Achebe 1958).

Another negative outcome of Western education in Africa has been the production of westernized Africans whose minds have become so occupied with western values and ways of life to the extent that they now deprecate their own indigenous ways of life in general. Rather than cooperate and seek the welfare of their own people based on ubuntu, most of them are obsessed with grotesque and voracious appetites for western goods and luxurious items; even if these would have negative consequences for their immediate environment. Of course, the controversies surrounding western science are not new. It began as the abuses of scientific practice increased from one generation to another starting from the Enlightenment era.

9.5 Science During and After the Enlightenment Period

Science was considered during the Enlightenment in the seventeenth century as the epitome of all knowledge. According to Trigg (1993), science was so elevated to the status of absolute truth and as the panacea to all human ills during the Enlightenment and that anything that could not be subjected to scientific objectivity was considered subjective and altogether "... beyond the scope of reason, and perhaps positively irrational" (Trigg 1993, p. 3). Skirbekk and Gilje (2001) summarized the tenets of the Enlightenment as follows:

- Humans are good by nature.
- The goal of human life is well-being in this world, not blessedness in the next.
- This goal can be achieved by humans alone through science (knowledge is power).
- The greatest obstacles in reaching this goal are ignorance, superstition, and intolerance.
- To overcome these obstacles, we need enlightenment (not revolution).
- Through more enlightenment, humans automatically become more moral.
- Through enlightenment, the world would move forward.
- Reason is possessed by all human beings.
- Natural law guarantees individual's rights against privileges and tyranny.

- The moral principle of enlightened self-interest holds that we should seek the best for ourselves.
- Sociologically, there is a harmony of self-interests: to fight for our own interests is to contribute to every one's welfare.
- An ideal state secures property rights and individual liberty and is efficient (i.e. state-protected capitalism, nationally; protectionism and colonization, internationally) (Skirbekk and Gilje 2001, p. 244).

Habermas (1971) considers the above tenets of humanism of the Enlightenment as tantamount to scientism. To him:

> modern science regardless of its mathematical rigour does not coincide with absolute knowledge in that its apparent narrow-mindedness; and to think otherwise is tantamount to scientism-i.e. "science's belief in itself: that is, the conviction that we can no longer understand science as one form of possible knowledge, but rather must identify knowledge with science" (p. 4).

The elevated view of science during the Enlightenment and afterwards soon began to wane however, as the abuses of scientific practice increased from one generation to another as exemplified for instance by the Slave trade, colonialism and the apartheid system of government all which resulted in the suppression of non-western peoples around the world. In view of this, scientists cannot simply busy themselves in their labs and turn deaf ears to public outcries against the abuses brought about by scientific, technological and industrial activities around the world. There is no denying the fact that the progress made by the western powers to conquer and colonize the diverse people groups around the world has largely depended on the application of the scientific (and technological) knowledge (Diamond 1999).

In the pursuit of alternative worldviews and to counter the narrow view of western science, some scholars have in the last two decades begun to turn their attention to the values and benefits of indigenous knowledge that have sustained the great bulk of human population for centuries without causing as much damage to the environment as has been the case with scientific practice in the last four centuries. The works of such scholars have stimulated interest among researchers in both developed and developing worlds to examine more closely the so-called ethno-science.

It is interesting to note that the needs and the aspirations of indigenous peoples for freedom, equity and social justice are very similar (e.g. Aikenhead and Elliott 2010; Cajete 1999; Ogunniyi 2004, 2007, 2016). In line with Fraser's (1998) social justice framework, the indigenous peoples worldwide are becoming more vociferous in their demand for: redistribution in terms of resources; recognition of their cultural heritage as unique and distinctly different from the dominant western culture with its assimilatory tendency; and participation in the sense that it affords them political rights as a people group towards self-determination (Nomlomo and Khuzwayo 2019). Having been denied of their human rights in the last four centuries, the indigenous peoples have begun to demand for the restoration of their lands which were unjustly and forcefully taken from them by the colonial governments and subsequent apartheid government. For instance in South Africa, the San-Koi indigenous communities have begun to demand that their lands and waterways be restored to them. They are also demanding their right to participate in the South African Parliament having been ignored in the past elections.

9.6 Nature of Science and Science Education

In its introductory remarks the Revised National Curriculum Statement Grades R-9 in South Africa contends that modern science had its roots in African, Arabic, Asian, American and European cultures as they search to understand the natural world through observation, careful investigation, data gathering, positing and testing hypotheses and drawing meaningful conclusions. The same curriculum Statement claims that science is a human heritage rather than belonging to any particular race or people group. As with other forms of knowledge, scientific knowledge is an ongoing process which usually involves individual and collective effort in the context of argumentation, discussions, debates, dialogues, the formulation, testing and validation of hypotheses and theories through a series of observations and carefully conducted experiments (DOE 2002). The new South African curriculum certainly deviates substantially from the old apartheid westernized science curriculum (e.g. DBE 2011; DOE 2002; Lee and Buxton 2010; Naidoo and Vithal 2014).

As stated earlier, science has been a blessing to humanity but there are copious examples of how science or its application has been used in an abusive way. Recent examples of this include: the atomic bombs unleashed on Hiroshima and Nagasaki in 1945 during the 2nd World War; the Bhopal disaster of 1984 in India; the mass deaths associated with asbestosis in South Africa for most of twentieth century; the nuclear accidents in Dauphin County, Pennsylvania, USA in 1979, the Chernobyl nuclear disaster in Ukraine in 1986 and the Fukushima Daiichi Plant disaster in Japan in 2011; and the phenomena of global warming, Greenhouse effects, erosion of the ozone layer, environmental pollution, just to mention a few. So, while we all applaud the beneficial effects of scientific practice we cannot ignore its glaring excesses.

A frequently mentioned objective of science education in the extant literature has been for science teachers and their students to understand the nature of science (Abd-El-Khalick et al. 1998). Today, science teachers and students are confronted with a massive knowledge explosion. The problem now is not whether science teachers and students are being sufficiently exposed to scientific information but whether or not they possess valid knowledge of NOS.

The renowned British philosopher, Dingle, nearly 70 years ago, observed that:

> It is one of the paradoxes of modern life that while-on the material and intellectual side, at any rate—it is dominated by science, very few people have any clear idea of what science is... This is true not only the rank and file, but the average scientist... It is conceivable, however, that this phenomenon indicates a defect in our educational system...We teach the practice of science well; we teach the understanding of science scarcely at all (Dingle 1952, p. 337).

After an extensive review of the literature McComas (1998) has concluded that while a lot of studies have attempted to enhance teachers' understanding of NOS through the study of the history and philosophy, very few have paid attention to the sociological component of scientific practice; thus presenting an incomplete image of the scientific modes of inquiry. Often, science teachers do not reflect the nature of science in their classrooms and consequently, produce students who are anything but scientifically literate.

Rudolph (2002) has also argued that the image of NOS presented in the school science curriculum and for whose understanding is sought for teachers and students is only a selective representation of what science is and lacks the historical dimensions of scientific practices. He contends further that current debates about the NOS in the school curriculum have centred on where the boundary between canonical school science and other forms of knowledge should be drawn. To him, what seems to be missing in these debates is a careful examination of how (given the diversity of scientific practices) what lies within the boundary of school science itself as how it has been determined. Certainly, the process of selecting what or not should be included in a science curriculum cannot be completely divorced from the socio-cultural and political forces involved in such selection. He concluded that the portrayal of science in the school curriculum cannot be fully understood without the due consideration of the socio-cultural-political contexts in which they are developed. If this is the case, who decides what or not should constitute a given science curriculum?

Increasingly, in recent times, dissenting voices are being heard within the academy and the society at large that the science being taught in our schools tends to ignore students' socio-cultural environment and as such it is not compatible with the postulates and aspirations of the communities that established the schools in the first place. This realization has further strengthened the position of the proponents of an inclusive science curriculum that also takes a full advantage of the wisdom and the funds of knowledge resident in the students' indigenous sociocultural environment (e.g. Aikenhead and Elliot 2010; DOE 2002; Lee and Buxton 2010; Ogunniyi 2016; Smith 1999).

Although a lot of commendable efforts have been directed at understanding the NOS in a societal context, these efforts seem to resemble the three blind men in the Hindustani fable attempting to describe an elephant. If anything at all, these efforts are based on the assumption that science can be represented in a single model rather than multiple perspectives. In reality, none has been able to provide a comprehensive image of NOS.

According to Ziman (2000) academic science is a complex multifaceted enterprise pursued in a variety of ways. It is a socio-cultural endeavour that has evolved among different groups of practitioners with shared traditions which in turn are articulated, transmitted and reinforced by members of these groups across political boundaries. Although academic science has no written code of conduct, all its members are constrained by their training to behave in certain predictable and morally responsible manner. But while a scientist is free to pursue his/her intellectual interest in an open manner, he/she is well aware that any claim he/she makes will not go unchallenged by other members of the scientific community.

Science educators have argued for decades that the science curriculum and instruction should reflect NOS to enhance students' scientific literacy. Several justifications have been made in support of this objective. Driver et al. (1996) identified five reasons for teaching NOS as follows:

- a utilitarian reason- an understanding of the NOS will help people to make sense of science in their daily lives;
- a democratic reason- an understanding of NOS will equip them with the necessary knowledge to make intelligent decisions on socio-scientific issues;
- a cultural reason-an understanding of NOS will help them to appreciate science as a critical element of contemporary culture;
- a moral reason-science as a critical aspect of modern culture and an understanding of its nature, particularly its ethical values, deserves a greater attention by the general public;
- an educational reason-to succeed in science a learner must understand NOS.

Of course, we could add to the list above such as the need to:

- prepare students for scientific careers;
- make them aware of the strengths and limitations of science;
- make them appreciate the tentative and revisionary nature of science;
- clarify their misconceptions about NOS and so on.

Whatever justifications we might adduce for teaching science, there are subtle assumptions we do not seem to pay much attention to such as: the belief that scientific knowledge is all we need to solve all our current multifarious local and global problems; that possessing such knowledge necessarily guarantees that we would make right decisions in the way we use or relate with our biophysical and socio-cultural environment; that by possessing such knowledge we could overcome the human tendencies and propensities to become cooperating victims of avarice, selfishness, greed, cupidity and other vices. But rather than despair, we must all admit that NOS is rather an elusive concept and all we need do is to make students to become aware of this reality. That would save them from falling into the error of scientism i.e. the belief that science is absolute truth and the panacea to all human ills (Habermas 1971).

9.7 Curriculum Reforms in South Africa

In her forward to the South African Curriculum Assessment Policy Statement, Angie Motshekga, the Minister of Basic Education, states the following about the new curriculum: "Our curriculum is the culmination of our efforts over a period of 17 years to transform the curriculum bequeathed to us by apartheid. From the start of democracy we have built our curriculum on the values that inspired our Constitution" (Act 108 of 1996).

The same curriculum statement emphasizes key issues such as democratic values, healing the divisions of the past, uniting the society, pursuing social justice, improving the quality of life, freeing individual potentials, building a democratic and open society, using the curriculum as the platform for change, and so on. These same issues dominated the students' uprising in 2017 regarding the lack of access to higher education by the majority Black students on account of their inability to pay the fees.

However, despite these curriculum reforms and students'protests, there has not been any significant change in the education system. In a current study involving five of the nine South African provinces, STEM and language education (STEMLE) teachers were asked to respond to a 20 open-ended questionnaire about their experiences in teaching and/learning these subjects during and after the colonial/apartheid period. An analysis of 98 responses received so far has indicated four major themes namely, that: STEMLE are still being used largely as a tool for propagating institutional racism; curricula are characterized by teacher-centred instruction which encourage rote learning; the schools in the disadvantaged communities still lack adequate human and material resources; and that the two foreign languages of instruction-English and Afrikaans are still the dominant modes of instruction. In light of these findings, it seems that the impact of colonilism/apartheid and associated ills such as racism, inequity and lack of social justice are still much around and that it will take a lot more time and effort than envisaged to redress the unsavoury situation.

9.8 Indigenization of the Science Curriculum

Although the call to indigenize the curriculum especially in Africa has been going on since the independent era of the 1960s, the recent call by South African university students to indigenize the curriculum during the "Fees Must Fall" protest has once again re-echoed the need for a visible change. One of the demands of students in this regard is that the curricula and instruction have not been indigenized. They therefore, want the language of instruction to be in English and the students' indigenous languages. The students decried the fact that the curricula have not changed much from the ones bequeathed to them by the colonial and apartheid governments. They wanted curricula that are relevant to their lives as well as free from the colonial/apartheid legacy. Another reason for the new emphasis on the indigenization of the curriculum is that South African students are under-performing in national and international assessments in science. In response to all this, some researchers have been looking for ways to indigenize the science curriculum and instructional practice in general (e.g. Beets and le Grange 2005; Naidoo and Vithal 2014; Ogunniyi 2004, 2007; Reddy 2006).

The Revised National Curriculum Statement Grades R-9 and the Curriculum and Assessment Policy Statement Grades 10–12 Physical Sciences (CAPS) further expect that students would be able to: (1) act confidently, become curious about natural phenomena, investigate relationships and solve problems in scientific, technological and environmental contexts; (2) know, interpret and apply scientific, technological and environmental knowledge; (3) demonstrate an understanding of the interrelationships between science, technology, society and the environment; and (4) appreciate the ubuntu-driven indigenous sciences and cultural values that have glued the indigenous communities together for centuries despite the political interferences and the imposition of the racist western education system on their traditional education (DBE 2011; DOE 2002).

9.9 Racism and Racist Education

More subtle and equally devastating throughout human history, has been the misuse of genetic science to justify slavery, racism and of indigenous peoples worldwide. The modern example of research in support of racial superiority of Whites over Blacks was that of Arthur Jensen in the last half of the twentieth century. Jensen attempted to show that IQ differences between White and Black children were as a result of some variations in their genetic constitutions. Although he denied ever implying that Black children were cognitively inferior to White children, he was convinced till the date of his death that genetic factors could not be ruled out completely.

Copious studies have consistently cast doubts on attempts to link the so-called intelligent quotient (IQ) with differences in school achievement on the basis of "race" or more correctly, people groups. Howard Gardner and his associates have shown unequivocally through the Harvard Project Zero that people are endowed with multiple intelligences rather than a single index known as IQ (Gardner 1993). It is a well-known fact much of our behaviours are not inborn but acquired from our forebears or the community in which we have been raised. The tragedies spurred by racism and misconstrued evolutionism during Hitler Germany in early twentieth century are still fresh in our minds even today.

An International Working and Advisory Group from Brazil, South Africa and the United States defined racism as:

> The denial of our shared humanity, violations of the human rights to which all people are entitled. It is a moral blight, source of festering injustice and serious economic problem…Racism is an invention, an utter and total fabrication that grew up as a justification for the military and legalistic takeover of land, labor, and people (Beyond Racism 2000, pp. 3–4).

Gill and Levidow (1987) paint a graphic picture of racism at work in the history of Britain during the colonial era up to the present period in the following way:

Racism is far more than conscious, or unconscious, kinds of personal prejudice, though these clearly perpetuate racist institutional practices. It is more, too, than the way institutions operate to the disadvantage of black people. The state racism implicit in British nationality laws, for example, clearly reinforce the personal prejudices of white people while making black people more vulnerable to economic exploitation… In the search for profits, Western powers have historically enslaved black people, conquered and impoverished much of the Third World, and induced millions of its people to immigrate to metropolitan countries. There they have been treated as cheap labour-an 'internal colony' to be exploited or expelled as expedient (pp. 1–2).

9.10 Personal Experience

I was born during Second World War and I can still remember how we sang and prayed for God's protection over King George VI and later, Queen Elizabeth II of Great Britain as Heads of the British Empire respectively. Of course, there is nothing wrong with that. As children marching to our classrooms after every morning assembly, we sang with gusto such songs as:

British Empire shall never perish (2ce)
Believe in God British Empire
Shall never perish, shall never fall (2ce)
We pray, we pray, we pray, we pray, we pray
British Empire shall never perish!

Empire Day, empire Day (2ce)
The day that we celebrate
The birthday of King of the White people
Glory, glory, glory!
Praise, praise, praise!
We give thanks to God for sparing our lives to see this day
May we celebrate it again next year!

Whatever sense one makes of these songs, some element of brainwashing of these innocent Black children cannot be justified for any reason. First of all their forebears were under a colonial rule. They had no rights to protest this act even if or when they became aware that what they were doing or being forced to do was an abuse of their human right. While a racist White reader's ego might be boosted by such songs, a non-racist White might find them repulsive. But whatever the case, the fact remains that after several decades, these and similar songs and similar acts of suppression have remained indelibly etched in my mind!

Fashesh (1997), a Palestinian scholar draws some parallels between the politics in Palestine with the occupation of their minds; meaning they can only think in a particular rigid way or what we call here in Africa people with a colonial mentality. In the same vein, Khuzwayo (2005) claims that the Western values that students have imbibed have instilled in them a passive acceptance of authority rather than providing them with the conceptual tool necessary for creative and independent thought.

In our science classes we had to abandon our indigenous knowledge about diverse phenomena and contended ourselves only with what was presented in the science classroom by the teacher or the textbooks. We had to sit for science examinations more suitable for British children than African children. Likewise, we had to do the same IQ tests, which were used to determine into what study streams we would be placed and so on. All these experiences are likely to result in the production of individuals with what is known as colonial mentality- i.e. an individual exhibiting an acute inferiority complex.

During my graduate studies in the US in the 1970s, I was often the only Black person in some advanced science courses. On entering some classrooms, I was often asked by both professors and students if I had missed the room I had intended to go to. I was even admonished by my supervisor and colleagues not to take some of the so-called killer courses. However, it did not take long before they changed their minds as they saw that I performed reasonably and even obtained higher scores than many of my White counterparts in such courses. My foolhardiness knew no limit when I told my supervisor that I wanted to attempt my prelim exam-an 8-h doctoral exam barely a year after my Masters programme. After much counselling by my supervisor he allowed me to take the exam and to everyone's surprise I made it! This incidence jolted many of my slumbering White colleagues who had been preparing for the same exam for a couple of years to attempt the same exam as soon as I succeeded. Of course, that experience in Madison was not the last time that I encountered all forms of prejudice. But in a way, such an experience helped to reinforce my sense of identity that had been eroded for so long.

In 1998, after a break at a UNESCO Conference on Higher Education, I entered a nearby restaurant for a quick lunch. A male waiter approached me and informed me that Black people are not welcomed. Some of my White friends were very angry about this but could do nothing. I thanked the waiter and moved to the next restaurant and had a good meal there. Though racism has become more sophisticated nowadays, it is still much alive judging from recent upsurge of the right-wingers in a number of Western countries. Of course, racism is not peculiar to White people; it is equally demonstrated among the so-called people of Colour. Racism is a psychological disease, a superiority complex and a self-defense mechanism caused by a sense of insecurity. An effective way to cure this blight early in life is to create a learning environment by using an inclusive pedagogy that provides ample opportunities for students to freely interact and work together as equals.

9.11 Anti-racist Science Curriculum and Pedagogy

The apartheid curriculum was based on the medical or deficit model and as such, suffered from such vices as structural racism. However, the new curriculum is based on social and inclusivist model and as such is liberatory, participatory and supportive of self-image (Johannes 2006). According to Mutegi (2011) curriculum reforms that are based on the medical model tend not to have sufficient dialogues with the target communities in that they are underpinned by the perceptions of White curriculum planners whose positionality is likely to influence their research findings.

In terms of the curricular demand for teachers to integrate science with students' indigenous knowledge, the Science and Indigenous Knowledge Systems Project

(SIKSP) which I coordinate has adopted a dialogical argumentation instructional model (DAIM) based on a social model of instruction. In other words, DAIM provides the participant teachers with ample opportunities to discuss the curricular materials in an atmosphere of dialogues, criticisms and collaborative consensus without being pressurized to accept the views of the curriculum planners or any one for that matter (Ogunniyi 2007, 2011).

9.12 Dialogical Argumentation Instructional Model

As stated in the previous section, DAIM is an innovative strategy that allows participating teachers to interrogate the content of the new curriculum before implementing it in their classrooms. This in a sense makes them to construe curriculum implementation as an evolving process rather than a completive enterprise or rhetoric of conclusions for which they can have no input. DAIM is underpinned by two theoretical constructs namely, the Aristotelian association of ideas and the contiguity argumentation theory (CAT) which draws ubuntu notion of communality and interdependence. DAIM enables students to tackle problems collaboratively in the way similar to how indigenous communities solve their problems by starting from individual effort, then small-group (family) effort, and finally whole-group (community) effort. It is at the community level that collaborative consensus is reached often with the teacher performing the role of a facilitator (Ogunniyi 2007).

The new curriculum demands that teachers integrate science and indigenous knowledge. In light of this, the SIKSP group has attempted to translate scientific terms into students' indigenous terms to facilitate their understanding of difficult scientific concepts. Nevertheless, this effort is still at the preliminary stages. Taylor (2016) has identified some obstacles to translating science into indigenous languages such as: the rudimentary nature of many African indigenous languages; the multiplicity of meanings assigned to the same terms; inadequate human and material resources; the power of English in the socio-political economy of the country; and so on.

The Concept Literacy Project (CLP) at the University of Cape Town in collaboration with Rhodes University and the University of KwaZulu-Natal produced a Multilingual Resource Book for mathematics and science in English, isiXhosa, isi Zulu and Afrikaans. In his forward to the book, Saleem Badat (2009), the Vice-Chancellor of Rhodes University reflecting on Harvey Kaye's notion of 'prophetic memory' attests to:

> the oppressive role that language has played in our history to thwart the potential of non-English or Afrikaans speaking learners and marginalize social groups; critique of the injustice of the continuing lack of effective educational opportunities for millions of learners; consciousness of the educational and social challenges of our present and that our history teaches that nothing is gained without creativity, boldness and determined endeavour; imagination to conceive of new ways of doing; and the desire to remake our schooling, induct learners into scientific literacy, and through these to help reshape our country (2009, p. 5).

Another attempt to indigenize science is that of the Language of Instruction in Tanzania and South Africa (LOITASA) Project (2001–2011) that translated several science concepts to isiXhosa and Kiswahili (Brock-Utne et al. 2006). Likewise, SIKSP (2004–2014) has trained many science teachers on how to implement a science-IK curriculum in their classrooms, produced instructional materials and translated some scientific terms in English to equivalent indigenous African languages (Ogunniyi 2007, 2011).

In cases where the same indigenous words are used to describe different phenomena, the SIKSP team coins new words or borrows new words from the same indigenous language or other close indigenous languages. However, in cases of a non-existing word or concept e.g. 'atom' becomes "i-athomu' and 'molecule', becomes 'i-molecule' in Zulu by coining new terms from the English terms. Alternatively, the Zulu word 'amandla' or 'power' (English) is substituted with the Setswana word 'maatla' which has a close intonation to 'amandla' instead of using 'Tsenya lefoku' (another Setswana term for power) and so on (Table 9.1).

According to Vygotsky (1986), language is a critical tool for thought. It is an indispensable tool "not only for the construction of the world of thought but also for the construction of the world of perception …Language is an energy, an activity, not only of communication and self-expression but of orientation in the universe (Anshen 2000, pp. 11–12). But when indigenous students' languages are relegated to the background, they are deprived of the most potent tool for reasoning, self-expression and sense of identity.

9.13 Knowledge and Moral Responsibility

The conundrum, 'knowledge is power' is not an idle cliché considering the way science has been used since the Enlightenment Period in the seventeenth century. In the same vein, our discussion on the nature of science for social justice will be incomplete without touching the issue of moral responsibility. The formerly colonized people around the world (particularly the indigenous communities in Africa) hold divergent views about science. While they are generally aware of the contribution of science to their overall welfare, they are equally aware of how the same tool

Table 9.1 List of translated scientific terms in English into indigenous African *terms*

English	Yoruba	Zulu	Shona	Swahili	Tigrigna
Square	onihamẹrin	Kubakaki	Sikweya	mraba	trb'ït
Energy	Agbára	Amandla[a]	Simba[a]	Nishati	x'ät
Cell	pádi	Iseli	Sero	Kiini	Wahyo
Speed	Iyara	Isivinini	Havukavu[a]	Kasi	ĥayli
Force	Ipá	Amandla[a]	Simba	Nguvu	ĥayli

[a]Same word for different scientific terms

has been used to subjugate them, pollute their environment and/or exploit their resources without compensation. In other words, we cannot talk about the nature of science for social justice without considering the conflicting views that different people groups hold about science.

Since the Enlightenment and the birth of modern science, scholars have been urged to embrace the rationalist doctrine of determinism and to reject the notion of free will and personal responsibility (Szasz 1996). Determinism is the doctrine that every fact in the universe is guided by natural laws and that these laws or the way they are applied are free of any human intentionality. This of course, contrasts sharply to the older doctrine, fatalism which asserts that all the facts in the physical universe and hence in human history, is absolutely dependent on and conditioned by their causes. In psychology, determinism is the doctrine asserting that the will is not really free but determined by contextual mitigating conditions (Runes 1975). So if a military pilot releases poisonous chemicals on assumed enemy territory resulting in the mass deaths of soldiers and civilians, the conjoining circumstances rather than the perpetrator of the act might be to blame! In the same vein, if the scientific knowledge is used to cause the deaths of millions people or destroy an environment as has been the case throughout human history, nobody should be blamed! We can even pull this argument further by asking, why are certain countries allowed to pile up weapons of mass destruction why others are not allowed to do so? What is the justification for this?

The doctrine of scientism which has pervaded the intellectual tradition in Europe since the Enlightenment era has attributed almost all of human behavior to the activities of genes and neurotransmitters; thus negating the issue of intentionality and culpability. The obvious dilemma of course is when can one attribute 'mentality' to an individual, treat his/her behaviour as an intentional act, and hold him/her responsible for it? (Szasz 1996). It should be obvious by now that we cannot abdicate our moral responsibility towards the knowledge we create or how it is applied. To do so would be to show a high sense of irresponsibility. It is in this sense that the scientific knowledge must be tempered with reason, reflection and wisdom not only in the way we see it but also how it affects the rights of others.

9.14 Conclusion

This chapter has examined different perspectives about the nature of science for social justice. For instance, while it has acknowledged the significant contribution that science has made to the socioeconomic development of many countries, it has also been used by Western Imperial colonial powers, and recently by international companies as a weapon of oppression and mass destruction of indigenous peoples around the world. Science has even been used to justify all forms of human atrocities and injustices such as: the Slave Trade; the exploitation of human and material resources; the atomization of society; and the bastardization of traditional beliefs and cultural practices to the extent that today many indigenous peoples around the world have lost their sense of identity. It is in light of this that ubuntu; a central African worldview, has been proposed in the chapter as a potential palliative to mitigate the abuses and the adverse effects of scientific and industrial activities since the industrial revolution in the seventeenth century.

Further, the chapter construes ubuntu as having the potential to tame scientific practice so to speak and to promote human virtues not only within the indigenous communities but the society at large. The contention in the chapter is that a person whose philosophy of life is governed by the ubuntu way of life would not use his/her knowledge, in this case, scientific knowledge to exploit or harm other humans or to destroy the environment they live in simply because he/she is in position of power. The call here is to give science a human face by using the scientific knowledge to promote human welfare rather than subjugate them. In conclusion, there is need to implement science curricula and instructional strategies that capitalize not only on the benefits of science but also the indigenous sciences or funds of knowledge that have contributed to the sustainability of the human race for several centuries. By harnessing the potentials of both science and indigenous knowledge many intractable problems bedeviling our society today could ultimately be overcome.

References

Abd-El Khalick, F., Bell, R. L., & Lederman, N. G. (1998). The nature of science and instructional practice: Making the unnatural natural. *Science Education, 82*, 417–436.

Achebe, C. (1958). *Things fall apart*. London: Penguin Random House.

Aikenhead, G. S., & Elliot, D. (2010). An emerging decolonizing science education in Canada. *Canadian Journal of Science, Mathematics and Technology Education, 10*(4), 321–338.

Anshen, R. N. (2000). *Preface to language and thought: The Frick collection* (Monograph Three). Wakefield: Moyer Bell.

Badat, S. (2009). *Forward to Understanding concepts in mathematics and science*. Cape Town: Maskew Miller Longman.

Beets, P., & le Grange, L. (2005). 'Africanising' assessment practices: Does the notion of Ubuntu hold any promise? *South African Journal of Higher Education, 19*, 1197–1207.

Beyond Racism. (2000). *A report of the international working and advisory group* (pp. 3–4).

Bishop, A. J. (1990). Western mathematics: The secret weapon of cultural imperialism. *Race and Class, 32*(2), 51–63.

Brock-Utne, B., Desai, Z., & Qorro, M. (Eds.). (2006). *Focus on fresh data on the language of instruction debate in Tanzania and South Africa*. Cape Town: African Minds.

Cajete, G. A. (1999). *Igniting the sparkle: An indigenous science education model*. Skyland: Kivaki Press.

Chaplin, K. (2006). The ubuntu spirit in African communities downloaded on the 29th of July, 2015. https://www.google.com/search?q=the+ubuntu+spirit+in+african+communities&ie=utf-8&oe=utf-8.

Department of Basic Education. (2011). *Curriculum and assessment policy statement grades 10–12* (Physical sciences). Pretoria, Government Printing Works.

Department of Education. (2002). *Revised national curriculum statement for grades R-9 (schools) – Natural sciences*. City to be named later: Department of Education.

Diamond, J. (1999). *Gun, germs and steel*. New York: W.W. Norton & Company.

Dingle, H. (1952). History and philosophy of science. *Universities Quarterly, 6*, 337–345.
Driver, R., Leach, J., Millar, R., & Scott, P. (1996). *Young people's images of science*. Buckingham: Open University Press.
Fashesh, M. (1997). Mathematics, culture, and authority. In A. Powell & M. Frankenstein (Eds.), *Ethnomathematics: Challenging eurocentrism in mathematics education* (pp. 273–290). Albany: State University of New York Press.
Fraser, N. (1998). *Social justice in the age of identity politics: Redistribution, recognition and participation*. Discussion Papers, Research Unit: Organization and Employment FS I 98–108, WZB Berlin Social Science Center.
Gardner, H. (1993). *Multiple intelligences: The theory in practice*. New York: Basic Books.
Gill, D., & Levidow, L. (1987). *Anti-racist science teaching*. London: Free Association Books.
Habermas, J. (1971). *Knowledge and human interest*. Toronto: Beacon Press.
Johannes, E. (2006). *Voice, disability and inclusion: A case study of biology learners with cerebral palsy*. Unpublished Thesis, University of the Western Cape.
Khuzwayo, H. (2005). A history of mathematics education research in South Africa: The apartheid years. In R. Vithal, J. Adler, & C. Keitel (Eds.), *Researching mathematics education in South Africa: Perspectives, practices and possibilities*. Cape Town: HSRC Press.
Lee, O., & Buxton, C. A. (2010). *Diversity and equity in science education: Research, policy, and practice* (Multicultural education series). New York: Teachers College Press.
McComas, W. F. (Ed.). (1998). *The nature of science in science education: Rationales and strategies*. Boston: Springer Academic.
Motshekga, A. (2011). Department of Basic Education (2011). In *Curriculum and assessment policy statement (CAPS) grades R-12 (schools)*. City to be named later: Department of Basic Education.
Mutegi, J. W. (2011). The inadequacies of "science for all" and the necessity and nature of a socially transformative curriculum approach for African American science education. *Journal of Research in Science Teaching, 48*(3), 301–316. https://doi.org/10.1002/tea.20410.
Naidoo, P. D., & Vithal, R. (2014). Teacher approaches to introducing indigenous knowledge in school science classrooms. *African Journal of Research in Mathematics, Science and Technology Education, 18*(3), 253–263.
Nomlomo, V., & Khuzwayo, B. (2019). Exploring teachers' experiences of mathematics, science and language education during and after apartheid: A case of the eastern cape and Kwazulu-Natal provinces. *Proceedings of the 5th international conference of AASIKS* (pp. 114–124).
Nussbaum, B. (2003). Ubuntu: *Reflections of a South African on Our Common Humanity. Reflections, 4*(4), 21–26.
Ogunniyi, M. B. (2004). The challenge of preparing and equipping science teachers in higher education to integrate scientific and indigenous knowledge systems for learners. *South Africa Journal of Higher Education, 18*(3), 289–304.
Ogunniyi, M. B. (2007). Teachers' stances and practical arguments regarding a science- indigenous knowledge curriculum, paper 1. *International Journal of Science Education, 29*(8), 963–985.
Ogunniyi, M. B. (2011). The context of training teachers to implement a socially relevant science education in Africa. *African Journal of Research in Mathematics, Science and Technology Education, 15*(3), 98–121.
Ogunniyi, M. B. (2016). Omoluabi: The goal of Yoruba traditional or indigenous education. *Proceedings of AASIKS* (pp. 133–141).
Reddy, V. (2006). *Mathematics and science achievement at south African schools in TIMMS*. Cape Town: Human Sciences Research Council.
Rudolph, J. (2002). *Scientists in the classroom: The cold war reconstruction of American science education*. New York: Palm Grave.
Runes, D. D. (Ed.). (1975). *Dictionary of philosophy*. Little Field, Adams & Co: Totowa.
Skirbekk, G., & Gilje, N. (2001). *A history of western thought*. London: Routledge.
Smith, L. T. (1999). *Decolonizing methodologies: Research and indigenous peoples*. Malaysia: Zed Books.

Szasz, T. (1996). *The meaning of mind*. Westport: Praeger.
Taylor, N. (2016) *Putting language into the mathematics and science equation*. http://www.zenex-foundation.org.za/downloads/Dinaledi%20booklet.pd. Accessed on 8 July 2016.
Trigg, R. (1993). *Rationality and science*. Oxford: Blackwell.
Venter, E. (2004). *Studies in the philosophy of education, 23*, 149–160.
Vygotsky, L. (1986). *Thought and language*. Cambridge, MS: The Massachusetts Institute of Technology.
Ziman, J. (2000). *Real science: What it is, and what it means*. Cambridge: Cambridge University Press.

Chapter 10
Teaching Robust Argumentation Informed by the Nature of Science to Support Social Justice. Experiences from Two Projects in Lower Secondary Schools in Norway

Stein Dankert Kolstø

10.1 Introduction

This chapter presents an analysis of two science projects where grade 8 students constructed evidence-based arguments informed by their developing understanding of the nature of science (NOS). In the first project, the students experimented on the toxicity of household chemicals. In the second project students constructed models for energy consumption related to a local transport issue. Based on the analysis, the chapter suggests a set of design principles for science curricula that will enable students to produce evidence-based arguments expressing views related to their own interests. Fundamental to this analysis, is the assumption that the ability to construct evidence-based arguments strengthens students' ability to promote their own views in the interest of social justice. This is of special importance for students not enculturated into such argumentation through their upbringing. To promote one's own views in a debate means to construct arguments that are resistent to criticism. It also requires critiquing others' arguments. Several insights about NOS, like the destinction between observations and possible inferences, can guide sound construction and evaluation of evidence-based arguments. Insights into NOS can therefore support students as they construct arguments related to issues of their own concern.

Issues that are relevant to students might differ considerably depending on local context, socioeconomic situations, and personal interests. Students in disadvantaged areas may experience more acute issues than students from wealthier areas. However, all students may experience the need to articulate their concerns and requirements to attract attention to their own situations. Students from different backgrounds might have experienced varying degrees of enculturation into ways of constructing robust argumentation. Arguments based on superficial inquiries or

S. D. Kolstø (✉)
Department of Physics and Technology, University of Bergen, Bergen, Norway
e-mail: Stein.Dankert.Kolstoe@uib.no

H. A. Yacoubian, L. Hansson (eds.), *Nature of Science for Social Justice*, Science: Philosophy, History and Education,
https://doi.org/10.1007/978-3-030-47260-3_10

incorrect interpretation of facts may be refuted and may weaken the student's position. To strengthen their cases, they could support their opinions with evidence-based arguments. To support social justice, it is therefore important that schooling develops the ability of all students to construct robust evidence-based arguments based on adequate understanding of the nature of science (NOS).

The core elements of NOS include basing claims on evidence, a distinction between observation and inference, and the roles of creativity, testing, critical thinking, and communication of arguments (Lederman 2007; Osborne et al. 2003). Insight into these elements, combined with experience of how they guide epistemic practices and the development of scientific arguments, might enable more students to construct robust evidence-based arguments.

Science education for social justice places emphasis on student agency and on enabling students to use science for their own agendas (Barton et al. 2003). Basu and Barton (2010) conducted empirical research on democratic educational practices to identify a model of democratic science pedagogy, that is, a science pedagogy aimed at empowering all students to use science in accordance with their own needs. This model has critical science agency, shared authority, and constructions of community as its key principles. Critical science agency implies that students have opportunities to influence the way science is used and to use science in accordance with their own values and perspectives. Shared authority implies that students are free to raise their voices and provide suggestions and critiques, and that students' knowledge is valued. Constructions of community imply the opportunity to work together in a supporting and caring environment that enables all students to learn.

This model emphasises how teaching for social justice should combine opportunities to learn with opportunities to bring forward the students' own perspectives and use science in accordance with their own needs and values. The present study expands this model by focusing on how students' use of science in accordance with their own perspectives can involve evidence-based argumentation, i.e., argumentation involving scientific or other factual knowledge about the material world. Moreover, the study identifies design principles for science teaching to provide experiences with construction of robust evidence-based arguments.

In science, argumentation is linked to knowledge justification, and claims should be justified with logical reasoning and empirical and theoretical evidence (Jiménez-Aleixandre and Erduran 2007). Lacking awareness of how scientific arguments can be scrutinised might lead to less robust argumentation. Experience and knowledge about scientific inquiry and critical thinking will empower students to construct arguments that better withstand criticism.

One of the central goals of teaching science for social justice, and of the two projects that are analysed in this study, is to 'position students as knowledge-constructors and critics, rather than passive recipients' (Thadani et al. 2010, p. 22). Increasing students' ability to construct knowledge and arguments through inquiry is therefore a necessary aspect of teaching for social justice (Garii and Rule 2009; Thadani et al. 2010).

Teaching for social justice includes 'tying the academic content to students' own lives, recognising that this will empower them within the contexts of their lives and communities' (Garii and Rule 2009, p. 491). In science education, this aim implies the provision of students with experiences of how science could be used for their own agendas (O'Neill 2010). This aim also presupposes the fostering of student ownership of content, student autonomy, and student mastery experiences (Barton and Tan 2009). The aim also requires the inclusion of students' knowledge as legitimate sources of introductory ideas and contributions to class dialogues (Barton and Tan 2009; Basu and Barton 2010).

The purpose of this chapter is the identification of design principles that enhance students' ability to construct robust evidence-based argumentation based on their own perspectives. That is, science teaching for social justice focusing on students' construction of arguments that are informed by NOS and critical thinking. To enable the identification of such design principles, this paper analyses two science projects based on two questions:

1. What practices did the students participate in that involved construction of evidence-based arguments reflecting aspects of NOS and critical thinking?
2. What underlying design principles might have contributed to student participation in learning activities where they used facts and scientific argumentation to support their points of view?

The suggested set of design principles are intended as a starting point for design-based research and a guide for curriculum design. The principles are general and do not describe specific teaching or learning activities. This set of principles might be denoted as a high-level conjecture. Sandoval (2014) has discussed design-based research and identified a high-level conjecture as 'a theoretically principled idea of how to support some desired form of learning, articulated in general terms and at too high a level to determine design'. In this chapter's discussion, the design principles are compared to Basu and Barton's (2010) key principles for democratic science pedagogy presented above.

10.2 Theoretical Framework

The theoretical framework for the analysis of the two science projects includes a conception of NOS and a conception of robust argumentation. The nature of science refers to the epistemology of science, science as a way of knowing, and the values and beliefs that are essential to the development of scientific knowledge (Abd-El-Khalick and Lederman 2000). Lederman (2007) formulated a set of seven NOS objectives for science education, which together represent a condensed formulation of the "consensus view" of NOS. According to this view, students should learn about the following characteristics of science: observation is distinct from inference; scientific (empirical) laws are distinct from (explanatory) theories; science is based on observations of the natural world and involves human imagination and creativity; scientific knowledge is subjective and theory-laden; science is as a human enterprise practised in the context of a larger culture; scientific knowledge is never absolute or certain; and NOS refers to the epistemological underpinnings of the activities of science.

Although the concepts are related, NOS may be seen as distinct from scientific inquiry, which refers to the methods and procedures of science (Lederman 2007). However, Grandy and Duschl (2008) warned that such a differentiation oversimplifies the nature of observation and theory, and emphasised that inquiry practices are guided by epistemological thinking. This comprehensive view of NOS is echoed in the well known Delphi study by Osborne et al. (2003), which identified nine key ideas about NOS to be taught in school science. The nine key ideas were categorised as ideas about the nature of scientific knowledge, the methods employed in science and institutions, and the social practices in science.

Several authors (Allchin 2011; Yacoubian 2012) have suggested that teaching NOS should be aimed at inculcating the use of epistemic understanding to guide inquiry and should focus on the dimensions of reliability in scientific practice. In line with this reasoning, this paper takes the perspective that students' ability to engage critically with science and scientific argumentation might profit from NOS teaching, which is integrated into student inquiry and related to topical issues. One major reason for this is that several aspects of NOS reflect values and practices employed by scientific communities to enable the development of reliable knowledge backed by robust argumentation. Examples in the above-mentioned Delphi study include asking questions and seeking answers; using experimental methods, hypotheses, controls, and critical testing; distinguishing between data and interpretations; using creativity and collaboration; and subjecting new developments to critique through activities such as peer review. Moreover, the Delphi study includes insights about the tentative and revisionary nature of scientific knowledge and the use of reports, for example, to communicate arguments and new developments. Compared to Lederman's conceptual core objectives, the Delphi study represents a somewhat broader picture of NOS and places more emphasis on epistemic practices in science.

Ideas about NOS guide epistemic practices in science and are therefore relevant for students aiming to construct and criticise evidence-based arguments. Using the core elements of NOS identified by Lederman (2007) and the Delphi study (Osborne et al. 2003), the arguments constructed by the students in the two projects, and their epistemic practices, will be analysed for their use of ideas about NOS.

The main purpose of this study is to discuss how to support the development of students' competences in critical thinking and construction of robust argumentation. The analysis therefore includes a focus on instances and characteristics of the evidence-based argumentation and critical thinking students practised and became more skilful in. An argument might be regarded as a justified claim, i.e., supported by data (Toulmin 1958). According to Toulmin (1958), arguments might also contain elements like warrants, qualifiers, and rebuttals. In a discussion involving several points of view, arguments might meet critique. A main purpose of critical thinking is to decide on what to believe (Bailin et al. 1999), e.g., to judge the quality of arguments. An argument can be viewed as more or less robust depending on its ability to withstand criticism. In the process of constructing an argument, the arguer might use critical thinking to evaluate the strength of different parts of the argument in order to identify element in need of improvement. Consequently, a crucial part of critical thinking is to identify, construct, and evaluate arguments (Facione 1990).

According to Golding (2011), a critical thinker needs a sophisticated epistemic understanding as, in addition to other elements, 'critical thinking is about constructing and evaluating reasoned judgments' (p. 358). Moreover 'interpreting the reliability of scientific claims requires a broad understanding of scientific practice' (Allchin 2011, p. 522). However, while critical thinking is an integral aspect of NOS, it is also a separate area of scholarly study. Bailin et al. (1999) have identified five kinds of intellectual resources that are necessary for critical thinking: background knowledge; operational knowledge of the standards of good thinking; knowledge of key critical concepts (e.g., being able to distinguish a value statement from an empirical statement); heuristics (strategies, procedures, etc.); and habits of mind (e.g., open-mindedness and fair-mindedness). Moreover, critical thinking often requires 'imagining possible consequences, generating original approaches and identifying alternative perspectives' (Bailin et al. 1999). Thus, creativity plays as important a role in critical thinking as it does in NOS and scientific inquiry.

Bailin and Battersby (2016) have argued for 'a conception of critical thinking as a practice—the practice of inquiry' based on the need to integrate skilled performance and the acquisition of the virtues inherent to the practice of critical thinking. The claim that critical thinking and inquiry are interlinked practices and competencies is also supported by the seminal Delphi Expert Consensus (Facione 1990), which has stated that 'critical thinking is essential as a tool of inquiry'. The consensus report goes on to state that 'critical thinking is a liberating force in education and a powerful resource in one's personal and civic life', which is consistent with one of the main purposes of teaching for social justice.

Several researchers have explored how understandings of NOS might support students' critical thinking in socio-scientific contexts (Bell and Lederman 2003; Khishfe 2012; Leung et al. 2015; Matthews 1994; Yacoubian 2015). Neither Bell and Lederman (2003) nor Khishfe (2012) have found any impact of personal decisions on socio-scientific issues after NOS instruction. However, Yacoubian (2015) has argued that the students need to not only *understand* aspects of NOS, but also be guided to *apply* their understanding in relevant contexts. Khishfe's (2012) study included such practices. She examined students' justifications and found that more students in treatment groups were referring to NOS aspects; the students were backing their views with empirical evidence and indicating greater awareness of the tentative nature of evidence.

This chapter assumes that experience with NOS in inquiry environments can improve students' abilities to construct and evaluate arguments in socio-scientific contexts. This assumption is supported by several studies that have documented how students' critical thinking could be positively influenced by inquiry-based science teaching (Gupta et al. 2015; Hand et al. 2018; Quitadamo et al. 2008) and topical socio-scientific contexts (Goeden et al. 2015; Merchan and Matarredona 2016; Wang et al. 2017). These topical socio-scientific issues provide opportunities for students to relate science to life outside school.

10.3 Data and Analysis

10.3.1 The Cress Project

The first of the two projects to be discussed was initiated by a male science teacher at a lower secondary school. He had two Grade 8 science classes and wanted to do an inquiry-based project as some of his students showed a lack of motivation for science. Also he thought the inquiry methods were appropriate for the next topic to be taught, NOS and scientific inquiry. The teacher felt somewhat unexperienced with the topic and with inquiry-based teaching and asked this author to collaborate.

The school is situated in a suburban area where students have varied socioeconomic background, but the average is above that of the local municipality. Varied socioeconomic backgrounds of students are typical of many schools in the area. At the national test in mathematics the year of the project, the school's score was one point below the national average, which was defined to be 50 (www.udir.no/in-english/). Of the five performance levels, only 22% of the students at the school were in the top two, while the national figure is 32%. There are no national tests in science. The students had conducted practical work in science before, but had not experienced inquiry-based teaching at this school.

The main goal in this project was to help students, in the context of an environmental issue, begin to understand how scientific claims are constructed and might be inspected for trustworthiness. Students were asked to imagine that a local environmental organisation wanted to advise people to use less harmful chemicals. The organisation needed reliable knowledge on what types of chemicals are harmful so that people would not lose trust in their advice. The students worked in groups of three and four to test the impact of a household or garage chemical of their choice on the growth of plants. All groups were given 10 pots with garden cress as a model plant to use in an experiment of their own design. Seven 45-min science lessons were allocated for the Cress project.

The data from the project consist of observational notes from the classroom lessons, including students' oral contributions during class discussions; student experiment plans and experimental reports; and students' written responses to tasks during lessons and to questions on an end-of-chapter test. Data also include planning documents and the teacher's and researcher's written reflections during the project.

10.3.2 The Energy Project

The second of the two projects was suggested by two university teachers and this author as part of a 3-year collaboration with a lower secondary school. The project was developed in cooperation with eighth grade teachers in science, social science, and mathematics. All the data were collected from this author's collaboration with the two female teachers running the project in one class. These two teachers showed great interest in the project and thought it was important to raise the relevance of the science teaching for their students.

The school is situated in a suburban area with students from similar mixed socio-economic backgrounds as the first school. At the national test in mathematics that year, the school's score was at the national average (www.udir.no/in-english/). Of the five performance levels, 33% of the students at the school were in the top two levels that year and 12% were at the lowest level, while the national figures were 30% and 11% respectively. Also in this class, the students had conducted practical work in science before, but had not used inquiry-based teaching.

The goal of this project was to increase students' confidence to use facts to build up their own arguments on a topical socio-scientific issue. There was a local debate on whether to expand an existing light rail system. Due to climate and pollution considerations, the municipality had decided that an existing line was to be extended past the students' school. Students were asked to develop a method to compare light rail and more roads in terms of energy consumption and CO_2 emissions. Then, the students were to apply this method to their own inquiry and write a scientific report presenting their exploration and conclusion. A set of introductory tasks was designed to introduce students to the concept of kilowatt hours and how energy can be measured using different concepts (e.g., gasoline consumption). The scientific report had to be written as a poster and each group presented their report orally to the rest of the class. All presentations were followed by a teacher-mediated class discussion that focused on critical questions and the identification of interesting points. Students worked in groups of two to six students, and the project involved 16 school lessons.

In this project, all 16 lessons were videotaped, using one whole class camera capturing the teacher and blackboard and five GoPro cameras capturing different groups. Relevant passages in the videos were transcribed. Data from this project also included students' written reports and posters; self-evaluations of learning gains, effort, and challenges; and the teacher's and researcher's videotaped oral and written reflections.

10.3.3 Method and Analysis of the Two Projects

The analysis of the two science projects focused on student experiences with NOS, evidence-based argumentation, and critical thinking. Such experiences might impact students' abilities to construct and criticise knowledge claims. In addition the projects will be inspected for presence of key aspects of democratic science education. All field notes, videos, and written work by students were carefully inspected and relevant sections, i.e., involving talk or writing of arguments, claims, critiques, inquiries, and NOS, were marked. Marked sections were further inspected, and the elements of NOS and critical thinking involved were described.

The first dimension of the analysis, the identification of situations where students experience NOS, included aspects related to dimensions of reliability in scientific practice. The analysis included instances where students expressed elements of NOS with their own words and where elements of NOS were reflected in the students' practices or where students were challenged by their teacher to do so. In specific, the analysis identified situations where students formulated research questions, designed and explicated methods of inquiry, used controls and critical testing, collected and interpreted observations, experienced relevant critique, discussed assumptions, based conclusions in reports on evidence, discussed the tentative nature of scientific claims, and exhibited other elements of NOS.

The second dimension of the analysis implied identification of students' experiences, discussions, and writing involving construction and critical thinking about arguments, including situations where students were challenged or supported by their teacher to do so. In specific, the analysis identified situations, discussions, and written work where students created or presented arguments; identified questions or comments on possibly weak aspects in plans, experiments, observations, interpretations, conclusions or justifications; suggested or commented on alternative interpretations or ways of testing the correctness of claims; or practised or commented upon dispositions associated with critical thinking (e.g., scepticism, open-mindedness, requests for grounds for factual claims or value judgements). This dimension also included instances where students or teachers questioned the relevance of a claim or expressed the importance of a critical attitude.

These two foci of analysis included descriptions of the tasks and scaffolding that are used in the identified situations and practices. Student participation in such situations was also noted. This inclusion enabled a discussion of design principles informed by challenges and successes in scaffolding students to acquire the kind of practices that are the focus of this study.

Finally, to judge the extent to which the projects supported social justice through providing experiences and insight into the construction of evidence-based arguments, three interlinked key aspects of democratic science education were examined. First, do the students bring their own perspectives and express their own views in discussions and reports? Second, do they use their voices and knowledge to construct their own arguments and critiques? Third, are students experiencing a supporting and caring community enabling all students to participate in activities?

Based on a comparison of results of the analyses along the described dimensions and the characteristics of the projects, a set of tentative design principles was identified.

10.4 Findings from the Cress Project

At the outset of the Cress project, the students were informed that they would learn more about NOS and how scientific claims are made. They were told that they would make a practical inquiry, formulate clear questions and plans, make detailed observations, and use those observations to formulate a concluding claim. As the teaching unfolded, this author and the teachers decided to focus on critical thinking as an important aspect of scientific practice. The final learning goals stated that students should learn the following:

- The nature of scientific research questions.
- The importance of identifying and controlling variables, nonbiased and systematic observation.
- How concluding claims need to be consistent with observations and results.
- How the introduction, method, results, and discussion (IMRaD)-structure of scientific reports and its strict division between empirical results and final claims enables critical inspection of a study.

The central pedagogical idea of the Cress project was to use students' experiences from different phases of the practical inquiry as starting points for discussions and mini-lectures on elements of NOS. Typically, each lesson started with a class dialogue on experiences from the previous lesson that sought to highlight students' reflections and articulate answers on NOS-related questions. Tasks and scaffolding were designed to guide students to use relevant practices, formulate their own ideas, and discuss interpretations of elements of NOS.

10.4.1 Practices That Reflect Ideas About NOS

The first lesson sought to engage students and provide an introduction to a body of ideas. The key idea was that in science, studies need to be designed carefully in order to make concluding claims reliable. Students were shown a provocative claim from a scientific study: 'If you sleep less than 7–8 h each night you are more likely to get a cold'. Following a short class discussion, students worked in small groups on the following prompts: 'Might this claim be correct? How should researchers have conducted their work for us to trust what they find out? Suggest some ideas!' The students had relevant ideas that were shared and written on the blackboard. The methods that were used by the researchers were then revealed and discussed in the class. This process signalled that students' ideas were valued, and provided students at all levels of abilities a reservoir of ideas for subsequent discussions and design of methods.

Students began to design their own methods of inquiry based on these introductory activities and a template. The students had the freedom to choose chemicals according to their interest. They brought substances like nail polish remover, engine oil, and dish soap to the classroom, and decided how to check whether these would harm cress plants. As an introduction to the planning phase, the teacher led a discussion about the importance of testing hypothesis about toxicity before making claims.

During the Cress project, the students experienced discussions of observations and how observations do not speak for themselves, but need to be interpreted. Before the students began to run their experiments, the teacher initiated a class discussion on whether some observations were possible, such as if the engine oil was toxic. Many students said 'no', many said 'yes', and there was a discussion over 'Why [the plant] might have become withered?' In the next lesson, before students went to observe the effects of their treatments, the teacher picked up again on that issue by asking 'Will you be able to observe whether your chemical is dangerous?' Student comments ranged from 'No!' to 'We might have given too small doses' and 'It might have got too little water'.

All the groups planned to use different amounts of their chosen chemical on the cress plants. This made the details of the inquiry methods different. In both classes, at least one group had the idea of leaving some pots untreated by chemicals as control, and this idea spread to most of the other groups and became implemented in their plans.

In the last lesson on the project, the concepts of variables and control was introduced and discussed in relation to students' experiences. The teacher invited students to discuss 'What could possibly vary in these inquiries?' Students' contributions resulted in the following list being written on the blackboard: '[Different] substances, amounts of the chemical, mixes of chemicals, amounts of watering, ways to add the chemical, numbers of cress plant, treatment times, light.' The teacher added the word 'Variables' as a heading for the list. The students were then asked to indicate the strategies their group had used to make sure that only the chemical was responsible for the withered plants. Next, the teacher conducted a discussion on the value of using control plants.

In several reports, students commented explicitly on how the use of control plants enabled comparison. The group that tested dish soap stated, 'We have several cress pots, some with ordinary water and some with dish soap. Then we can see the difference.'

About half of the 15 reports made comments indicating awareness of the value of controlling variables, as in the following example: 'After [treatment was added] they got a bit of water in order not to dry out. We gave them exactly the same amount.'

A final activity stimulated students to reflect on their resulting insights about how to ensure quality of methods and reasoning when using inquiry to produce scientific claims. The students were asked to work in pairs and 'write down four things we need to take into account when investigating a research question'. The 37 responses in one of the classes included the following: 'Have a good plan, thorough method', 'They must vary a lot to see what really affects the plant', 'They observe, one must observe only what is being measured', 'Accurate observation, orderly table', 'Discuss the reasons why this happened'. Each group shared two responses in class, which were written under the appropriate heading on the blackboard (introduction, method, results, and discussion) to support a final class discussion.

The end-of-chapter test included questions about observations, such as 'Why do researchers first write down exactly what they see, and then ask how it might be explained'. In their answers, 22 of 44 students explained that there might be other causes from the one anticipated. Some answers were detailed while others were simpler, as the following two examples illustrate:

> Because then you can decide for yourself if the conclusion is true. You can see how it might fit the hypothesis and whether you will believe it because they explain how they have achieved the results and what they have seen.
>
> It's important that they first write down what they saw, so they did not forget it. And then they can begin to find an explanation and compare.

Seven students gave answers that were not relevant to the question.

10.4.2 Practising Argumentation and Critical Thinking

Students practised scientific argumentation and critical thinking when drafting experimental plans and reports, taking the end-of-chapter test, and in responses to oral and written tasks. Twelve of the sixteen experimental plans included the key practice of using the control group to make fair tests. Ten experimental plans also included one or more arguments about qualities or potentially weak aspects of their method. The following is an excerpt from a plan that included both (text from the template in italics):

> *Suggested method:* Different dosage of diesel. Two plants with one drop, two plants with 4 drops of diesel, two plants with 7 drops of diesel and two with a lot of diesel and finally one without diesel to enable comparison. *We will observe:* Does it wither, change colour or become smaller or larger. *The method may be unreliable because*: If we had had too much diesel it could have been drowned because of shortage of oxygen.

All the groups made a plan, started their own experiment, logged observations, and wrote a report. As illustrated by the following example, all reports included a discussion (counting between seven and 181 words) with a justified claim about the harmfulness of the tested chemical:

> In our opinion the answer is that plants do not tolerate juice because the juice has substances that the plant cannot withstand. The observation supported us because the plants withered and we thought the plants do not tolerate the juices. We are sure that this answer is correct because the cress withered and then we have something that can prove that the plants didn't withstand juice.

The structure and logic of the scientific reports, i.e., providing a claim backed by observations and a sound method, were discussed several times during the project. All 15 reports separated results from inferences in the discussion. All but one of the reports made explicit reference to empirical results that supported the concluding claim. None of the conclusions in the reports were judged to be unacceptable. Seven reports demonstrated open-mindedness by stating a different conclusion from the one hypothesised, as in the following example: 'We don't think that charcoal lighter fluid is very dangerous for plants. This we believe because only those plants that received very much fluid [sic] were clearly injured.'

Six of the students' reports included eight examples of creative critical thinking by identifying possible weaknesses in methods, alternative interpretations, or additional tests, as in the following example:

> To make sure they [the cress plants] were really dead, we gave them some water to see if they become fresh and alive again, but that did not happen and therefore we believe that petrol is very harmful to plants.

In the end-of-chapter test, eight students suggested possible reasons why their conclusion might be wrong, including issues related to quality of observations, control of independent variables, and the doses used in the treatment.

10.4.3 Ownership and Critical Science Agency

The analysis above reveals that all groups of students picked a chemical of their own interest, made experimental plans and reports, and constructed a concluding argument backed by their own observations. Within the common general questions, students formulated their own research question, hypothesis, and interpretations. This suggests that students brought some of their own interests and perspectives, although within the constraints set by the project. These findings also show that the students took the fictive context seriously. According to the teacher, most students showed more interest in the writing of the reports than he had seen before, and many groups were actively discussing as they were writing. This indicates that the context involving an environmental concern and the issue of trustworthiness of claims was meaningful for most students, resulting in a situation where argumentation and critical evaluations became natural.

This conclusion is also supported by comments made by the students in a rubric the students were asked to complete to evaluate the project. All groups used word like *fun, interesting, exciting,* or *worked fine* in their comments. Five groups added a comment reflecting the environmental focus of the context of the project. For example, one group commented that 'It was very interesting to work with this. It was very disappointing that environmentally friendly gasoline is not so environmentally friendly.'

Throughout the Cress project students were challenged to discuss in small groups and share their own ideas in class, and often these were written on the blackboard. Such sharing implies valuing of students' ideas. Also, students' ideas were not assessed but used by the teacher as starting points for presenting ideas about NOS. As this sharing and valuing of students' ideas was related to their inquiry projects, it also implied an establishing of an epistemic culture in the classroom where students' epistemic agency was practised and valued.

The use of group discussion and sharing of ideas also implied scaffolding the development of ideas needed to create experimental plan and understand key points of NOS. Most pairs of students contributed in class with ideas as part of think-pair-share activities, and all groups designed an experiment and logged observations. However, fewer students contributed with comments in class discussions that focused on conceptual knowledge. Key points related to NOS and critical evaluation of ideas was in focus in several activities and discussions throughout the project.

This repetitive focus probably contributed to positive results at the end-of-chapter test. The teacher did not include activities that stimulated reflections based on examples of critical thinking in students' plans or reports.

To support learning and mastery for all, the overall research question for the project, 'What chemicals might impact the growth of plants?', was deliberately designed to allow for inquiry projects at many levels of complexity. When students were asked to make a research question, design a method using the provided cress plants, do the experiment, and write a report, they solved these challenges differently. An easy solution designed by one group was to put the chosen chemical on some plants and see what happened. Other students used different doses of their chemical on different plants and used each dose on a minimum of two plants for increased reliability, and some additional untreated plants for control. One of these groups also wanted to check the conditions where the plants were stored between lessons, to ensure equal light conditions for all plants. This freedom in complexity of methodology, combined with different types of scaffolding, probably explains why all groups managed to produce plans, experiments, evidenced-based claims, and reports. Thus all students had experiences enabling them to participation in discussions.

10.5 Findings from the Energy Project

The specific learning goals identified for the Energy project included the following:

- Experience how to back up claims with facts and clear reasoning so that those claims are not criticised or ignored.
- Practice the ability to collect the necessary facts, build a model to compare measures in a structured way, and put forward a fact-based argument.
- Understand how energy can be used as a common yardstick for comparison across different contexts.
- Understand the concept of energy consumption per passenger kilometres, and use this concept for fair comparisons.

The project also emphasised how to help the students gain insight on how a report is structured, why the method is explained in scientific reports, and why the results and discussion of those results are presented in separate sections. The purpose was to provide an introductory awareness of the use of environmental impact assessments reports in management and how these might be used to find and critique arguments that relate to issues of interest.

The central elements of the pedagogical thinking behind the Energy project were to engage students, provide for the sharing of ideas, and support continuous improvement of ideas. A real-life context and a driving question were designed to engage students and to allow all groups to develop their own specific research question and method of inquiry.

10.5.1 Practices That Reflect Ideas About NOS

In the Energy project, discussions of aspects of NOS were restricted to ideas that were embedded in the structure and logic of scientific reports and characteristics of scientific methodological thinking. The different sections of a report were presented as the project developed, and examples from students' tentative descriptions of methods, results, and discussions were shared and discussed. During model development and the writing of final reports, students in the Energy project were challenged to explicitly articulate their methods. The following extract from a report is a typical example of how the students did this.

> The method we used was comparison. We mostly compared numbers in [kilowatt hours] Kwh. What we compared was the difference in figures between car and light rail. How did we do that? We first found out how many people took the light rail. If they had not taken the light rail they had most probably driven a car. Then we found out how many percent (%) of those who would have taken a bus (2100), electric car (990), diesel car (5614), petrol car (4400), and so vi multiplied with 1.4. 1.4 is average number of persons in a car.

When discussing how to explicate methods, the teacher explained how the method supports an implicit claim about the quality of data presented and enables criticism of possible weak points. In discussions following presentations of reports, the teacher often challenged the rest of the class to comment, thus stimulating students to apply their developing ideas about the characteristics of scientific reports:

Teacher: What makes this a good research report?
Student 1: There's a lot of order, so we can see where the introduction is and where the method is, and.
Student 2: They have explained very good what they want to investigate and how they did it. And, very well explained, and also they explain at the end, eh, what they could conclude, in a way.

All group reports explicated methods, separated data and interpretations, and formulated an evidence-based argument in the concluding discussion. However, explanations for these practices were not explicitly articulated by students during discussions or presentations.

Although students were not challenged to articulate their understanding of NOS, students were challenged to explain the difference between methods and results, and between results and concluding claims.

The importance of argumentation, quantification, correctness, and explanation of assumptions was discussed on several occasions. As illustrated in the following dialogue, the teachers repeatedly explained how scientific arguments needed to be convincing.

> *Teacher*: If you want to compare, you may want to have some numbers to compare. Just saying 'a bit much', and 'a little more', makes it difficult to compare, in a way. *Student*: Should we make an average, kind of? *Teacher*: … The question is, in a way, how may I know that what you claim is correct? You have to make it convincing. *(The discussion goes on)*

The idea of fair comparison often came up when teachers supervised groups. Upon being asked by the teacher what they had found so far, the students in one group stated 'It takes nine times as much energy with the city rail.' The teacher then asked if they have thought about the number of passengers. One student replied, 'Yes, so there are many more who take it, right'. Figures were given in energy per person in the final report of this group.

The key NOS idea exemplified here, the importance of designing one's methods in ways that makes the results and conclusion reliable and robust against criticism, was a recurrent theme in discussions.

10.5.2 Practising Argumentation and Critical Thinking

Evidence-based argumentation was evident during supervision of groups and in all reports. Such argumentation ranged from short statements to more elaborate discussions, as the following two examples indicate:

> *From conclusion in report from group two*: It is better to use the light rail lane, and not cars, because the light lane uses only 0.62 kWh per person from Lagunen to Flesland, while the car (petrol car) uses about 3.2 kWh on the same stretch.

> *From conclusion in report from group seven*: Our evaluations are, that when you drive a car in and out of the city, you spend a lot less than the light rail use, but if you think about it, the light rail is actually better because it carries 212 passengers and a car max 5. Early in the morning there are only 1 max 2 in the car while in the light rail there are maybe 70 people, so in the long run the light rail is much better than cars. It uses a lot more kwh than a car but also it carries many more.

Throughout the project, the teachers reminded students of the need to use correct facts to justify claims:

Student: The light rail uses more energy than cars.
Teacher: In order to justify that claim, you need figures.

The video data and written reports made it evident that critical discussions during the project typically involved practical reasoning, fact-based argumentation, and critical discussion of figures used by peers.

The practical reasoning was characterised by students using their everyday knowledge about energy, environmental issues, and transport-related needs to make arguments, as exemplified in the following excerpt from an early discussion in group 2:

> *Student 1*: We could still just use cars [and not build light rail] and build a lot of bicycle and car roads. *Student 3*: But then we will be using more and more energy. *Student 1*: But think about the fact that more and more people use electric cars. *Student 3*: But think also of what kind of source the energy is from. *Student 2*: Hydropower? *Student 3*: And? ... *Student 1*: We do not have enough hydro power if all the cars are going to use it. *(Continuous discussion about hydropower, coal, and wind energy from Denmark)*

This excerpt also exemplifies how critical discussions sometimes involved the critical skill of considering alternatives.

Their fact-based argumentation typically included figures about travelling distances and energy consumption of different types of vehicles. These critical discussion of figures included an example of a group being asked by a peer to explain a figure for the number of cars travelling a stretch every day: 'How can there be 2.5 million cars on that road when there are 5 million cars in all of Norway?'. This critical question focusing on a possible inconsistency was resolved when it was explained that the first figure concerned the number of cars passing a counter during the preceding year.

Following the presentation by another group, a question was raised about their use of the figure of 212 seats in the light rail as a basis for calculating energy consumption per passenger. Some students remembered that the light rail was said to often be very full, while others had read that it had 217 passengers on average. One student pointed out that not all passengers take the rail the entire route. The teacher stated 'It's very good you are so awake'. She summarised by highlighting the importance of stating the figures and assumptions used, and that assumptions need to be checked when comparing reports.

Critical discussions among students were most often observed when teachers contacted groups. Typically, the teacher asked for an update and then asked a challenging question, which led to a discussion. Some of these discussions were also initiated by students.

10.5.3 Ownership and Critical Science Agency

In the Energy project, all student groups formulated different research questions and methods of comparison. For example, two groups calculated energy consumption for each person who travelled a certain distance by light rail and by car. The consumption rates were then compared, considering statistics on the number of persons in cars and light rail at a comparable distance. Another group included energy costs of the materials used in construction and CO_2 emissions involved. Yet another group found statistics on types of preferred transport and calculated how many people would use the new light rail. The group also calculated the distance the light rail had to go before the energy construction costs were lower than the energy saved by shifting from cars to light rail. The groups' concluding arguments also varied. While some focused on energy costs only, other groups emphasised other aspects, e.g., that 'We don't think the light rail is needed, as there are buses on that stretch already'. The teachers' positive comments to all groups during supervision and presentations signalled to the students that their diverse perspectives were valued.

The diversity of research question and methods suggest that students made choices based on their own perspectives and competence levels. On several occasions students continued working in the classroom after lessons had ended, and the two teachers commented that the students' engagement was higher than normal.

This indicates that the context involving a local topical issue and the focus making trustworthy reports to inform politicians was meaningful for many students, resulting in a situation where argumentation and critical thinking became natural.

On several occasions, the teachers stated that the students were supposed to come up with ideas, find facts, find ways to do relevant calculations, and construct their own conclusions:

> You will come up with ideas on whether city railways are smarter than cars. How might we find out? There are many ways, you should develop your own ideas, develop a method, compare with and without the city rail.

The teachers did not evaluate students' ideas or indicate that there were correct forms of thinking. Instead, the teachers facilitated sharing and mutual evaluation of ideas. One example is how the teacher structured an introductory task: 'Work in groups for 6 min: Make suggestions as to what might affect how much energy it costs to carry people by car and by city rail.' After 6 min, the teacher called for attention and stated 'I want one idea from each group!' She noted ideas on the blackboard for easier sharing among groups. She wrote two headings on the blackboard, 'Road' and 'Light rail', to enable structured comparison of the ideas.

These teacher practices implied valuing and support of students' ideas and funds of knowledge. A reoccurring activity in the Energy project was the mutual evaluation of students' ideas and critical discussions related to their inquiry projects. These practices constituted an epistemic culture in the classroom where students' epistemic agency was stimulated and practised. Together with several introductory tasks and the simple template for the report, this implied scaffolding of students' practices and learning opportunities for all students.

Throughout the project, the teachers signalled trust in students' ability to practice agency:

Teacher: Now we share ideas. No expectations about finished ideas, but start sharing. Can you start, group two?

Student: We talked about where the power comes from. That the power must come from a source with no pollution.

Typically, teachers would follow up with remarks such as 'Any comments or questions to this group? Any ideas for you to take on-board?'

During the final presentations of reports, the teachers expressed trust in students' abilities to construct a model and a justified conclusion: 'Speak louder, Marit, because what you're saying is smart, so just go for it!'

The use of scaffolding and signals of trust in students' abilities suggested a caring and supportive learning culture in the classroom. Although at different levels of complexity, all groups constructed models, made calculations, and wrote justified conclusions. The presentations and the students' written self-evaluations indicate that the project was an experience in mastery for all groups and for most students. One of the teachers summarised the evaluations by teachers from all the classes as follows:

The teachers express that they would like to have this project again. The students learned a lot, new concepts, collaboration, and mastered things they did not think they could handle.

10.6 Didactic Principles for Teaching for Robust Argumentation and Critical Thinking

The two projects reveal that it is possible to design science projects for students to practice ownership and use their growing insights in NOS to construct and present arguments and carry out critical thinking. Although this is not a sufficient basis for establishing design principles, it is possible to use the two projects to identify design principles to be used as working hypotheses for further exploration.

Both projects included a social context, and the tasks were designed to have some relevance for participation in such contexts. Moreover, students could make personal choices and engage with issues within their own interests and abilities. Thus, the tasks and situations bear a resemblance to real-life situations for students engaged in issues and discussions where evidence-based arguments are relevant. Differences in opinions on such issues might trigger argumentation from diverse fields of knowledge. However, a focus on trustworthiness of evidence-based arguments can make science, NOS, and critical thinking relevant. The following three design elements are therefore suggested as important to the development of social relevance and student ownership, which are in turn necessary for the motivation and engagement needed for lasting participation in classroom activities:

1. Identify a real-life context that might be meaningful to the students and includes evidence-based arguments with potentially disputed trustworthiness to trigger students' engagement.
2. Design situations where trustworthiness is at stake to give students a natural need to construct evidence-based arguments and to inspect all arguments critically.
3. To enable ownership, mastery, and autonomy for all students, design driving questions and scaffolding tasks which can be interpreted to any level of complexity and adapted by the students to their interests and abilities.

School teaching always aims to support students' competency development. Science teaching for social justice must therefore support intended learning while allowing for ownership and activities that resemble real-life situations. The students involved in the cress and the energy projects have gained introductory insights into NOS and critical thinking as well as experiences in using such insights in their own projects. This indicates that the projects to some extent enabled planned learning while involving the students in practices that resembled real-life activities.

The Cress and the Energy project had few and interlinked learning goals. Moreover, teachers used tasks, presentations, and discussions to challenge students to create, share, and improve ideas and to repeat key ideas and practices. The following three design elements are suggested for their contribution to intended learning through active knowledge construction in science classrooms that are characterised by real-life context and student autonomy:

4. Formulate learning goals which are interlinked, manageable, and represent aspects of NOS and critical thinking that are relevant for evaluating the robustness of arguments.
5. Cultivate an epistemic culture in the classroom that resembles epistemic values in science: all students have a legitimate voice, all ideas are welcomed and explored, and the goal is the evaluation and improvement of ideas.
6. Students repeatedly encounter important ideas and situations to support their development of deep understanding and new knowledge-based habits.

The six principles are not meant for inquiry-based teaching in general, but more specifically for guiding teaching aimed at increasing all students' capacity to construct robust evidence-based arguments, with the ultimate aim of promoting social justice.

10.7 Discussion

This chapter analysed two science projects where students constructed evidence-based arguments related to a real-life issue. The analysis revealed that all student groups constructed arguments and that many students participated in critical discussions of observations, figures, or arguments. Moreover, students participated in practices reflecting key aspects of NOS and focused on the reliability of scientific practices and arguments.

The analysis of students' practices and written reports in the Cress project revealed that the students expressed awareness of the difference between observations and possible inferences and articulated a range of ideas related to scientific methods. In the end-of-chapter test, half of the students applied the distinction between observations and inferences. Furthermore, in their scientifically structured reports, all groups communicated evidence-based arguments. The students also discussed, designed, and implemented methods, tested hypotheses, interpreted data, and made use of, and discussed, the concepts of variables and the control of variables.

In the Energy project, all groups distinguished between data and inference in their reports and communicated evidence-based arguments. The students also discussed, designed, articulated, and implemented their own methods. Several students formulated differences between methods, results, and concluding claims in their own words. Moreover, students participated in discussions about the importance and characteristics of scientific argumentation, the argumentative structure of scientific reports, fair comparison, articulation of assumptions, and the importance of checking information for correctness.

In the two projects, students experienced how using elements of NOS, such as basing arguments on evidence, separating results and inferences, and using their own creativity and thinking when developing methods and inferences, resulted in quality arguments. Moreover, the use of control plants in the Cress project and evidence-based calculations in the Energy project provided an experience of how to make arguments less susceptible to criticisms. Consequently, the two projects showed the students how an awareness of the elements of NOS supports the construction of defensible evidence-based arguments. Increasing students' ability to construct such arguments might enable more students to provide robust support for own views in issues of interest. This suggests that awareness of NOS in the context of argument construction might have a role in supporting social justice.

Practices involved in the two projects were compared with Basu and Barton's (2010) model for inclusive science teaching aimed at empowering all students to use science in accordance with their own needs. Within the chosen contexts, students constructed arguments based on their own perspectives and choices. Their ideas, perspectives, and knowledge were valued, and they experienced support and opportunities to learn.

The present study expands Basu and Barton's model by including characteristics of science teaching that empower students to construct and critique evidence-based arguments related to real-life issues. These characteristics are formulated in a set of six didactical principles. In their model, in order to promote social justice, science teaching needs to ensure access to science for all students and enable the use of science for students' own purposes. Our results imply the possibility of specifying their concept of 'critical science agency', i.e., opportunities for students to engage with science in accordance with their own perspectives, to include opportunities for students to use science in class to construct evidence-based arguments in accordance with their own needs. Moreover, the importance of empowering students to be able to construct robust arguments related to issues of interest implies that epistemological autonomy is an important practice in science class. An explication of this aspect involves a specification of the key element of 'shared authority' in their model.

It is likely that increased insights into issues of reliability will enhance students' ability to construct robust evidence-based arguments. As reliability is a main concern in science, scientific practice has developed appropriate methods, values, and ways of thinking. In the two science projects analysed, key elements of NOS were dominant in one and less prominent in the other. The Energy project had less emphasis on increasing students' ability to explicate key elements of NOS. The development of students' epistemic habits might gain from explicit attention to relevant elements of NOS (Lederman 2007). At the same time, experience with critiques of evidence-based arguments and discussions of how to make arguments more robust is probably necessary. Autonomous application of abstract tenets of NOS is demanding and needs to be developed through experiential learning. The present study indicates that a combined focus on NOS and critical thinking might help to create classroom environments where students' construction of evidence-based arguments is guided by key ideas about NOS and critical thinking. However, more research seems needed to understand how NOS should be included to support the development of students' abilities to construct robust arguments.

A basic assumption in this study is that science teaching for social justice requires the development of all students' abilities to construct robust evidence-based arguments. The analyses of the Cress project and the Energy project indicate that insights into NOS and scientific practices involving critical thinking might support students in constructing such arguments, thus supporting social justice. However, the ultimate aim is students' autonomous construction of arguments in issues related to their own interests outside school. This probably presupposes trust in their own abilities to inquire into issues and construct evidence-based arguments that are robust to some extent. Consequently, critical science agency does not only depend on the students' scientific knowledge and desire to learn. It also depends on the student's trust in their own abilities to construct knowledge claims and critical comments that are based on their inquiries into issues of concern. As self-efficacy refers to a person's belief that he or she can do what is necessary to successfully achieve a specific goal or task (Bandura 1997), this might be denoted as *epistemic self-efficacy*. In addition to psychological and affective states, self-efficacy is influenced by mastery experience, observing others' experiences, and social support and feedback (Bandura 1997). This implies that experience is necessary, but might prove insufficient if this is unsuccessful. An emphasis on mastery for all, sharing and discussions between groups, and a supporting and caring teacher thus seems paramount.

Acknowledgments This work was supported by the the Norwegian Research Council [grant number 275835].

References

Abd-El-Khalick, F., & Lederman, N. G. (2000). The influence of history of science courses on students' views of nature of science. *Journal of Research in Science Teaching, 37*(10), 1057–1095. https://doi.org/10.1002/1098-2736(200012)37:10%3C1057::AID-TEA3%3E3.0.CO;2-C.

Allchin, D. (2011). Evaluating knowledge of the nature of (whole) science. *Science Education, 95*(3), 518–542. https://doi.org/10.1002/sce.20432.

Bailin, S., & Battersby, M. (2016). Fostering the virtues of inquiry. *Topoi-an International Review of Philosophy, 35*(2), 367–374. https://doi.org/10.1007/s11245-015-9307-6.

Bailin, S., Case, R., Coombs, J. R., & Daniels, L. B. (1999). Conceptualizing critical thinking. *Journal of Curriculum Studies, 31*(3), 285–302. https://doi.org/10.1080/002202799183133.

Bandura, A. (1997). *Self-efficacy: The exercise of control*. Macmillan.

Barton, A. C., & Tan, E. (2009). Funds of knowledge and discourses and hybrid space. *Journal of Research in Science Teaching, 46*(1), 50–73. https://doi.org/10.1002/tea.20269.

Barton, A. C., Ermer, J. L., & Burkett, T. A. (2003). *Teaching science for social justice*. Teachers College Press.

Basu, S. J., & Barton, A. C. (2010). A researcher-student-teacher model for democratic science pedagogy: Connections to community, shared authority, and critical science agency. *Equity & Excellence in Education, 43*(1), 72–87. https://doi.org/10.1080/10665680903489379.

Bell, R. L., & Lederman, N. (2003). Understandings of the nature of science and decision making on science and technology based issues. *Science Education, 87*(3), 352–377. https://doi.org/10.1002/sce.10063.

Facione, P. A. (1990). Critical thinking: A statement of expert consensus for purposes of educational assessment and instruction: Research findings and recommendations (the Delphi report). Retrieved from ERIC Document Reproduction Service No. ED315423: https://eric.ed.gov/

Garii, B., & Rule, A. C. (2009). Integrating social justice with mathematics and science: An analysis of student teacher lessons. *Teaching and Teacher Education, 25*(3), 490–499. https://doi.org/10.1016/j.tate.2008.11.003.

Goeden, T. J., Kurtz, M. J., Quitadamo, I. J., & Thomas, C. (2015). Community-based inquiry in allied health biochemistry promotes equity by improving critical thinking for women and showing promise for increasing content gains for ethnic minority students. *Journal of Chemical Education, 92*(5), 788–796. https://doi.org/10.1021/ed400893f.

Golding, C. (2011). Educating for critical thinking: Thought-encouraging questions in a community of inquiry. *Higher Education Research & Development, 30*(3), 357–370. https://doi.org/10.1080/07294360.2010.499144.

Grandy, R., & Duschl, R. (2008). Consensus: Expanding the scientific method and school science. In R. Duschl & R. Grandy (Eds.), *Teaching scientific inquiry: Recommendations for research and implementation* (pp. 304–325). Rotterdam: Sense.

Gupta, T., Burke, K. A., Mehta, A., & Greenbowe, T. J. (2015). Impact of guided-inquiry-based instruction with a writing and reflection emphasis on chemistry Students' critical thinking abilities. *Journal of Chemical Education, 92*(1), 32–38. https://doi.org/10.1021/ed500059r.

Hand, B., Shelley, M. C., Laugerman, M., Fostvedt, L., & Therrien, W. (2018). Improving critical thinking growth for disadvantaged groups within elementary school science: A randomized controlled trial using the science writing heuristic approach. *Science Education, 102*(4), 693–710. https://doi.org/10.1002/sce.21341.

Jiménez-Aleixandre, M. P., & Erduran, S. (2007). Argumentation in science education: An overview. In S. Erduran & M. P. Jiménez-Aleixandre (Eds.), *Argumentation in science education: Perspectives from classroom-based research* (pp. 3–27). Dordrecht: Springer Netherlands. https://doi.org/10.1007/978-1-4020-6670-2_1.

Khishfe, R. (2012). Nature of science and decision-making. *International Journal of Science Education, 34*(1), 67–100. https://doi.org/10.1080/09500693.2011.559490.

Lederman, N. (2007). Nature of science: Past, present, and future. In S. Abell, K. Appleton, & D. L. Hanuscin (Eds.), *Handbook of research on science education* (pp. 831–880). New York: Routledge. https://doi.org/10.4324/9780203824696.

Leung, J. S. C., Wong, A. S. L., & Yung, B. H. W. (2015). Understandings of nature of science and multiple perspective evaluation of science news by non-science majors. *Science & Education, 24*(7–8), 887–912. https://doi.org/10.1007/s11191-014-9736-4.

Matthews, M. R. (1994). *Science teaching. The role of history and philosophy of science*. New York: Routledge.

Merchan, N. Y. T., & Matarredona, J. S. (2016). Contributions of intervention teaching using socioscientific issues to develop critical thinking. *Ensenanza De Las Ciencias, 34*(2), 43–65. https://doi.org/10.5565/rev/ensciencias.1638.

O'Neill, T. B. (2010). Fostering spaces of student ownership in middle school science. *Equity & Excellence in Education, 43*(1), 6–20. https://doi.org/10.1080/10665680903484909.

Osborne, J., Collins, S., Ratcliffe, M., Millar, R., & Duschl, R. A. (2003). What "ideas-about-science" should be taught in school science? A Delphi study of the expert community. *Journal of Research in Science Teaching, 40*(7), 692–720. https://doi.org/10.1002/tea.10105.

Quitadamo, I. J., Faiola, C. L., Johnson, J. E., & Kurtz, M. J. (2008). Community-based inquiry improves critical thinking in general education biology. *Cbe-Life Sciences Education, 7*(3), 327–337. https://doi.org/10.1187/cbe.07-11-0097.

Sandoval, W. (2014). Conjecture mapping: An approach to systematic educational design research. *Journal of the Learning Sciences, 23*(1), 18–36. https://doi.org/10.1080/10508406.2013.778204.

Thadani, V., Cook, M. S., Griffis, K., Wise, J. A., & Blakey, A. (2010). The possibilities and limitations of curriculum-based science inquiry interventions for challenging the "pedagogy of poverty". *Equity & Excellence in Education, 43*(1), 21–37. https://doi.org/10.1080/10665680903408908.

Toulmin, S. (1958). *The uses of argument*. Cambridge: Cambridge University Press.

Wang, H. H., Chen, H. T., Lin, H. S., Huang, Y. N., & Hong, Z. R. (2017). Longitudinal study of a cooperation- driven, socio-scientific issue intervention on promoting students' critical thinking and self-regulation in learning science. *International Journal of Science Education, 39*(15), 2002–2026. https://doi.org/10.1080/09500693.2017.1357087.

Yacoubian, H. A. (2012). *Towards a philosophically and a pedagogically reasonable nature of science curriculum*. Doctoral dissertation, University of Alberta, https://era.library.ualberta.ca/items/143c1900-551c-434c-9763-d8559349b5e6

Yacoubian, H. A. (2015). A framework for guiding future citizens to think critically about nature of science and Socioscientific issues. *Canadian Journal of Science Mathematics and Technology Education, 15*(3), 248–260. https://doi.org/10.1080/14926156.2015.1051671.

Chapter 11
Social Images of Science and of Scientists, and the Imperative of Science Education for All

Agustín Adúriz-Bravo and Alejandro P. Pujalte

11.1 Introduction

Fifty years of research on the images of science and of scientists that circulate in Western societies have shown that naïve and distorted images prevail. Many authors have pointed out that these *folk* images are scarcely adequate from an educational point of view, since they can become genuine obstacles in achieving a good-quality science education for all. This is seen in many enquiries on students of different educational levels (Mead and Metraux 1957; Beardslee and O'Dowd 1961; Brush 1979; Chambers 1983; Leslie-Pelecky et al. 2005; Pujalte et al. 2012b) and on science teachers (Hodson 1998; Chen et al. 1997; Adúriz-Bravo 2001; Manassero and Vázquez 2001; Hugo and Adúriz-Bravo 2003; Vázquez et al. 2006; Demirbaş 2009).

Naïve images of science and of scientists are at odds with the idea that knowledge *of* and *about* science is indispensable for people to exercise full citizenship. Results of empirical research reveal a broadly extended view of scientific activity and its products that is positivistic, empirical-inductivist, a-historical, individual, decontextualized, and value-free (Gil et al. 2001; Fernández et al. 2002).

Science is very often believed to be a neutral, objective, almost infallible endeavor with no recognizable human aims –apart from just "knowing"– and aiming at the sheer discovery of truth. The vast majority of social actors that have been studied (students, teachers, citizens, even scientists) share a portrayal of science as an elitist and excluding activity, mainly conducted by white middle-aged males who follow "the" scientific method and are above the rest of humanity due to their intellectual skills (Aikenhead 1984; Gagliardi and Giordan 1986; Hodson 1992; Newton and Newton 1992; Fernández et al. 2002).

A. Adúriz-Bravo (✉) · A. P. Pujalte
Instituto CeFIEC, Facultad de Ciencias Exactas y Naturales, Universidad de Buenos Aires, Ciudad Autónoma de Buenos Aires, Argentina
e-mail: aadurizbravo@cefiec.fcen.uba.ar; alejandropujalte@gmail.com

H. A. Yacoubian, L. Hansson (eds.), *Nature of Science for Social Justice*, Science: Philosophy, History and Education,
https://doi.org/10.1007/978-3-030-47260-3_11

There is also an accumulated corpus of literature on the transmission and strengthening of these images of science and of scientists through formal science education, especially in levels K-12. Studies suggest that these views on the so-called "nature of science" (NOS) that circulate at school may constitute a relevant "didactical obstacle" for scientific literacy, and that this might be particularly relevant in the case of students coming from a socio-economically vulnerable background.

In this chapter we engage the question: why teach science in compulsory education in the twenty-first century? We adhere to the idea that the currently proclaimed aim of a quality science education *for all*, with equity and social justice, crucially needs: (1) explicitly addressing the social imaginary of science in a critical way, and (2) designing and implementing pedagogies that are founded on, and present students with, a more democratic conception of the nature of science.

11.2 Images of Science and of Scientists as an Object of Research

Although the term "image" is rather polysemic, there is a sense employed in different disciplinary fields that makes reference to the idea of a (primary visual) representation that is activated in the mind of subjects when they are driven to think of something concretely. In this sense, Cassirer (1957) points out that we humans represent reality in images produced from the multiple relationships that we establish with the world.

With regard to these images that Cassirer studies, there is an expression coined within the framework of Gestalt psychology *–pregnancy*. Pregnancy is a quality that establishes the degree of impact of an image on an observer due to the characteristics that make it easily memorable. According to Chacón and Sánchez-Ruiz (2009), pregnancy is a mechanism that allows a reduction of ambiguities or distortions, aiming at the simplest or most constant forms in order to construe significant units, or "figures". This "synthesizing" character of the image in terms of Gestalt theory can also be somehow found in the notion of *social imaginary*, understood as the construction of cultural representations through which we make sense of the world, often with a "displacement of meaning": symbols already available are invested with other meanings than their "normal" or canonical (Castoriadis 1987). Thus, it is assumed that the imaginary is separated from the real.

The synthesizing character that has been very often given to the term "image" in different disciplinary fields is what we will be using here when referring to what society thinks or believes about science and about the people who do science. In this way, investigating these images has been a traditional concern of the sociology of science (Collins and Pinch 1993) and of the social studies of science and technology, particularly within an area of research on *public understanding of science* (Ryan and Aikenhead 1992; Bauer and Schoon 1993; Locke 1999) that is focused on elucidating the reasons for the existing "gap" between science and society.

A pioneering study around the public perception of science is that of Robert Davis in 1957 (Cortassa 2012). This was a nation-wide survey in the USA for the NASW (National Association of Science Writers) that included, among its dimensions of analysis, people's images of the scientific profession and their predisposition towards it. From that study on, researchers have concluded that societies' ignorance of science not only is wide-spread but also comes associated with negative attitudes towards, and lack of interest in, science (Cortassa 2012).

In the same year of Davis's study, Margaret Mead and Rhoda Metraux published a survey (Mead and Metraux 1957) on a sample of 35,000 American high school students; they asked them what they thought about science and how they saw scientists. Figures 11.1 and 11.2 show some of the main results from that piece of research.

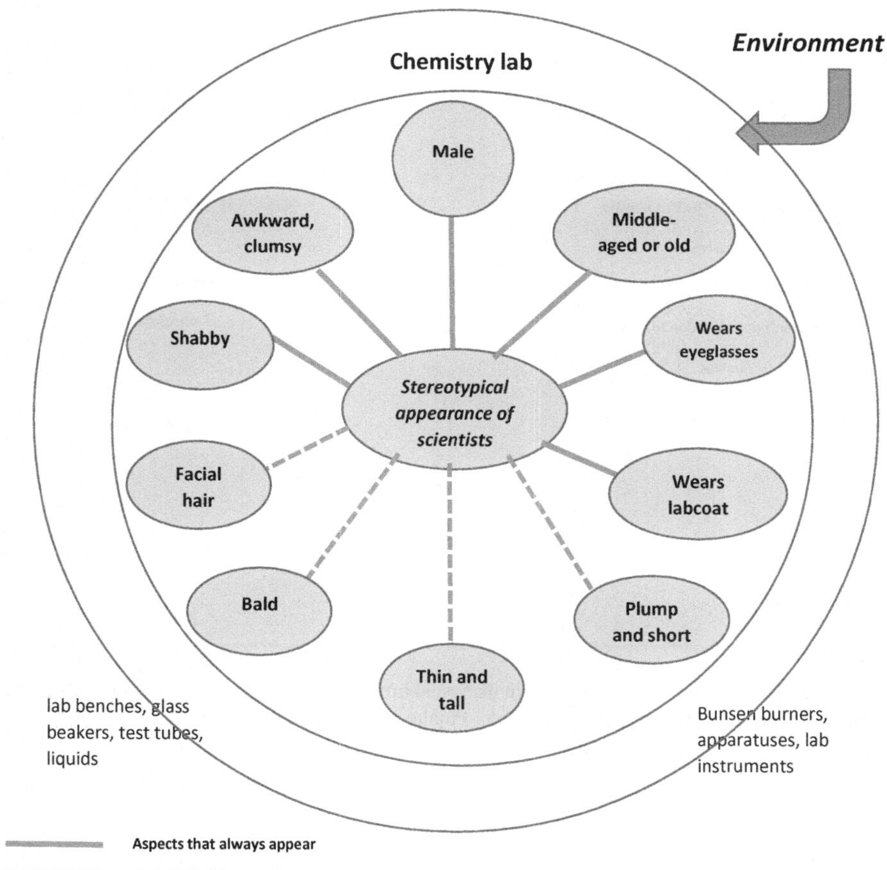

Fig. 11.1 Recurring traits in people's conceptions on the appearance of scientists and on the characteristics of their workplace, based on the results of the study by Mead and Metraux (1957). (Source: Pujalte et al. 2012b: 262)

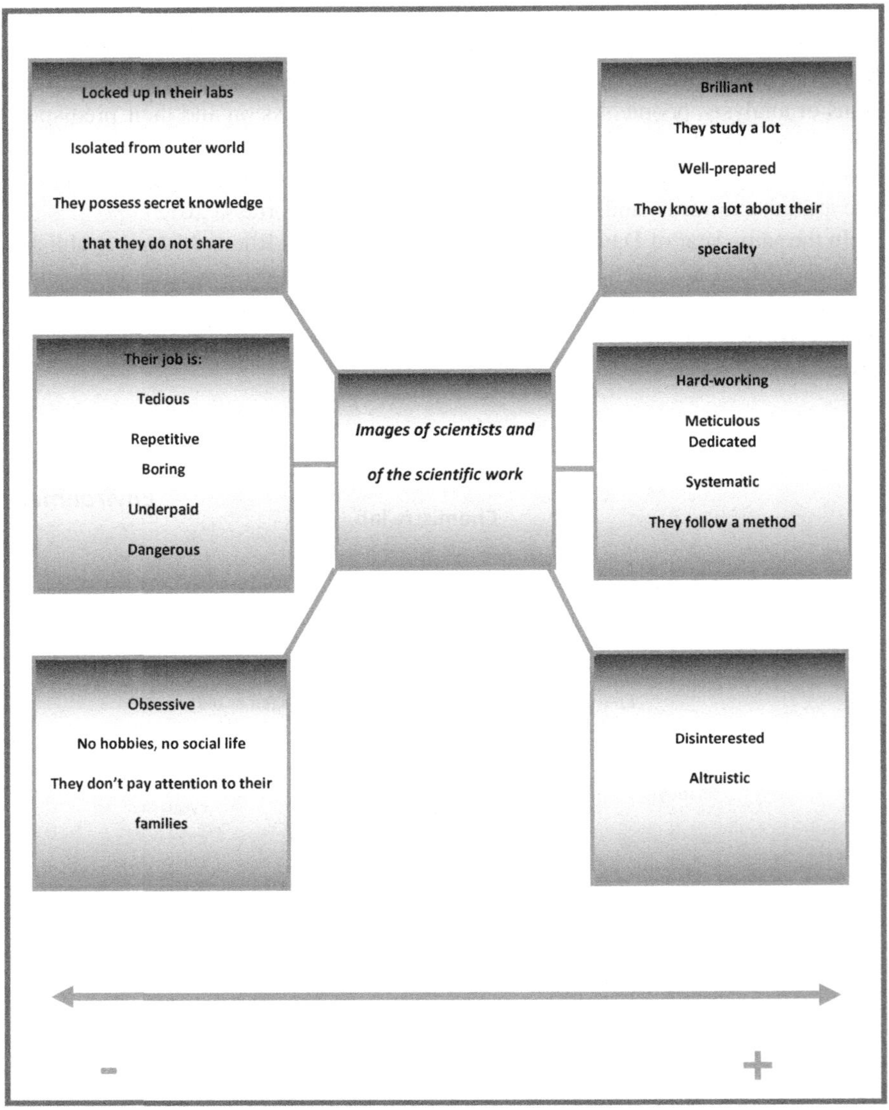

Fig. 11.2 Positive and negative aspects in the social image of scientists, based on the results of the study by Mead and Metraux (1957). (Source: Pujalte et al. 2012b: 263)

11.3 Images of Science and of Scientists in Didactics of Science: "Nature of Science" as a Research Line

Inquiries about the "folk" image of scientists are part of the line of research in didactics of science known as nature of science (NOS), which, for almost three decades now, has been examining ideas about science both in science students and in science teachers (McComas 1998; Adúriz-Bravo 2005).

In fact, the interest of NOS research in exploring these images arises from its main concern: achieving high-quality scientific literacy for all through new curricula that include content of science and *about* science (i.e. "meta-scientific" content). That is, in addition to grasping some scientific content, citizens should be able to answer questions about what science is, how it changes over time, and how it relates to society and culture (Adúriz-Bravo 2005, 2009). This would require the incorporation of new content: a critical, meta-discursive reflection on science as a process and as a product. Such content would ostensibly come from the classical meta-sciences –the philosophy, history and sociology of science.

Exploratory studies show that the populations examined do not generally have adequate views on the nature of science; accordingly, much has been written about the importance of incorporating meta-sciences into science education. Rosalind Driver and her colleagues (1996) proposed a series of arguments to defend the inclusion of the nature of science in science teaching:

1. *Utilitarian argument.* NOS gives meaning to the scientific endeavor and allows action with technological objects and processes.
2. *Democratic argument.* NOS provides the foundations for informed decision-making in socio-scientific issues.
3. *Cultural argument.* NOS helps appreciating the value of science as part of contemporary culture.
4. *Moral argument.* NOS helps understanding the norms of the scientific community, which imply commitments of moral value.

NOS also helps learning science content and can increase the "supply of qualified scientists to maintain and develop the industrial processes on which national prosperity depends" (Driver et al. 1996: 11).

Following Norman Lederman (2006), we could say that research in NOS has progressively dived into the following focuses of attention:

1. students' conceptions about the nature of science,
2. curriculum aimed at improving students' ideas about the nature of science,
3. teachers' conceptions about the nature of science,
4. proposals for the improvement of teachers' conceptions, and
5. relative effectiveness of various instructional practices.

In that same review, Lederman recognizes some results that arise from the vast number of studies already conducted on the nature of science:

1. Secondary students' conceptions on NOS are naïve or distorted.
2. Secondary science teachers hold views on NOS that are scarcely adequate for a teaching aligned with the new curricula.
3. Views on NOS are better learnt through "explicit and reflective" teaching than implicitly through "doing science".
4. Teachers' conceptions on NOS do not automatically translate into classroom practices. (This is why, in this chapter, we distinguish between "discursive" and "enactive" conceptions on NOS.)
5. Science teachers do not consider NOS (as teaching content) to be of equal status with "traditional" science content.

The next subsections are devoted to reviewing studies on the images of science and of scientists in students and teachers.

11.3.1 Inquiry into Students' Images of Science and of Scientists

After the foundational anthropological work by Mead and Metraux, many other investigations converged in the same negative conclusions on young people's images of scientists (e.g. Beardslee and O'Dowd 1961; Brush 1979). Soon a fruitful perspective of inquiry was introduced in this line of research to go deeper into those negative images –using drawings.

When students are required to draw a scientist and its work environment, results show high uniformity, with independence of their age, gender, socio-cultural background, ethnicity and other context variables. Most drawings depict white males in their lab coats working alone in a laboratory that resembles a school lab for chemistry. The first systematic analysis of this kind of drawings was due to David Chambers (1983); with his DAST (Draw-a-Scientist Test) he obtained drawings from 4800 American and Canadian boys and girls aged 5–12, over the period 1966–1977. Other pieces of research that followed confirmed the pervading presence of this cliché of scientist in different cultures and different educational levels (Maoldomhnaigh and Hunt 1988; Parsons 1997; Räty and Snellman 1997; She 1998; Brosnan 1999; Song and Kim 1999).

In these investigations, drawings of scientists are usually accompanied by questions that ask students to express some of the scientists' characteristics in writing. Descriptions that are obtained by these means are also very coincident: the typical scientist is distracted, absorbed by his work, not caring about social life, occupied in things that only he can understand, without family or friends, without other interests or motivations.

All these investigations show a *stereotype*; we support the working hypothesis that such a stereotype of a scientist featuring in the drawings is an "epiphenomenon" of the drawer's particular image of science. We assume that students personify in their scientist their own conceptions on the scientific activity.

The "stereotypical scientist" still prevails in the science classes after more than five decades of research and interventions; such stereotype does not constitute a model to which the majority of students want to adhere:

> While most people express respect and admiration for scientists, the dominant stereotype discourages those who do not fit the stereotype from seeing themselves as scientists. Children's attitudes toward science and their participation in science are strongly defined, highly gendered and formed as early as kindergarten. These attitudes affect future education and career choices, as well as attitudes toward, for example, public support of scientific research. (Leslie-Pelecky et al. 2005: 173)

Such disidentification of students with science and scientists in many cases arises from the fact that they do not consider themselves able to understand the science that is taught to them (Pujalte et al. 2012a, b).

Many studies have indicated that this negative image of science and of scientists is formed at very early ages (Newton and Newton 1998; Adúriz-Bravo et al. 2013). Another issue examined in those studies is the gender in the drawings –almost every research subject draws a male scientist.[1] There are nevertheless some nuances: younger children sometimes represent their own gender, and, for developmental causes, girls are more able to do drawings that can be recognized as females. In boys' drawings such distinction is sometimes hard to establish, and they adhere far more usually to the stereotype than girls (Losh et al. 2008).

There are also large-scale studies that intend to compare students' ideas and attitudes towards science across countries; in a study called "Science and Scientists", directed by Svein Sjøberg of the University of Oslo (Sjøberg 2000), children from emerging economies presented a much more positive view on science and technology than those in rich countries. In these countries, children (especially boys) represent the scientist as a mad and cruel individual; children from developing regions consider scientists as idols and heroes.

It has been repeatedly pointed out that, in developed economies, young people hold a very poor public image of science and are not attracted towards careers of science and technology. This contradiction –given the fact that these people live in societies based on scientific and technological knowledge– can be explained as the result of traditional science teaching and a bad image of science circulating in the media.

Other publications on this issue include an investigation on Catalan children's images of scientists (Escalas Tramullas et al. 2009). This study analyzed a selection of 250 drawings made by Catalan girls and boys aged 6–18, who participated in a "Draw a Scientist" contest within the framework of the Project "Discover the Researchers' Facets" funded by the European Commission. In the contest, the participants were asked to draw a male or female scientist without further specifications; the idea was to condition the results as little as possible. The drawings were analyzed –using the canonical DAST variables– by a team of experts in didactics of

[1] Most probably a cis-, heterosexual male drawn under a heteronormative canon.

science from the Observatory of Science Dissemination of the Universitat Autònoma de Barcelona.

DAST as a test and its application have suffered reformulations following academic critique. For instance, Manzoli et al. (2006) point out that this test: (1) does not show how and where the stereotype is constructed, (2) provides a static image that does not allow students to give details of the processes of science, and (3) might not be eliciting students' conceptions, but rather icons purposefully shaped to be recognizable by those who have asked for them. Symington and Spurling (1990) have also cast doubt on the drawings being the "true" images of scientists of the drawers.

Studies that go beyond the classical DAST require, in addition to the drawing, writing a little story having the scientist as its main character (Manzoli et al. 2006; Reis and Galvão 2006, 2007; Reis et al. 2006). Results of these more "sophisticated" studies lead the authors to two main conclusions: (1) children are capable of producing representations of scientists that go beyond the stereotype, which is only a "skeleton" intended to capture key features that they have taken from the media, and (2) children show surprising awareness of the social, ethical and political aspects of the scientific activity and its methods.

In Argentina, research into the images of science and of scientists has constituted, for more than ten years now, a fruitful line of our group GEHyD (Group of Philosophy, History and Didactics of Science) at the Universidad de Buenos Aires. In one of our studies (Pujalte et al. 2012b), we analyzed literary materials –science fiction produced by high-school students in the framework of the contest "La ciencia en los cuentos" ("Science in tales"). The corpus is interesting insofar participants were motivated by a genuine interest in producing original narratives; this permits assuming that the images of science and of scientists in their tales are more "spontaneous".

Another relevant antecedent is a piece of research that intended to reformulate DAST instructions; this was conducted by a team of primary teachers in Uruguay (Adúriz-Bravo et al. 2013). Around 700 drawings by students aged 5–11 were analyzed. In this investigation, several wordings of the instruction for drawing were tested. Additionally, drawing was complemented with interviews to the children; the use of interviews proved crucial in order to interpret the drawings of younger subjects (kindergarten and early primary students).

All the accumulated research shows that this stereotyped image of scientists is formed at very early stages; as schooling progresses, its most characteristic features are stressed (Dibarboure 2010; Pujalte et al. 2011a, b). This process is accompanied with increasing lack of interest in science subjects among children, and the consequent reduction in their enrollment in science careers.

This exclusion usually has an important gender component: girls are strongly deterred from studying science. In this sense, Jones et al. (2000) suggest that school –despite being an institution that is far from being free of gender bias– should be the place where students can shape the best values and attitudes towards science. They also point out that science teachers *cannot elude the responsibility of presenting science as equally appropriate for boys and girls*, generating equally capable participation in the use of the scientific instruments, and fostering reflection on scientific activity in all children. They also conclude that maintaining the sexist *status quo* and not aiming at the transformation of cultural practices relegates female students to the periphery of science.

Along the same line, Mary Wyer (2003) claims that, for women not to abandon scientific careers, determining factors are positive images of scientists, positive attitudes regarding gender equity, and positive experiences in the science classrooms. The widespread installation of stereotypes that "dictate" which groups can succeed in science leads to many young people disidentifying themselves with science and taking for granted their lack of capacities for it. This is particularly the case with minorities, as Steele (1997) well points out.

Given the situation reported, many research and innovation projects have been developed to evaluate the impact of specific teaching interventions that engage students' images of scientists. The aim of such interventions is to generate more representations of science *that aim at including a diversity of people*. Most of those studies agree that interventions are more effective the earlier they are conducted (i.e. in the first years of schooling, when girls and boys have their first formal approach to science content). For example, Bodzin and Gehringer (2001) studied the effect of involving real scientists in didactical units designed to teach science to primary school students. Through pre- and post-tests they showed significant improvement in the recurring aspects of the stereotype of scientist.

Cakmakci and colleagues (2011) suggest that both results of empirical research on public images of science and of scientists and theoretical analyses on the nature of knowledge, cognition and teaching can be used to design instructional activities aimed at improving students' ideas about science. They recognize that, in general, these strategies are not sufficiently used or they are applied too late.

11.3.2 Genesis of the Stereotype of Scientist

It is highly significant that drawings of scientists present such uniformity across ethnicity, age, gender, socio-economic status and cultural background. Not only students draw stereotyped scientists; the same image appears in teachers and in the general public. It is generally argued that this image of male scientists with a white lab coat, eyeglasses and disheveled hair is omnipresent in the media, constituting itself in mandatory reference. Thus, comics (Gallego Torres 2007), cartoons (Vílchez-González and Perales 2006), cinema (Moreno Lupiáñez 2003; Weingart et al. 2003; Guerra Retamosa 2004; Weingart 2006) and TV (Long and Steinke 1996; Steinke et al. 2008; Long et al. 2010) have portrayed such a scientist, endowing him with ambition and madness.

A study in *New Scientist* (Hills and Shallis 1975) surveying scientific researchers and laypeople found differences in their images of scientists. Scientists characterized their colleagues as accessible, sociable, open, unconventional, and with varied interests; non-scientists saw scientists as aloof, retracted, reserved, highly conventional, with few interests and markedly unpopular. Both groups converged, nevertheless, in portraying scientists as males. With the input "when I think of a scientist, I think of…", non- scientific respondents gave the name of a famous scientist (most popular were Archimedes, Marie Curie, Darwin, Galileo, Newton, Louis Pasteur and Linus Pauling, coming from popular films of the period). With the same input, the surveyed scientists answered that "they thought about their colleagues". Science novices show, in their answers to surveys on scientists, a conception of them working alone and best represented by the "great discoverers" (Frayling 2005).

Returning to the stereotype of the "mad scientist", appearing in many fiction films as well as in other cultural productions, it is worth asking from where mass media have taken this emblematic figure. Answering this question requires going back to the alchemists. Although these characters are present in Antiquity all over the world, it is Medieval European alchemists, with their hermetic and magic traits, that will be perpetuated in the stereotype of scientist. Alchemy at that time presented itself as a promise of fabulous riches and longevity: the philosopher's stone would allow transmuting metals into gold and the elixir of youth would end the threat of disease and death (Eliade 1983). This figure of the alchemist persisted until the Renaissance and even Modernity (Vickers 1984). Practitioners of alchemy worked in secret laboratories, almost always underground, and in the middle of the night. Always moving in the gray zone between the natural and the supernatural, they used obscure language, full of symbols, as a way of preserving their knowledge.

Literary fiction in the nineteenth century was responsible for revamping the figure of those proto-scientists (Haynes 2003). Goethe's Faust and Mary Shelley's Frankenstein are two examples of how natural and mystical forces combine in the pursuit of an obsession: to become gods, to be immortal, to create life. Particularly suggestive is the explanation proposed by Joachim Schummer (2006) about why this image is used in literature: it would be a literary response to the emergence of the new chemistry, the prototype of experimental science seen as a threat to the unity of knowledge and linked to atheism, materialism and nihilism. In this emerging figure of the mad scientist there appears an additional element: Greek *hubris*, the excess and arrogance. According to Schummer's explanation, writers resort to this thoughtless scientist in order to highlight the negative aspects of science and attack the doctrines of Enlightenment. The construed image is that of a man who lives as a hermit, wasting his own life in pursuit of his grandiose dreams, detaching himself from all things mundane, estranged from his family and friends. It is sometimes the figure of a rogue or a charlatan, who promises to reveal miraculous techniques and robs his victims of their possessions. As expressed by Schummer:

> In order to make hubris a morally convincing accusation for their readers, nineteenth-century authors created the mad scientist. Transformed from the mad alchemist already established in the medieval literature, the mad scientist combines hubris with all the moral perversion that nineteenth-century writers could imagine. Borne out of the need for serious arguments, this literary figure has dominated the public view of science ever since. (Schummer 2006: 124–125)

This nineteenth-century fiction literature on the mad scientist has been extensively adapted to the cinema and has thus become massive. From there, the figure of the stereotypical scientist has turned into a regular character on television and in comics, cartoons and advertisements. The rescue of the character by mass media brings it closer to students: reference to science and to scientists proliferate in those media, but strict information is loaded with lack of rigor, sensationalism, exaggeration of the "triumphs" of science and omission of its provisional character. An activity that is conceived as the blatant announcement of discovered truths is coherent with this conception of scientists as wizards, magicians or heroes (Schäffer 2011).

The significant number of hours that children, adolescents and young people are exposed to TV series and movies, cartoons and videogames no doubt makes an impact in their images of science and of scientists. Research shows that there are very few programs in which *actual* scientists are shown, the screen is dominated by fictional scientists who are almost always white males endowed with superior intelligence. When women appear, they are put in a secondary role and their opinions are underestimated, naturalizing the common-sense idea that they have less ability and preparation for science (Long et al. 2010).

11.3.3 Teachers' Images of Science

As we have indicated in previous paragraphs, research coincides in pointing out that, as schooling progresses, students' representations about science and scientists become more and more stereotyped. Not only that, they become increasingly similar to their teachers' representations (Fung 2002). Accordingly, in parallel with research on students' images of science and of scientists, investigations on the images sustained by pre- and in-service teachers were conducted. Such investigations talk about "deformed", "distorted" or "inadequate" views (Chen et al. 1997; Hodson 1998; Gil et al. 2001; Manassero and Vázquez 2001; Fernández et al. 2002; Vázquez et al. 2006; Demirbaş 2009). In some studies it is contended that such images are transmitted to students during science teaching (e.g. Gil et al. 2001; Fernández et al. 2002); hence the urgent need to diagnose teachers' conceptions in order to intervene on them (Abd-El-Khalick 2001).

Researchers agree that inadequate images of science arise from a markedly empirical-inductivist view that considers science as an ahistorical, individualistic, value-independent, a-problematic and decontextualized enterprise, and therefore neutral, objective and infallible. At the same time, science is shown as an elitist and excluding activity, founded on a scientific rationality centered on a single method. Its cryptic and hermetic character is often highlighted: science can only be "deciphered" by "initiated" people. Thus, the whole "ethos" of science as a cultural activity is shaped by non-inclusive, elitist and discriminatory constraints.

International literature on this issue is vast, but there is still paucity of interventional studies to test materials and activities that promote teachers' understanding of the nature of science (Acevedo-Díaz 2007a, b). In relation with this, the ineffectiveness of isolated courses on NOS has been highlighted; authors suggest "explicit and reflective" teaching of the nature of science that should be linked to the scientific content that teachers must teach in their classroom (Buaraphan 2009) and also to didactical considerations on how to teach it (Adúriz-Bravo 2007).

On the other hand, it is necessary to ask whether certain "profiles" of conceptions and beliefs are "activated" in certain contexts. According to Barnett and Hodson (2001), the same teacher teaches different content in different ways to students, depending on the particular conditions of those students and of the school. Teaching about science would then be highly "context-specific" and therefore, teachers' decisions in the classroom would strongly depend on the specific social and cultural conditions of students. But such an adaptation to the context would not always result in quality science education, especially when it comes to students from socially disadvantaged contexts (Gómez et al. 2004), who enter their secondary school at a disadvantage with respect to the "cultural capital" that school legitimizes (Bourdieu and Passeron 2003).

The important number of studies that inquire on teachers' images of science and of scientists make use of a wide variety of instruments, which examine their discourse and practice. Among such instruments, we can mention the following written questionnaires: *Teacher's Beliefs about Science-Technology-Society* (TBA-STS, Rubba and Harkness 1993), *Views on the Nature of Science* (VNOS, Lederman et al. 2002), *Views on Science-Technology-Society* (VOSTS, Aikenhead and Ryan 1992; Aikenhead et al. 1989) and its Spanish adaptation, COCTS (Manassero et al. 2001, 2003), and *Views on Science and Education Questionnaire* (VOSE, Chen 2006). This last questionnaire also asks teachers about NOS teaching. Despite the differences between all these instruments, the results obtained greatly coincide: teachers of all educational levels adhere to a conceptualization of science that can be more or less related to traditional philosophies of science –positivism, empiricism, naïve realism. Even so, teachers' ideas about science are far from homogeneous; they usually appear hybridized with some more "contextualist" and "historicist" conceptualizations, which can be considered akin to the so-called new philosophy of science of the 1950s and 1960s.

All the surveys that we have listed above access what we call the "declarative" image of science, emerging when teachers are explicitly posed questions (usually accompanied with a set of fixed answers to choose upon) on typical aspects of science (method, validity, values). The problem is that teachers' declarative image of science is not constituted only be their epistemological positioning; such image also includes beliefs about science teaching, students and aims of science education. It is almost always found that the declarative image of science sustained by teachers is "democratic" and "inclusive": it explicitly states the aim of a science education for all.

But such an image coexists in teachers with what we call their "enactive" image of science, the one put into action in their classroom practices. Our hypothesis here is that this latter image is usually inadequate when teachers operate in front of students from disadvantaged socio-economic contexts: teachers tend to apply a "deficit" view on the capacities of those students and consequently to teach in ways they think may "assist" them (Pujalte 2014). This "remedial" image of school science would contribute to the exclusion of students from appreciating and understanding science and would thus hinder their proper insertion as citizens in society.

11.3.4 Analyzing Teachers' Images of Science Through Models

The diagnosis that teachers' ideas on NOS are far from recent and contemporary conceptualizations from the philosophy of science is relevant insofar traditional views on what science is and how it works can become "epistemological obstacles" when planning and executing science teaching that is adjusted to the proclaimed aim of a high quality science education for all, with equity and social justice (Furió et al. 2001; Acevedo-Díaz 2004). Teachers' ideas (labeled as "deformed", "mythical", "distortive", "common-sense", "folk", etc.) are "inadequate" in relation with science curricula in the twenty-first century, which aim at the preparation of citizens. Such ideas, nevertheless, form a more or less coherent image of science (Fernández et al. 2002), they behave as conceptual schemes or implicit models to organize discourse and guide action.

Such internal coherence allows us to model teachers' ideas about science as having a "family resemblance" with traditional models from the philosophy of science (i.e. those produced during the first half of the twentieth century in schools such as logical positivism, critical rationalism and the "received view"). It is our contention in this chapter that teachers use a deficit model with disadvantaged students and formulate an "assistential" teaching proposal because of their empirical-positivistic conception of science.

Logical positivism of the Vienna Circle in the 1920s (represented by philosophers such as Schlick and Carnap) intended to elaborate an epistemological reconstruction of science detached from the conditions of production of scientific knowledge. As Palma and Wolovelsky (2001) point out, scientists as human subjects were not considered in this picture of science, as contextual determinations could only explain, in the best case, the errors of science. This highly idealized reconstruction of science resorts to a "categorical" rationality based on formal logic: science objectively seeking the truth is erected to epitome of rational thinking (Glavich et al. 1998).

Karl Popper's early critique to logical positivism reclaims recognizing the mediation of subjects in knowledge construction, but his reconstruction of the scientific activity is still based on the *norm* that subjects should stick to deductive logic and seek objectivity through intersubjectivity: scientists should establish sound logical relations between propositions but at the same time interact with one another to reach rational consensus (Popper 1935/Popper 1992). As Javier Echeverría (1998: 82) points out, Popper's position is based in a moral imperative on how science *should* be conducted that is part of the Zeitgeist: Robert K. Merton's CUDOS (communism, universalism, disinterestedness and organized skepticism) was the orthodox, idealized, normative model for the goals and methods of science.

Teachers' stereotype of scientist summarizes to a large extent the set of "positivistic" qualities that collective imaginary attributes to people who do science: they are good observers, they study a lot, they dedicate themselves to their jobs, they are neat, methodical, careful and thorough, honest, disinterested, etc. But this set of qualities is just one of the facets of the scientist that appears in the imaginary; the other is apparently opposite: scientists are characterized as imaginative, creative, brilliant, and at the same absent-minded and a little mad (*à la* Albert Einstein). This two-faced image, already identified in Mead and Metraux's seminal study, is explained with Schummer's hypothesis on the genesis of the mad scientist and is compatible with the heavily analytic-syntactic reconstruction of the scientific activity propounded at the beginning of professionalized philosophy of science.

Modeling science as an observation-based, rigid, algorithmic, precise, exact and infallible activity leads teachers to adhere to an individualist and elitist view. As science is seen as the domain for especially gifted individuals, teachers implicitly transmit negative expectations towards students belonging to minorities; there is clear discrimination on the basis of gender, ethnicity and social status (Fernández et al. 2002). Elitism is operationalized in an extremely formal, heavily mathematical presentation of content that does not seek to be "accessible" to students, as we have hinted at the beginning of the chapter; thus, all human aspects of science as an activity are veiled or suppressed (Fernández et al. 2002).

One of the conceptions on the nature of science among teachers is then that the structure of scientific knowledge has a specific and elaborate syntax that requires logical and analytical rigor to address it; this would lead them to believe that there are some students with the appropriate skills to understand science and some others who do not possess such skills. For the latter, teachers would present a science of lower quality, of a superficial and anecdotal nature, adjusted to their possibilities.

Explicit or implicit adherence of teachers to this elitist view of science is transmitted to students: they assume the characterization of science as a "naturally" difficult subject and they internalize (i.e., make their own) the discourse in relation to their lack of the "necessary" capacities to learn science. This is particularly acute with students from disadvantaged contexts, as we have shown in a study carried out with 96 students (14–16 years old) in a school attended by teenagers from lower socioeconomic background (Pujalte et al. 2012a). In such study we asked the students to draw a scientist and to provide five keywords to characterize scientists and five keywords for science. In their drawings, only 11 students show more than one scientist; only 12 draw women, and only 10 locate science outside the lab. In their lists of keywords, the most frequent terms for science are: difficult/complicated, interesting/entertaining, experimental, requiring effort/dedication/study, exhausting, time-consuming, requiring complete dedication, risky, boring.

Keywords that they use to characterize scientists are: smart/intelligent/bright/wise; they study, work and experiment a lot; able/skillful; dedicated/careful/responsible; boring. Other words less frequent: ingenious, elderly, antisocial, weird, crazy, lonely, commonplace, uncool/bitter.

Along with this characterization, an overwhelming total of 86 out of 96 students manifested that they did not picture themselves doing science. The reasons they give for this show how they are excluded: "I don't like science", "it's too difficult", "you have to study a lot", "it's not for me", "it's only for clever people", "I am not good enough", "I am not able", "I don't understand it".

11.4 Cultural Perspectives in Science Education: What Science for Students from Disadvantaged Contexts?

Some recent studies, aimed at understanding inequity in education, emphasize the importance of incorporating a sociocultural perspective to address the emotional commitment put into action in science teaching and the impact of emotions on student performance. It has been found that, when the cultural capital of students is not valued in the science classroom, they perceive a gap between their own knowledge, values and dispositions and the school science curriculum (Elmesky 2001; Seiler 2002; Olitsky and Milne 2012). When this happens, negative emotions appear that interfere with learning. Some authors recommend that science programs should change in order to be more relevant for students' interests, especially in low-income urban areas. In other words, instead of focusing on why a particular student stops relating to, and participating in, the science classes, efforts should be made to involve the class as a whole, using knowledge of the students' culture in order to increase curriculum relevance of science content. In this way, students would begin to feel positive emotions about their participation in science, which would lead to greater commitment (Olitsky and Milne 2012).

As Emdin (2012) points out, some very relevant innovative perspectives in science teaching, such as constructivist approaches, nature of science and pedagogical content knowledge can prove inapplicable in urban science classrooms unless they are first focused specifically on the needs of these more marginalized students. Our didactical knowledge needs to take into account the specificity of these audiences in a way that the science that is taught *makes sense for their lives*.

From this approach that takes into account cultural perspectives in relation to the teaching of science, Nancy Brickhouse (2013) also focuses on epistemological aspects. She adheres to the premise that learning takes place in a sociocultural context that impresses knowledge with culturally specific structures; it might be the case, then, that scientific knowledge taught at school has been generated in contexts so different from those in which many students live that it makes access to it very difficult for them. Brickhouse adds that the contexts in which scientists work can be understood as a very particular "culture", in which scientists share repertoires of ideas, values and practices. In that sense, Brickhouse says that, in many traditional school environments, such repertoires are presented in a highly stereotyped and decontextualized way to the point that they reinforce an elitist image of science.

From this perspective, the solutions suggested from standards-based reforms will never be enough to address inequality in science learning. The fact of just providing equal opportunities for learning will not automatically give the expected results; in addition, the *epistemological* and *cultural* dimensions of scientists' science and school science that must mediate in order to provide access to science for all must be taken into account (Brickhouse 2013: 47).

11.4.1 Self-Fulfilling Prophecy and Labeling Theory

The expression of "self-fulfilling prophecy" refers to the increase of the probability of occurrence of a situation in someone's life when it is anticipated by them on the basis of their beliefs. The expression was used by Merton (1968) in his book *Social Theory and Social Structure*. In education, this idea is dubbed as "Pygmalion effect", from the well-known article by Robert Rosenthal and Lenore Jacobson (1968) where they presented research they carried out with teachers and students of primary school that allowed them to conclude that the expectations on the performance of the students -intentionally induced by the researchers- impacted on the students' grades. Those students that the teacher believed would achieve the best results (based on the information received from the researchers) ended up obtaining the best grades and vice versa.

According to Jussim et al. (1996), social scientists have been interested in stereotypes for many years as a particular source of expectations that can contribute to social inequalities and injustices. In relation to the fact that the effects of teachers' expectations have a higher incidence on students of stigmatized or disadvantaged social groups, the authors argue that the most likely explanation would be in the stereotypes that teachers carry. In this regard, Ray C. Rist (1999) proposes to resort to the "labeling theory", used in various studies from social sciences, as a legitimate explanatory framework from which to address the social processes that will influence educational experience and how they contribute to success or failure at school. According to Rist, all agents in charge of social control -parents, teachers, authorities, etc.- assign "labels" to individuals, with which they rate their individual behaviors and attitudes and classify them socially. And this is consubstantial to school practice (Rist 1999: 615).

Rist, based on Good and Brophy (1973), describes the process that takes place in the classroom according to the following scheme:

1. The teacher expects specific behavior and performance from each specific student.
2. Due to the different expectations, the teacher has a different behavior with the students.
3. The way in which the teacher treats the students indicates to them behavior and performance that the teacher expects from them, and affects the concept of themselves, the motivation for performance and the level of aspirations.

4. If the teacher's treatment is maintained over time, and if the student does not actively resist or modify it in any way, it will tend to shape their achievements and attitude. Students with high expectations will be led to achieve high performance, while the achievements of students with low expectations will decrease.
5. Over time, students' performances and attitudes will get closer and closer to what was expected of them.

11.4.2 Public Understanding of Science: The "Deficit Model"

The pioneering studies by Davis and by Mead and Metraux inaugurated a long tradition aimed at characterizing and explaining a "gulf" between science and society:

> There is an increasing need to develop public understanding of science and technology. The fruits of science and the products of technology continue to shape the nature of our society and to influence events which have a world-wide significance. Yet the gulf between the daily lives and experience of most people and the complexity of science and technology is widening. Remarkably few individuals are familiar with the details of the industrial processes involved in their food, their medicine, their entertainment or their clothing. (Oppenheimer 1968: 206)

From the 1950s and 1960s on, there has been growing international recognition of the need to prepare citizens capable of understanding and participating in the implementation of scientific and technological policies, citizens who can participate in the debates and in decision-making in socio-scientific issues. The recurrence to the expression 'scientific literacy' increases, in a clear analogy with general literacy: just as people must know how to read and write in order to be inserted into the work force, they should also acquire minimum knowledge to be able to function and act in a world that is increasingly dependent on science and technology (Fourez 1994).

In the 1970s the expression "functional illiterates" is introduced to refer to those individuals who have not acquired the minimum competencies to adequately perform in such highly scientific and technological society (Carullo 2002). As Polino et al. (2003) state, it is in this context that the lack of a scientific culture is portrayed as an ignorance that must be remedied.

In the 1980s, such approach will be called, within the Anglo-Saxon tradition of studies on public communication of science, the "deficit model". Scientific knowledge is conceptualized as corpus of information; therefore, it is possible to measure how much of that information individuals possess and to establish their level of deficit of understanding. Along the same line, Ziman (1992) argues that traditional practices in scientific communication tend towards diagnosing the cognitive shortcomings of the public and then trying to solve them.

Deficit models define a unidirectional and vertical relationship with the audience: the general public is considered ignorant in scientific matters. Then, communication of science and technology should be aimed at getting people out of such ignorance (Polino et al. 2003). In contrast with this deficit approach, democratic models consider an active, engaged participation of the audiences; the general public assumes a more protagonic role in the practices of social appropriation of knowledge. This second view comes from the critical approaches in social studies of science and technology (Hermelin 2011).

As Carina Cortassa (2012) states, associating the problem of the science-society gap with a cognitive deficit results in a reassuring, self-exculpating explanation to the problem. If it is assumed that the obstacle in the relationship between the audiences and science lies in the difficulties of the former to understand content, then the problem can be solved in an acritical and patronizing way. The gaps are bridged using a "therapeutic" mechanism: the absence of knowledge suffered by individuals is faced with the aim of "curing" them of their ignorance and apathy (Cortassa 2012).

11.5 From Discourse to Classroom Practice

In their discourse, there are teachers who conceive science and technology from humanistic and contextual perspectives, where the scientific and technological activity appears intertwined with their purposes and values, and these change over time. Nevertheless, in their classroom practice positions are enacted that show an elitist conception of science: there would be a "first class" science to be taught to those students who have certain basic conditions and a "second class" science for those who do not comply with these supposed standards.

In principle, there would be two conclusions to highlight from the research that we have conducted (Pujalte 2014). Firstly, a declarative image of science that is predominantly contextualist and democratic would not constitute a predictive indicator of an enactive image of science that is democratic and inclusive. Secondly, teachers who present a declarative profile mostly empirical-positivistic show in their practice elements of an image of deficit and assistentialism, at odds with the imperatives of "science education for all" (as characterized in Fensham 1985).

It should also be noted that many traditional or dogmatic views on how to teach science, not being "deficit" or "assistential" images per se, contribute to the emergence of enactive images of science that discriminate and exclude (in Brickhouse's terms, as we explained above).

Such elements are functional to heavily traditional pedagogies for science teaching, characterized by the following traits:

1. The teacher as the "owner of knowledge", the voice of authority, the enunciator of truths, monopolizing discourse in the classroom.
2. "Sanctioned" or "official" textbook in the same status as teachers, being the authorized reference around which the class is structured.
3. Closed, stereotyped structures for the dialogue between teacher and students, very often in the template of an IRE (initiation – students' response – teacher's evaluation) sequence. Through this, students' interventions are typified as "correct" or "incorrect" answers.

4. Teachers understanding scientific language merely as a labeling system, and not as system for the interpretation of natural phenomena (Sutton 1996/2008), thus emphasizing activities that "name", "classify" and "define" objects, properties and processes. In accordance with this, a style of evaluation where students are required to retrieve this language in a non-heuristic and non-explanatory fashion.
5. Related to the previous point, classroom activities fall far from posing genuine problems: what prevails are questionnaires asking for literal answers extracted from the textbook or "enquiries" that are reduced to collecting factual information on a topic.
6. Experimental, laboratory or field practices that only serve the purpose of "demonstrating" the theoretical principles previously presented.

These traditional pedagogies are broadly extended for a multitude of reasons: prevailing teaching traditions that are heavily institutionalized, very tight curriculum constraints, low-quality materials, teachers' own educational history (providing traditional teaching models to follow). This adds up with little explicit reflection on the nature of science. The result ends to be absolutely consistent with the folk views on science that we have described: true, absolute, universal, univocal, permanent knowledge that is best transmitted under the form of formally compact, decontextualized formulations.

This very much extended conception of the nature of science that reigns in the classroom constitutes a major obstacle towards a high quality science and technology education that empowers and emancipates students and aims at social justice in the classrooms. We believe that the key to a change in the inadequate images of science and of scientists among teachers would be directly related to an intervention in pre- and in-service science teacher education. Such intervention should include a carefully planned selection of meta-scientific content and discussion of why it is necessary to integrate the nature of science in science curricula of all educational levels.

The aim of such formative interventions around an epistemologically well-founded nature of science would be that teachers make their own the need for a science learning that enable students to think critically, to actively participate in crucial debates in society, and to make informed decisions in relevant science and technology issues.

References

Abd-El-Khalick, F. (2001). Embedding nature of science instruction in preservice elementary science courses: Abandoning scientism, but... *Journal of Science Teacher Education, 12*(3), 215–233.

Acevedo-Díaz, J. A. (2004). Reflexiones sobre las finalidades de la enseñanza de las ciencias: educación científica para la ciudadanía. *Revista Eureka sobre Enseñanza y Divulgación de las ciencias, 1*(1), 3–16.

Acevedo-Díaz, J. A. (2007a). Una selección de artículos sobre decisiones tecnocientíficas y enseñanza de las ciencias (I). *Revista Eureka sobre Enseñanza y Divulgación de las Ciencias, 4*(1), 195–201.

Acevedo-Díaz, J. A. (2007b). Una selección de artículos sobre decisiones tecnocientíficas y enseñanza de las ciencias (II). *Revista Eureka sobre Enseñanza y Divulgación de las Ciencias, 4*(2), 358–363.

Adúriz-Bravo, A. (2001). *Integración de la epistemología en la formación del profesorado de ciencias* (Ph.D. dissertation). Bellaterra: Universitat Autònoma de Barcelona.

Adúriz-Bravo, A. (2005). ¿Qué naturaleza de la ciencia hemos de saber los profesores de ciencias?: Una cuestión actual de la investigación didáctica. *Tecné, Episteme y Didaxis*, Extra Issue, 23–33.

Adúriz-Bravo, A. (2007). La naturaleza de la ciencia en la formación de profesores de ciencias naturales. In Gallego Badillo, R., Pérez Miranda, R. y Torres de Gallego, L. N. (Eds.), *Didáctica de las ciencias. Aportes para una discusión*. Bogotá: Universidad Pedagógica Nacional.

Adúriz-Bravo, A. (2009). La naturaleza de la ciencia "ambientada" en la historia de la ciencia. *Enseñanza de las Ciencias*, Extra Issue, 1178–1181.

Adúriz-Bravo, A., Dibarboure, M., & Ithurralde, S. (Eds.). (2013). *El quehacer del científico al aula: Pistas para pensar*. Montevideo: Fondo Editorial Queduca.

Aikenhead, G. S. (1984). Teacher decision making: The case of Prairie High. *Journal of Research in Science Education, 21*, 167–186.

Aikenhead, G. S., & Ryan, A. G. (1992). The development of a new instrument: "Views on Science-Technology-Society" (VOSTS). *Science Education, 76*(5), 477–491.

Aikenhead, G. S., Ryan, A. G., & Fleming, R. W. (1989). *Views on science-technology-society: Form CDN.mc.5*. Saskatoon: Department of Curriculum Studies, University of Saskatchewan.

Barnett, J., & Hodson, D. (2001). Pedagogical context knowledge: Toward a fuller understanding of what good science teachers know. *Science Education, 85*(4), 426–453.

Bauer, M., & Schoon, I. (1993). Mapping variety in public understanding of science. *Public Understanding of Science, 2*(2), 141–155.

Beardslee, D. C., & O'Dowd, D. D. (1961). The college-student image of the scientist. *Science, 133*(3457), 997–1001.

Bodzin, A., & Gehringer, M. (2001). Breaking science stereotypes. *Science and Children, 38*(4), 36–41.

Bourdieu, P. Passeron, J. (2003). *Los herederos: los estudiantes y la cultura*. Buenos Aires: Siglo XXI.

Brickhouse, N. W. (2013). Conceptions of inequality in the era of Bush/Obama. In *Moving the equity agenda forward* (pp. 39–51). Dordrecht: Springer.

Brosnan, M. J. (1999). A new methodology, an old story? Gender differences in the "draw-a-computer-user" test. *European Journal of Psychology of Education, 14*(3), 375–385.

Brush, L. R. (1979). Avoidance of science and stereotypes of scientists. *Journal of Research in Science Teaching, 16*(3), 237–241.

Buaraphan, K. (2009). Preservice and inservice science teachers responses and reasoning about the nature of science. *Educational Research and Reviews, 4*(11), 561–581.

Cakmakci, G., Tosun, O., Turgut, S., Orenler, S., Sengul, K., & Top, G. (2011). Promoting an inclusive image of scientists among students: Towards research evidence-based practice. *International Journal of Science and Mathematics Education, 9*(3), 627–655.

Carullo, J. C. (2002). *La percepción pública de la ciencia: el caso de la biotecnología* (No. 660.6 C329). Instituto de Estudios Sociales de la Ciencia y la Tecnología, Universidad Nacional de Quilmes, Buenos Aires, Argentina.

Cassirer, E. (1957). *The philosophy of symbolic forms. Volume three: The phenomenology of knowledge*. New Haven: Yale University Press. (German original: 1929).

Castoriadis, C. (1987). *The imaginary institution of society*. Cambridge: Polity Press. (French original: 1975).

Chacón, P., & Sánchez-Ruiz, J. (2009). La estructura familiar de Los Simpsons© a través del dibujo infantil. *Revista Mexicana de Investigación Educativa, 14*(43), 1129–1154.
Chambers, D. W. (1983). Stereotypic images of the scientist: The draw-a-scientist test. *Science Education, 67*(2), 255–265.
Chen, S. (2006). Development of an instrument to assess views on nature of science and attitudes toward teaching science. *Science Education, 90*(5), 803–819.
Chen, C. C., Taylor, P. C., & Aldridge, J. M. (1997). *Development of a questionnaire for assessing teachers' beliefs about science and science teaching in Taiwan and Australia.* Paper presented in annual meeting of the National Association for Research in Science Teaching, Oak Brook.
Collins, H., & Pinch, T. (1993). *The Golem: What everyone should know about science.* Cambridge, MA: Cambridge University Press.
Cortassa, C. (2012). *La ciencia ante el público.* Buenos Aires: Eudeba.
Demirbaş, M. (2009). The relationships between the scientist perception and scientific attitudes of science teacher candidates in Turkey: A case study. *Scientific Research and Essays, 4*(6), 565–576.
Dibarboure, M. (2010). La naturaleza de la ciencia como contenido escolar. *Quehacer Educativo*, 100.
Driver, R., Leach, J., Millar, R., & Scott, P. (1996). *Young people's images of science.* Bristol: Open University Press.
Echeverría. (1998). *Filosofía de la Ciencia.* Madrid: Akal.
Eliade, M. (1983). *Herreros y alquimistas.* Madrid: Alianza Editorial.
Elmesky, R. (2001). *Struggles of agency and structure as cultural worlds collide as urban African American youth learn physics (Ph.D. dissertation).* The Florida State University.
Emdin, C. (2012). Reality pedagogy and urban science education: Towards a comprehensive understanding of the urban science classroom. In *Second international handbook of science education* (pp. 59–68). Dordrecht: Springer.
Escalas Tramullas, M. T., Ruiz Mallén, I., & Zorrilla Pujana, J. (2009). *El científic dibuixat.* Barcelona: Generalitat de Catalunya.
Fensham, P. J. (1985). Science for all: A reflective essay. *Journal of Curriculum Studies, 17*(4), 415–435.
Fernández, I., Gil, D., Carrascosa, J., Cachapuz, A., & Praia, J. (2002). Visiones deformadas de la ciencia transmitidas por la enseñanza. *Enseñanza de las Ciencias, 20*(3), 477–488.
Fourez, G. (1994). *Alphabétisation scientifique et technique : Essai sur les finalités de l'enseignement des sciences.* Brussels: De Boeck Université.
Frayling, C. (2005). *Mad, bad and dangerous?: The scientist and the cinema.* London: Reaktion Books.
Fung, Y. Y. (2002). A comparative study of primary and secondary school students' images of scientists. *Research in Science & Technological Education, 20*(2), 199–213.
Furió, C., Vilches, A., Guisasola, J., & Romo, V. (2001). Finalidades de la enseñanza de las ciencias en la Secundaria Obligatoria. ¿Alfabetización científica o preparación propedéutica? *Enseñanza de las Ciencias, 19*(3), 365–376.
Gagliardi, R., & Giordan, A. (1986). La historia de las ciencias: Una herramienta para la enseñanza. *Enseñanza de las ciencias, 4*(3), 253–258.
Gallego Torres, A. (2007). Imagen popular de la ciencia transmitida por los cómics. *Revista Eureka sobre Enseñanza y Divulgación de las Ciencias, 4*(1), 141–151.
Gil, D., Fernández, I., Carrascosa, J., Cachapuz, A., & Praia, J. (2001). Para uma imagem não deformada do trabalho científico. *Ciência & Educação, 7*(2), 125–153.
Glavich, E., Ibáñez, R., Lorenzo, M., & Palma, H. (1998). *Notas introductorias a la filosofía de la ciencia. I. La tradición anglosajona.* Eudeba: Buenos Aires.
Good, T., & Brophy, J. (1973). *Looking in classrooms.* New York: Harper & Row.
Gómez, S., Franchi, M., Tarantini, M., Bonan, L., Adúriz-Bravo, A., & Meinardi, E. (2004). Escuelas con poblaciones en riesgo social: Proyecto de intervención e investigación en el área de ciencias naturales. In N. Bocalandro, M. Mateu, E. Ibarra, & J. Botto (Eds.), *La educación*

en biología: Para una nueva relación entre ciencia, cultura y sociedad (pp. 161–163). Córdoba: ADBiA.

Guerra Retamosa, C. (2004). Laboratorio y batas blancas en el cine. *Revista Eureka sobre Enseñanza y Divulgación de las Ciencias, 1*(1), 52–63.

Haynes, R. (2003). From alchemy to artificial intelligence. *Public Understanding of Science, 12*(3), 243–254.

Hermelin, D. (2011). Un contexto para la comunicación pública de la ciencia y la tecnología en Colombia: de las herencias euro-céntricas a los modelos para la acción. *Co-herencia, 8*(14), 231–260.

Hills, P., & Shallis, M. (1975). Scientists and their images. *New Scientist, 67*(964), 471–474.

Hodson, D. (1992). In search of a meaningful relationship: An exploration of some issues relating to integration in science and science education. *International Journal of Science Education, 14*, 541–562.

Hodson, D. (1998). Science fiction: The continuing misrepresentation of science in the school curriculum. *Curriculum Studies, 6*(2), 191–216.

Hugo, D., & Adúriz-Bravo, A. (2003). Algunos elementos teóricos para la investigación del conocimiento profesional del profesorado de ciencias naturales acerca de la naturaleza de la ciencia. In A. Adúriz-Bravo, G. A. Perafán, & E. Badillo (Eds.), *Actualización en didáctica de las Ciencias Naturales y las Matemáticas* (pp. 23–42). Bogotá: Magisterio.

Jones, M. G., Howe, A., & Rua, M. J. (2000). Gender differences in students' experiences, interests, and attitudes toward science and scientists. *Science Education, 84*, 180–192.

Jussim, L., Eccles, J., & Madon, S. (1996). Social perception, social stereotypes, and teacher expectations: Accuracy and the quest for the powerful self-fulfilling prophecy. *Advances in Experimental Social Psychology, 28*, 281–388.

Lederman, N. G. (2006). Research on nature of science: Reflections on the past, anticipations of the future. *Asia-Pacific Forum on Science Learning and Teaching, 7*(1), 1–11.

Lederman, N. G., Abd-El-Khalick, F., Bell, R. L., & Schwartz, R. S. (2002). Views of nature of science questionnaire: Toward valid and meaningful assessment of learners' conceptions of nature of science. *Journal of Research in Science Teaching, 39*(6), 497–521.

Leslie-Pelecky, D. L., Buck, G. A., & Zabawa, A. (2005). Broadening middle school students' images of science and scientists. *Journal of Materials Education, 27*, 173–178.

Locke, S. (1999). Golem science and the public understanding of science: From deficit to dilemma. *Public Understanding of Science, 8*(2), 75–92.

Long, M., & Steinke, J. (1996). The thrill of everyday science: Images of science and scientists on children's educational science programmes in the United States. *Public Understanding of Science, 5*(2), 101–119.

Long, M., Steinke, J., Applegate, B., Lapinski, M. K., Johnson, M. J., & Ghosh, S. (2010). Portrayals of male and female scientists in television programs popular among middle school-age children. *Science Communication, 32*(3), 356–382.

Losh, S. C., Wilke, R. A., & Pop, M. (2008). Some methodological issues with 'Draw a Scientist' tests among young children. *International Journal of Science Education, 30*(6), 773–792.

Manassero, M., & Vázquez, A. (2001). Actitudes de estudiantes y profesorado sobre las características de los científicos. *Enseñanza de las Ciencias, 19*(2), 255–268.

Manassero, M. A., Vázquez, A., & Acevedo, J. A. (2001). La evaluación de las actitudes CTS. En Sala de Lecturas CTS+I de la OEI. Disponible en: http://www.oei.es/salactsi/acevedo11.htm

Manassero, M. A., Vázquez, A. M., & Acevedo, J. A. (2003). *Cuestionario de Opiniones sobre Ciencia, Tecnología y Sociedad (COCTS)*. Princenton: Educational Testing Service.

Manzoli, F., Castelfranchi, Y., Gouthier, D., & Cannata, I. (2006). *Children's perceptions of science and scientists: A case study based on drawings and story-telling*. Paper presented at 9th international conference on Public Communication of Science and Technology, Seul.

Maoldomhnaigh, M. O., & Hunt, A. (1988). Some factors affecting the image of the scientist drawn by older primary school pupils. *Research in Science & Technological Education, 6*(2), 159–166.

McComas, W. (1998). *The nature of science: Rationales and strategies*. Dordrecht: Kluwer.
Mead, M., & Metraux, R. (1957). Image of the scientist among high-school students: A pilot study. *Science, 26*, 384–390.
Merton, R. K. (Ed.). (1968). *Social theory and social structure*. New York: Simon and Schuster.
Moreno Lupiáñez, M. (2003). Cine y ciencia. *Revista Quark*, 28/29.
Newton, D. P., & Newton, L. D. (1992). Young children's perceptions of science and scientist. *International Journal in Science Education, 148*(3), 331–348.
Newton, D. P., & Newton, L. D. (1998). Primary children's conceptions of science and the scientist: Is the impact of a national curriculum breaking down the stereotypes? *International Journal of Science Education, 20*(9), 1137–1149.
Olitsky, S., & Milne, C. (2012). Understanding engagement in science education: The psychological and the social. In *Second international handbook of science education* (pp. 19–33). Dordrecht: Springer.
Oppenheimer, F. (1968). A rationale for a science museum. *Curator: The Museum Journal, 11*(3), 206–209.
Palma, H., & Wolovelsky, E. (2001). *Imágenes de la racionalidad científica*. Buenos Aires: Eudeba.
Parsons, E. C. (1997). Black high school females' images of the scientist: Expression of culture. *Journal of Research in Science Teaching, 34*(7), 745–768.
Polino, C., Fazio, M. E., & Vaccarezza, L. S. (2003). Medir la percepción pública de la ciencia en los países iberoamericanos: Aproximación a problemas conceptuales. *CTS+ I: Revista Iberoamericana de Ciencia, Tecnología, Sociedad e Innovación, 5*, 1.
Popper, K. R. (1992). *The logic of scientific discovery*. London: Routledge. (German original of 1935).
Pujalte, A. (2014). *Las imágenes de ciencia del profesorado: De la imagen declarativa a la enactiva*. Tesis doctoral. Universidad Nacional de Quilmes.
Pujalte, A., Porro, S., & Adúriz-Bravo, A. (2011a). Las imágenes de ciencia del profesorado: Su relación con una educación científica de calidad para todas y todos. *Tecné, Episteme y Didaxis*, Extra Issue, 410–415.
Pujalte, A., Gesuele, C., Márquez, M., & Adúriz-Bravo, A. (2011b). ¿Qué nos imaginamos al pensar en la gente que se dedica a la ciencia?: Implicaciones para una educación científica escolar de calidad para todas y todos. In *Avances en Educación en Ciencia y Tecnología: Enfoques y Estrategias: Año 2011* (pp. 352–354). San Fernando del Valle de Catamarca: Universidad Nacional de Catamarca.
Pujalte, A., Gangui, A., & Adúriz-Bravo, A. (2012a). "La ciencia en los cuentos": Análisis de las imágenes de científico. *Ciencia Ergo Sum, 19*(3), 261–270.
Pujalte, A., Porro, S., & Adúriz-Bravo, A. (2012b). *"Yo no sirvo para esto": La desidentificación con la ciencia de un grupo de estudiantes de secundaria: Perspectivas de análisis y propuestas superadoras*. Paper presented in X Jornadas Nacionales/V Congreso Internacional de Enseñanza de la Biología.
Räty, H., & Snellman, L. (1997). Children's images of an intelligent person. *Journal of Social Behavior & Personality, 12*(3), 773–784.
Reis, P., & Galvão, C. (2006). O diagnóstico de concepções sobre os cientistas a través da análise e discussão de histórias de ficção científica redigidas pelos alunos. *Revista Electrónica de Enseñanza de las Ciencias, 5*(2).
Reis, P., & Galvão, C. (2007). Reflecting on scientists' activity based on science fiction stories written by secondary students. *International Journal of Science Education, 29*(10), 1245–1260.
Reis, P., Rodrigues, S., & Santos, F. (2006). Concepções sobre os cientistas em alunos do 1° Ciclo do Ensino Básico: Poções, máquinas, monstros, invenções e outras coisas malucas. *Revista Electrónica de Enseñanza de las Ciencias, 5*(1), 51–74.
Rist, R. (1999). Sobre la comprensión del proceso de escolarización: Aportaciones de la teoría del etiquetado. In M. FernandezEnguita (Ed.), *Sociología de la Educación*. Ariel: Barcelona.
Rosenthal, R., & Jacobson, L. (1968). Pygmalion in the classroom. *The Urban Review, 3*(1), 16–20.

Rubba, P. A., & Harkness, W. L. (1993). Examination of preservice and in-service secondary science teachers' beliefs about science-technology-society interactions. *Science Education, 77*(4), 407–431.

Ryan, A. G., & Aikenhead, G. S. (1992). Students' preconceptions about the epistemology of science. *Science Education, 76*(6), 559–580.

Schäffer, M. (2011). Sources, characteristics and effects of mass media communication on science: A review of the literature, current trends and areas for future research. *Sociology Compass, 5*(6), 399–412.

Schummer, J. (2006). Historical roots of the 'mad scientist': Chemists in nineteenth-century literature. *Ambix, 53*(2), 99–127.

Seiler, G. (2002). *A critical look at teaching, learning, and learning to teach science in an inner city neighborhood high school (Ph.D. dissertation)*. University of Pennsylvania.

She, H. C. (1998). Gender and grade level differences in Taiwan students' stereotypes of science and scientists. *Research in Science & Technological Education, 16*(2), 125–135.

Sjøberg, S. (2000). Science and scientists: The SAS-study: Cross-cultural evidence and perspectives on pupils' interests, experiences and perceptions: Background, development and selected results. *Acta Didactica, 1*.

Song, J., & Kim, K. S. (1999). How Korean students see scientists: The images of the scientist. *International Journal of Science Education, 21*(9), 957–977.

Steele, C. (1997). A threat in the air: How stereotypes shape intellectual identity and performance. *American Psychologist, 52*, 613–629.

Steinke, J., Long, M., Johnson, M., & Ghosh, S. (2008). *Gender stereotypes of scientist characters in television programs popular among middle school-aged children*. Paper presented in annual meeting of the Association for Education in Journalism and Mass Communication, Chicago.

Sutton, C. (2008). Beliefs about science and beliefs about language. *International Journal of Science Education, 18*(1), 1–18. (Originally published in 1996).

Symington, D., & Spurling, H. (1990). The 'draw a scientist test': Interpreting the data. *Research in Science & Technological Education, 8*(1), 75–77.

Vázquez, A., Acevedo, J., Manassero, M., & Acevedo, P. (2006). *Creencias ingenuas sobre naturaleza de la ciencia: consensos en sociología interna de ciencia y tecnología*. In Actas del iv Seminario Ibérico de cts en la Educación Científica: Las relaciones cts en la Educación Científica. [cd-rom], Universidad de Málaga, Málaga.

Vickers, B. E. (1984). *Occult and scientific mentalities in the renaissance*. Cambridge: Cambridge University Press.

Vílchez-González, J. M., & Perales, F. J. (2006). Image of science in cartoons and its relationship with the image in comics. *Physics Education, 41*, 240–249.

Weingart, P. (2006). Chemists and their craft in fiction film. *HYLE-International Journal for Philosophy of Chemistry, 12*(1), 31–44.

Weingart, P., Muhl, C., & Pansegrau, P. (2003). Of power maniacs and unethical geniuses: Science and scientists in fiction film. *Public Understanding of Science, 12*(3), 279–287.

Wyer, M. (2003). Intending to stay: Images of scientists, attitudes toward women, and gender as influences on persistence among science and engineering majors. *Journal of Women and Minorities in Science and Engineering, 9*, 1–16.

Ziman, J. (1992). Not knowing, needing to know, and wanting to know. In B. V. Lewenstein (Ed.), *When science meets the public* (pp. 13–20). Washington: American Association for the Advancement of Science Washington.

Chapter 12
Images of Scientists in Textbooks Aimed at Students in Need of Supplemental Support – An Analysis of Adjustments

Lena Hansson and Lotta Leden

12.1 Introduction

Social justice perspectives urge us to focus on issues such as *who* science is for (Carlone 2004; Zacharia and Barton 2004), as well as on the empowering of students in relation to their daily lives and as citizens. Hackman (2005) argues that "social justice education requires an examination of systems of power and oppression combined with a prolonged emphasis on social change and student agency in and outside of the classroom" (p. 104). In science education literature, scholars point out the necessity of nature of science (NOS) perspectives in such a citizenship and/or activism oriented curriculum (Aikenhead 2006; Hodson 2008, 2014; Kolstø 2000; Yacoubian 2015). Thus, one important reason for NOS to be taught to *all* students in compulsory school is to meet the goals of increased social justice, including the empowering of students and the breaking of patterns of oppression of student groups that have traditionally been marginalized in the teaching of science. However, NOS has sometimes been seen as an add-on by teachers, considered only suitable for "clever" students or if there is time left (Henke and Höttecke 2015; Leden et al. 2015; Mulvey et al. 2016). This way to handle NOS could be viewed as a hindrance for different student groups in their citizen education.

When NOS learning is viewed as important from a social justice perspective it becomes central to examine how NOS is taught to different student groups. Swedish schools follow a policy of inclusion, which means they are obliged to offer supplemental support (Swedish 'extra anpassningar', "additional adjustments") for students who run the risk of not meeting the knowledge requirements of the curriculum:

L. Hansson (✉) · L. Leden
Department of Mathematics and Science Education, Kristianstad University, Kristianstad, Sweden
e-mail: lena.hansson@hkr.se; Lotta.leden@hkr.se

H. A. Yacoubian, L. Hansson (eds.), *Nature of Science for Social Justice*, Science: Philosophy, History and Education,
https://doi.org/10.1007/978-3-030-47260-3_12

> If … it can be expected that a student will not reach the minimum knowledge requirements, the student shall promptly be supported through supplemental support within the frame of ordinary teaching. (The Education Act, 3chp 5a§, our translation)

The art of supplemental support is decided by the teacher. Support can be of various types, such as teachers offering other kinds of explanations, clearer instructions, extra equipment or adjusted teaching materials (Swedish National Agency of Education 2014). In this chapter, we analyze adjusted textbooks which can be provided to students in need of supplemental support. The publishers offer both general and adjusted versions of the textbooks. Thus, the books can easily be used by different students in the same classroom. The publishers describe the books on their webpages. One of them describes the adjusted book as targeting "students with reading and writing difficulties/dyslexia, students with Swedish as a second language as well as other students who find the general books too hard…". The other publisher describes their adjusted book as a book for "students who want to have a slightly easier course and a smaller amount of text". Thus, the adjusted books have quite an unspecified target group.

Including NOS in science education means highlighting other aspects of science than the traditionally taught science concepts. One aspect of teaching NOS means shedding light on the *scientists* – emphasizing science as a human enterprise. Science as a human enterprise is, in one way or another, part of most NOS frameworks (e.g. Allchin 2011; Erduran and Dagher 2014; Lederman 2007) and highlights the need to not only focus on the products of science, but also on the processes of science in which people are involved. Thus, an inclusion of scientists communicates that scientific knowledge is not just out there, but that it has to be developed by humans. However, not only is the presence of humans important but the types of images that are communicated are too. Previous research has shown that such images, in textbooks for instance, are often stereotypical in relation to scientists' personal interests, characteristics and engagement in activities. Narrow and stereotypical images of scientists could constitute a hindrance for different student groups to identify with and thereby gain access to science and science education.

Therefore, how NOS, including images of scientists, is communicated to different student groups is of central importance when science teaching aims to increase social justice. This chapter focuses on variations and adjustments in regards to the descriptions of scientists in general science textbooks and textbooks written for this diverse group of students who are offered supplemental support in terms of adjusted textbooks. The analyzed textbooks are Swedish science textbooks for lower secondary school (grades 7–9, ages 13–16).

12.2 NOS and Scientists in Textbooks

Images of scientists and NOS presented in the media are often mythical and stereotypical (Allchin 2003; Van Gorp et al. 2014; see also Adúriz-Bravo and Pujalte, Chap. 11 in this volume). Certain prototypes are described in Van Gorp et al. (2014)

as genius (sometimes misunderstood), wizard, nerd, puzzler, adventurer, and mad scientist. These prototypes make up a collection of good heroes, helpers, enemies, and almost always "loners". The lone scientist has shown to make the profession particularly unattractive, especially to girls (Rommes et al. 2007). The study by Van Gorp et al. (2014) further shows that the products of science are emphasized much more than the processes of science.

One way to broaden the picture from the media is to include richer NOS descriptions in the teaching of science. There is, however, a reported lack of teachers' pedagogical content knowledge (Demirdöğen et al. 2015) for NOS as well as traditions to fall back on regarding NOS teaching. Therefore, the ways that NOS is communicated in science textbooks becomes crucial, since textbooks can function as a support for teachers. Furthermore, research shows that the textbook is a well-used teaching resource among science teachers (Abd-El-Khalick et al. 2017; Skolinspektionen [Swedish Schools Inspectorate] 2010).

However, previous research has reported that science textbooks often show a limited emphasis on NOS (McDonald and Abd-El-Khalick 2017), or coincides with the stereotypical ways that NOS is communicated in the media. Several studies have shown that many textbooks are focused on factual knowledge – that is, the end products of science and not how the knowledge is built or on the scientists and other individuals behind the knowledge (Brigham et al. 2011; Clough and Olson 2004; DiGiuseppe 2014; Knain 2001; Mason and Hedin 2011). Human actors, and their values, priorities, negotiations and controversies, are usually left out and when they are not, a stereotype picture of the scientist is common. A study by Yacoubian et al. (2017) found that the scientists in Lebanese science textbooks were mostly non-western white males, while Lebanese and/or Arab scientists were uncommon. Moreover, stereotypical images of scientists and science were further reinforced through scientists who discovered truths, while working alone and following the scientific method in a lab milieu. Clough and Olson (2004) argue that the common picture of a scientist who "discovers" truths hides the picture of the complicated processes that surrounds the interpretation of data.

In line with the above described international research on NOS in science books, previous Nordic studies describe a low emphasis on NOS in textbooks and that anecdotal descriptions of historical, western male characters are frequent, while there are very few examples of contemporary non-western or female scientists (Hedrén and Jidesjö 2010; Svennbeck 2003; Vesterinen et al. 2013). Such images neither contribute to the empowering of students to provide them with the understanding or engagement needed to take part in societal debates, nor do they contribute to students' possibilities of identifying with science (Aikenhead 2006). In line with these arguments, Knain (2001) highlights how the ideologies in science textbooks play a role for "which students are going to develop what relations to science and technology as future citizens" (Knain 2001, p. 319).

One reason as to why stereotypical images of NOS are not dealt with in science textbooks is described in DiGiuseppe (2014) who studied the publisher versus author impact on the choice of NOS in science textbooks. DiGiuseppe (2014) showed that an important part of the considerations over how to represent NOS in textbooks had to do with ideas of appropriateness. These ideas concerned both students' learning difficulties and difficulties for students and teachers to involve with explicit NOS content. Similarly, previous NOS research has shown that teachers sometimes argue that some NOS aspects are too complex for young students or for students who have been positioned as having learning difficulties (Henke and Höttecke 2015; Leden and Hansson 2017; Leden et al. 2015; Mulvey et al. 2016).

12.3 NOS and Students with Learning Difficulties

There is little research on science teaching in general for the heterogeneous group of students with various kinds of learning difficulties (Brigham et al. 2011; Therrien et al. 2011). There is even less research concerned with NOS teaching for these groups (however, some examples are discussed below).

A recent study by Mulvey et al. (2016) analyzed the effects of a professional development intervention where special education teachers were taught about the teaching of NOS. These teachers' classroom practices were later investigated. The authors argue that even though some scholars have suggested that NOS, due to its abstraction, ought to be avoided for students with special needs, "NOS may be empowering to students by encouraging them to explore more creative aspects of doing science" (Mulvey et al. 2016, p. 556). An important conclusion from the study was that teaching NOS to students with special needs changed the teachers' beliefs about students' capabilities and led to greater expectations on the students. Such findings could support NOS teaching for social justice and democratic citizenship where *all* students are regarded as individuals who are expected to take part in a democratic society. This concurs with the arguments in Lederman and Stefanich (2006), who claim that teaching focused on learning about NOS and "processes of reasoning rather than the products" (p. 57) contributes to an inclusive education where all students have the chance to develop abilities to make choices and express themselves. In line with this argument, Brigham et al. (2011), in a study of science education and disabled students, discuss NOS as an important goal of science education and question the typical science textbook focusing on facts rather than on NOS.

12.4 Textbooks and Students in Need of Supplemental Support

The textual structures in science textbooks have been characterized as descriptions that lack characters, story structures, narratives, argumentation and reflexivity (Knain 2001; Mason and Hedin 2011). Mason and Hedin (2011) discuss the "conceptual density" of the text, which is commonly very high in science textbooks. Studies on reading show that the higher density of a text the greater effort to understand and make meaning of it (Gibbons 2006; Hajer and Meestringa 2014; Mason and Hedin 2011; Wikman 2004): "For students with poor vocabulary and limited prior knowledge /.../ text density plays a critical role in intensifying comprehension difficulties" (Mason and Hedin 2011, p. 215).

The context of a text is of major importance both for the possibilities to interpret it and for its potential to create interest (Gibbons 2006; Hajer and Meestringa 2014). However, studies have shown that many texts are adjusted by being shortened through removing details and contexts and thus making it even denser. What should be done is to make the text richer in examples as it would make it both more interesting and easier to interpret (Hajer and Meestringa 2014; Wikman 2004). One part of enriching the context could be to show the knowledge processes instead of only presenting facts (Wikman 2004). Still, a recent Swedish study has shown that science teachers might be more inclined to reduce complexity (such as in assignments and texts) and focus more on "basic skills training" than teachers in other subjects (Jönsson 2018).

12.5 Research Question

A social justice perspective urges us to problematize notions of NOS as being too hard or complex for some student groups since NOS has a vital role in science teaching for citizenship. Thus, concerns for different student groups, for example those who are experiencing learning difficulties of various kinds, encourages us to look more closely at the communication of NOS images to various student groups (including marginalized groups).

The study presented in this chapter focuses on how the images of scientists in science textbooks (school years 7–9, ages 13–16) are adjusted between general textbooks and textbooks aimed at students in need of supplemental support. The research question guiding the study is: *What adjustments are made regarding images of scientists between general and adjusted textbooks?* These adjustments will be discussed from a social justice perspective.

12.6 Design of the Study

12.6.1 Context

Swedish students attend compulsory school for 9 years (ages 7–16). During this time they all take the same science courses (biology, chemistry and physics), where NOS, according to Swedish policy documents, should be included. Swedish school is goal steered. This means that the government provides aims, core content and knowledge requirements. However, the responsibility for how teaching is organized or what teaching materials are to be used is decided by the individual schools and teachers. In line with this structure there are no authorized textbooks, but there are several different science textbooks provided by different publishers for the Swedish market.

As mentioned previously, Swedish schools follow a policy of inclusion, which means that students from diverse backgrounds with a wide range of needs (such as clearer instructions, language support, extra equipment or adjusted teaching materials) take general classes in the regular school environment. The policy of inclusion means that schools are obliged to offer supplemental support, but the art of the support is often decided by the subject teacher in dialogue with the students and his or her parents.

12.6.2 Data and Analysis

As an approach to meet the need for differentiation within the same classroom, two textbook publishers in Sweden chose to publish both general textbooks and adjusted versions (aimed at compulsory school students aged 13–16). Such books are an example of supplemental support in terms of adjusted teaching material that teachers can offer students (see Swedish National Agency of Education 2014).

The present study builds on the analysis of general and adjusted versions of chemistry and physics textbooks published by the above mentioned publishers (in total 8 books).[1]

A content analysis (Hsieh and Shannon 2005) was performed on the textbooks. The texts were searched for instances where scientists were mentioned or illustrated in different ways. The unit of analysis was a specific section. A section was characterized as being a clearly defined paragraph often marked by a heading. Most pictures were judged as part of a text section since they were directly related to the contents of the text section.

In a first step all sections in the general books and the adjusted books were analyzed for the mentioning of scientists (or other individuals involved in science activities), characteristics (such as gender and nationalities), activities and contexts.

[1]The following books have been analyzed:

Andréasson, B. (2011). *Kemi: för grundskolans år 7–9. Fokus*. Stockholm: Natur & kultur.
Andréasson, B. (2011). *Kemi: för grundskolans år 7–9*. Stockholm: Natur & kultur.
Nettelblad, F. & Nettelblad, K. (2013). *Kemi Light*. Stockholm: Liber.
Nettelblad, F. & Nettelblad, K. (2013). *Kemi*. Stockholm: Liber.
Sjöberg, S. & Ekstig, B. (2011). *Fysik: för grundskolans år 7–9. Fokus*. Stockholm: Natur & kultur.
Sjöberg, S. & Ekstig, B. (2011). *Fysik: för grundskolans år 7–9*. Stockholm: Natur & kultur.
Undvall, L. & Karlsson, A. (2013). *Fysik Lightbok*. Stockholm: Liber.
Undvall, L. & Karlsson, A. (2013). *Fysik*. Stockholm: Liber.

These analytical choices were made due to previous reports on stereotypical images related to these aspects and their consequences for social justice.

All scientists mentioned in a section were noted. The gender of the scientist and any countries/nationalities mentioned in relation to them were also noted. Scientists are referred to in different ways in the textbooks. Sometimes the references are explicit in referring to specific scientists – mentioned by their name or referred to as "chemists", "researchers", "scientists", and so on. In other cases, the references are more implicit, such as "one, we, humans…". These words are not gendered in Swedish which means that we can only identify the gender if the book provides us with a name or a picture. The kind of activities that the scientists took part in, as well as the context for the activities (historical, general/uncertain or contemporary), were also noted. Activities in this analysis did not only mean descriptions of experiments or investigations, but also descriptions of feelings, states of mind, and disappointments (see Fig. 12.1).

In a second step, the section in the general book was compared to the section in the adjusted book in search for specific adjustments. The different types of adjustments were categorized in accordance with differences in relation to the references to scientists. This resulted in six different types of adjustments described in this chapter.

Examples from the textbooks are translated from Swedish by us.

Textbook section	Analysis
RESEARCH **An unexpected discovery** Magnetism is a property of metals. That is at least what **we** have believed so far. But now several **physicists** around the world have discovered that pure carbon can become magnetic. Can ordinary carbon in a pencil become magnetic? Yes, it sounded unbelievable and the discovery was so debated that it divided the **physicists** into two camps. But after 10 years of debate and research it is now proved that carbon during certain circumstances can become magnetic at room temperature. Why do we need magnetic carbon? Well, **scientists** believe that we can use it for medicines, biomechanics and for fast data storage in new kinds of computers. Therefore, it is possible that this discovery in the future might lead to an entirely new industry with carbon based electronics.	**References to scientists:** we, physicists, scientists **Context:** contemporary **Activities:** believe, discover, debate, research **Named scientists:** none **Gender:** not specified **Nationalities:** not specified (from around the world)

Fig. 12.1 Analysis of one section of a page (our translation) in the general version of a physics textbook (the entire section has been removed in the adjusted version). References to scientists are made in bold, and references to activities are underlined

12.7 Results

Scientists are mentioned to different extents in the different books, but there are also differences between chapters within the same book. Altogether, between 16% and 29% of the sections in the four general books include scientists. Sometimes the scientist is mentioned without much information or context, but there are also instances in the textbooks with more information about, for example, the activities the scientist is engaged in. Previous studies have reported that textbooks often show scientists as discovering truths. Such descriptions are common in the Swedish science textbooks reported on in this chapter. However, there are instances where more complicated research processes and/or more complex pictures of the scientists are provided. An example, from one of the general physics textbooks describes the research of Hans Christian Ørsted, and a less well-known scientist Jean Daniel Colladon, on the connection between magnetism and electricity. Verbs such a "discover" and "experimentation" are part of this description. However, the scientists' thoughts, Colladon's failures, and his bad luck, are also part of the description. Such descriptions challenge the image of scientists as "heroes" and give a more complex insight into the many people involved in science.

Another example, from a contemporary context, is the description of Dan Shechtman's discovery of quasiperiodic crystals. Even though his research is labelled as a discovery, other parts of the section describe how his ideas gradually changed other chemists' ways of viewing crystal structures. Furthermore, it describes how Shechtman, at first, was laughed at and bullied before he was able to convince other researchers to repeat his experiments. This description has traits of "hero" images of the scientist (people laughed but he was right in the long run). However, the description challenges the image of the research process characterized by experiments that are followed by easy, objective conclusions. This is an example of how a certain section can both challenge and reinforce stereotypes, and thus cannot simply be regarded as either stereotypical or non-stereotypical.

The following sections focus on the adjustments made between general and adjusted book versions regarding references to scientists. For instance, what happens to the sections with descriptions of Colladon and Shechtman? The analysis is only directed towards adjustments made concerning information about and references to scientists. Thus, other adjustments could very well have been done too (for example, concerning science concepts and models or concerning the language used), but this is not part of the present analysis.

12.7.1 Adjustments Between Book Versions

In general, the adjusted textbook versions are shorter (fewer pages) and each page contains less text than the general versions. These adjustments mean that, to different extents in the two book series, whole or parts of sections are removed or rewritten. Sections are also sometimes merged or divided. We have identified the following adjustments regarding scientists in the texts:

A. Remove an entire section (and therefore, scientists mentioned in the section are also removed)
B. Remove scientist from the section
C. Remove information about the scientist (such as characteristics or activities they are involved in)
D. Add scientists to the section
E. Add or emphasize information about scientists (such as characteristics or activities they are involved in)
F. No adjustment in respect of references to scientists (scientists and information about scientists are left (more or less) unchanged)

Below, the adjustments (A–E) are described and exemplified. As category F means no adjustment, this category is not further described in this chapter. Table 12.1 shows the frequencies of the different adjustments.

As Table 12.1 shows, there are many instances where no adjustments are made in respect to how references to scientists are made. One example is the section about Shechtman (described above) that is available in both books. However, oftentimes some kind of adjustments *are* made. The most frequent adjustments are *Remove section* and *Remove scientist.* All adjustments are described below.

A. ***Remove an entire section***

Using this adjustment, the authors exclude an entire section that made reference to scientists in the general book. This is a frequent adjustment and different types of sections that include a scientist can be removed. In one of the book series all special boxes with the headings "research" (see Fig. 12.1), "history" or "deeper knowledge" are systematically removed. Due to this adjustment, some chapters of the adjusted books show no references to scientists at all. When sections with scientists are removed it results in less references to research activities and the characteristics of scientists. Sometimes the excluded descriptions are rather limited in the general book, but other times they include rich examples of research activities and various

Table 12.1 Frequencies of different types of adjustments (% is given in relation to the total number of sections in the general books which refer to scientists). The coloured column shows the most frequent adjustment.

Remove section	Remove scientists	Remove information about scientists	Add scientists	Add/emphasize information about scientists	No adjustment in respect of references to scientists
72 (16%)	160 (37%)	42 (10%)	20 (5%)	11 (3%)	133 (30%)

scientists. When, for example, the section "Research in development" in one of the books is removed, the activities such as *Change ideas, believe, thought, consider, realize, discovered, invented, did not know, showed, tried to find out, put an hypothesis, ask for help, found* are also removed. Furthermore, the same section contains six drawings of scientists, such as an "old Egypt" and a fictional, future scientist (a girl). All of these activities and pictures of scientists are excluded from the adjusted version.

B. ***Remove scientist from section***

When the text is adjusted in this way the section in the general book can be identified in the adjusted version of the book. However, references to one or more of the scientists in the general book are excluded in the adjusted version. This sometimes means that information about the characteristics of the scientist and/or activities that the scientist is/has been involved in are also removed. This kind of adjustment could be described, to different extents, as dehumanizing the section and is the most frequent adjustment (see Table 12.1).

One example of this adjustment is a section about the sun (see Fig. 12.2). The general book includes information about size and distance concerning the sun, but it also mentions people's curiosity about the sun and names two scientists as well as the activities in which they were engaged (i.e., pondered, studied, discovered, found out, received the Nobel prize, and so on). In the adjusted version only factual information about size and distances are included. Everything else is excluded.

The sun is a small star

As long as we people have existed on Earth we have pondered over what the sun consists of and how it can be so warm and shine so bright. It is not until presently in the history of man that we have gained the answers. The first discovery was made in 1868 by the Frenchman *Pierre Jules Jansen*. When he studied light from the sun he discovered a new element that no one had known before. The element got the name Helium, from the Greek name for the sun (*helios*). Later, in the 1930ties, the physicist *Hans Bethe* figured out how helium reacts inside the sun so that enormous amounts of energy is built. He received the Nobel Prize in 1967.

The sun is a common type of star in the Universe. The *sun* is a star, our closest one. Compared to many other stars the sun is quite small. But compared to the Earth the sun is huge. One million Earths could be fitted into the sun. The distance to the sun is 150 million kilometers. We could also say that the distance to the sun is approximately eight *light-minutes*. That is, the time that passes when light travels from the sun to the Earth.

Fig. 12.2 One section from the general physics book (translation). Parts marked in grey constitute the remaining contents in the adjusted version

In another example – a section about the moon – the general book describes what you can see looking at the moon with binoculars. It also states that the most common mountains on the moon are called 'impact craters' (in Swedish, a specific concept 'ringberg' ('ring mountain') is used). This first part of the section is similar to the section in the adjusted version. The general book then continues by describing what humans previously believed about the moon (that plateaus were seas). This is not included in the adjusted version of the book. Furthermore, the general book describes how scientists have gained the knowledge we have today (*astronauts* have *brought home* material that has been *studied* by *scientists*, who have become *convinced* that there is no life on the moon), and that research is continuing (China has succeeded in landing a spacecraft equipped with a small cross-country car on the moon and plans to send humans there). This information has been removed from the adjusted book.

A slightly different, more unusual way to use this adjustment is exemplified in the introduction to a chapter about electricity. In the general book, Ørsted's and Colladon's struggles are described (see previous example, Sect. 12.7). In the adjusted version physicists are mentioned, but Colladon is not specifically mentioned. In this example, the main points remain but details are lost as are the human dimensions connected to a specific scientist's struggles and failures.

C. ***Remove information about scientist***

This adjustment means that even though the scientists in the general book is included in the adjusted version, some of the information concerning them has been removed. This could result in including less information about the characteristics of the scientist and/or include less information about the activities the scientist was involved with.

An example where characteristics of scientists are removed is a section in the general book that mentions the Swedish astronomer Anders Celsius, who lived and worked in Uppsala (Sweden) at the beginning of the eighteenth century. The adjusted version only mentions that he was a Swedish astronomer. Moreover, the general book includes a picture of Celsius that is not part of the adjusted version.

Another example shows how characteristics and activities of scientists are reduced in a section about free fall. In this section both the general book and the adjusted book describe Galileo Galilei's contribution to physics' understanding of free fall. Both books describe what Galilei thought about free fall, but that he could not do an experiment to investigate it due to the fact that they did not know how to empty a container of air at the time. The general book (but not the adjusted one) continues by describing that Galilei did not have a good enough watch, instead "he counted the number of water drops that fell before an object hit the ground. These experiments made Galilei understand why a leaf falls slower to the ground while a cone falls quickly". Finally, both books conclude that Galilei thought that if one took away the air then both objects would fall with the same speed.

In yet another example about the discovery of oxygen, the general book says:

> "In Sweden, we argue that it was the Swedish chemist Carl Wilhelm Scheele who discovered oxygen. He found the gas in 1773, but it was not until 1777 that he reported it in a book. In the research community it is very important that scientists inform others about their discoveries in a book or journal. Before Scheele had had the time to describe his discovery, two other chemists had also found the gas. Thus, in other countries they argue that other scientists discovered oxygen".

The adjusted version of the book only says: "Several chemists discovered oxygen at about the same time. The Swede Carl Wilhelm Scheele was one of them". These examples illustrate how some of the complexities of research activities are reduced.

D. ***Add scientist***

This adjustment is only used in one of the books. It means adding scientists to a section. Thus, the adjustment is the opposite to adjustment B (remove scientist). In some cases, the added scientist is not a specifically named scientist but is instead referred to, for instance, as "one". In an example from the introductory section to physics, the general book says: "this chapter is about water, air, and other substances, since they are examples of matter", while the adjusted version states: "In physics, *one tries to understand* how the world works. This particular chapter is about matter and different substances, such as water and air" (p. 6, italics added). Adding "one tries to understand" could be a way to humanize the science (see Wikman 2004).

In other (rare) cases a named scientist is added. For example, in one section about Galilei the adjusted version mentions that another scientist, Bruno (first name never mentioned in the textbook), had recently been burnt at the stake as an explanation as to why Galilei took back what he had said about the Earth moving around the Sun. Bruno is not mentioned in the general book.

E. ***Add or emphasize information about scientists***

Using this adjustment means adding/emphasizing information about scientists. In one unique example a whole paragraph is added. This is a special box describing how "our view of the universe has developed". In most cases the added information is about characteristics of the scientists (such as origin, gender or occupation) or about activities that the scientists are involved in. One example of adding characteristics is a section where Alessandro Volta is mentioned. The adjusted book adds that he was an *Italian physicist*. An example where activities are added is a section about how scientists learnt about the backside of the moon. The adjusted version adds a very small piece of extra information by mentioning that *photos* were taken and thus provide insight into how information can be collected.

The only example of where information is emphasized is in a paragraph about nuclear reactions where, among other things, the research of Lise Meitner and her colleague Otto Hahn is described. The general book mentions that "Otto Hahn was (alone!) in receiving the Nobel prize in chemistry in 1944". In the adjusted version the unfairness is emphasized both in the main text: "Otto Hahn was alone in receiving the Nobel prize for discovering fission. It was unfair, according to many people", and in the caption of a picture of Meitner saying: "Her boss, Otto Hahn, received the Nobel prize in chemistry in 1944, but Lise was forgotten!".

In summary, it is uncommon to add scientists or information about scientists in the adjusted versions (see Table 12.1). Among the adjustments, Removing scientists is the most common. The next section discusses what the adjustments mean concerning *who* the mentioned scientists actually become in the adjusted versions.

12.7.2 Who Does the Scientist Become in the Adjusted Books?

In cases where scientists are mentioned by name,[2] all four of the general books present a large majority of men and only a few women (see Fig. 12.3). An extreme example in this respect is one of the books in which 50 scientists are mentioned by name and only two of them are women. In all of the adjusted books one or more of the few female scientists from the general books are removed, even though the number of female scientists was already low to begin with (see Fig. 12.3). The exclusion has been made through the adjustments "Remove section" and "Remove Scientist" (see Table 12.1).

Even though the relative frequency of male and female scientists in the general and adjusted versions of the books do not differ significantly, the already low total number of female scientists becomes reduced by half in the adjusted books. This means that there are very few named female scientists left in the adjusted versions. In one of the adjusted books, no female scientists are mentioned by name. This means that the possibility for students to find female role models in the textbooks is limited, but even more limited in the adjusted versions.

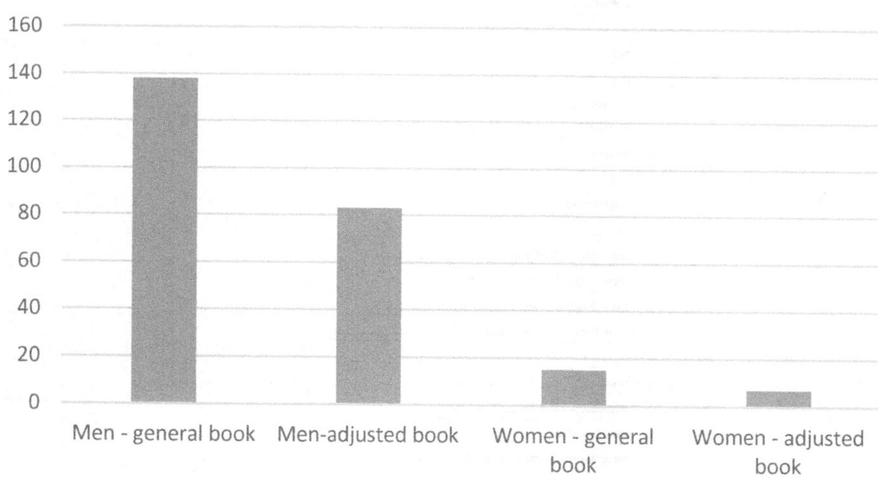

Fig. 12.3 Number of scientists (men and women) mentioned by name in the textbooks

[2]Often, the name is the only way to tell whether the individual is a female or male scientist. Therefore, we focus on named scientists. Except for a few instances where pictures are included without mentioning the name of the scientist.

Very few of the named scientists (in both the general and adjusted versions) are non-western. In an examination of countries and nationalities that are explicitly mentioned in connection to the scientists (named or non-named), it is clear that European countries and the US are most frequently mentioned in the general books (Fig. 12.4). It should also be noted that references to Greece and Egypt are almost exclusively from the early history of science, while recent references are mostly to Western Europe and the US. What is interesting is a large number of references to Sweden. For both the general and adjusted versions of the books such references are almost twice as frequent as references to the second most frequently mentioned country (see Fig. 12.4).

The Western perspective of the science textbooks means that most knowledge developed by scientists from other countries is excluded. As an example, there are almost no references to African or South American countries. An extreme (and unique) example of the Eurocentric perspective is how the student in a section of one of the textbooks is positioned as a "physics missionary" going to "the interior of a hot continent" to "tell those who live there what the world really looks like". This section is present in both versions of the book. The topic deals with how it can be proved that the Earth is a sphere. Even though the intention to focus on how we know things is laudable, the framing raises many questions.

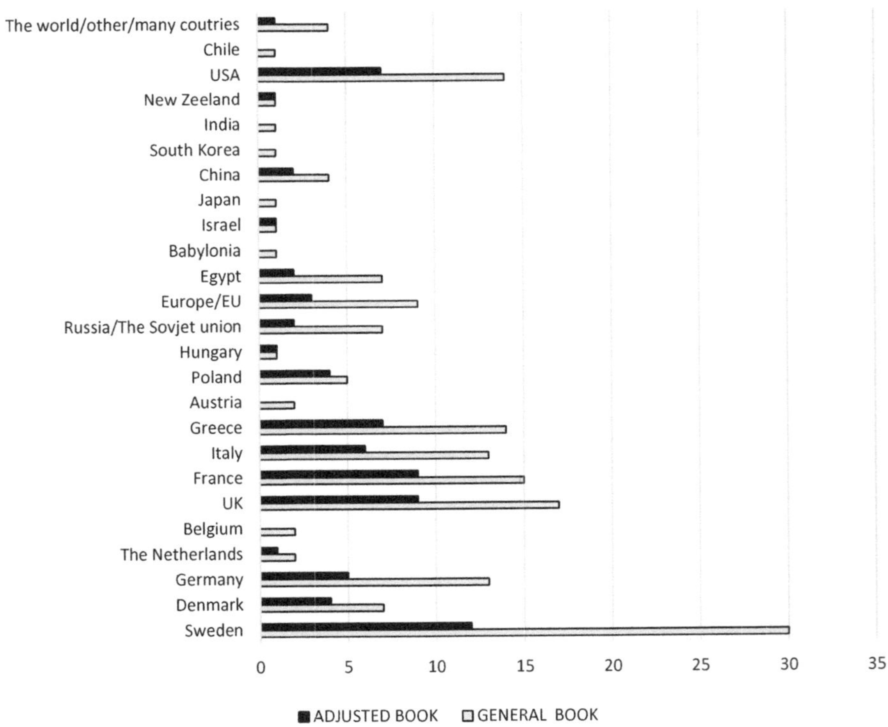

Fig. 12.4 Countries/nationalities mentioned in relation to scientists in the general books (N = 4) and adjusted books (N = 4)

The adjustments mean that the number and frequencies of different countries/nationalities associated with scientists are reduced (see Fig. 12.4). For example, references to scientists working in Russia/The Soviet Union are reduced from seven to two, and references to scientists from Asian countries are reduced from nine to three. References to the only African country – Egypt – are reduced from seven to two. Thus, when some of the few references to countries outside Europe and the US are removed the Eurocentric perspective is further strengthened.

12.8 Discussion

In this study we have analysed textbooks that are directed towards a heterogeneous group of students – described by one publisher as "students with reading and writing difficulties/dyslexia, students with Swedish as a second language, as well as other students who find the general books too hard…". The results show how the images of science and scientists to some extent, are narrowed in the adjusted textbooks – fewer examples are given that challenge the stereotype of the scientist as a western male. In addition, the number of complex and messy images of research processes that scientists are engaged in are lower than in the general textbooks.

The results presented in this chapter show that many scientists become excluded in the adjusted versions. This can be seen, to some extent, as a natural adjustment – if you are supposed to write a book with less content (and a reduced amount of text) then things have to be removed. Thus, the fact that scientists disappear from the adjusted version due to the adjustment "Remove section" is not surprising. Still, questions about *which* sections to remove are important. This is brought to the head in one of the book series where all sections that deal specifically with ongoing or historical research and its processes are removed.

We have also seen that the adjustment "Remove scientists" is frequently used. In these cases, the science content has been kept while one or more of the scientists have been removed. However, the publishers' and authors' decisions to remove scientists and thus limit the references to individuals and more narrative elements concerning what these individuals (scientists) are doing, changes the "voice" of the text. Much research points to the fact that such adjustments are counterproductive in respect of possibilities for conceptual learning. Greater conceptual density (Mason and Hedin 2011) and diminished contextual support (Hajer and Meestringa 2014) reduce students' possibilities to both make sense of it and to take an active interest in the content of the text (Knain 2001; Wikman 2004). Thus, such adjustments could be problematic in respect of students' possibilities to learn about scientific concepts and models. Furthermore, questions have to be asked concerning whether it is more important to learn science concepts and facts than learning about the individuals and the processes that lead to the knowledge. Through examples such as the section about the sun (see Fig. 12.2) where only facts about the sun remained in the adjusted version, we can see that references to NOS are probably still viewed as an add-on (c.f. Clough 2006), not only by teachers but also by authors and publishers. An add-on that could be excluded for students in need of an adjusted science textbook.

In addition, when adjustments result in removing one of very few women, or one of the few scientists connected to a non-western country, questions must be asked about whether this could have been avoided if a social justice perspective had been present in the adjustment process. The reduction of women and non-western nationalities mentioned in the adjusted books is especially notable since one of the publishers states that the books could be used by "students with Swedish as a second language", and we know that many of these students are born or have parents born in non-western countries.

The problem with removing female and non-western scientists could be ameliorated fairly easily if publishers and textbook authors become aware, and take action against, a reproduction of stereotypical images, and then include this perspective in both general and adjusted textbook versions. Such actions would mean including female scientists as well as non-Western male and female scientists who have often been excluded in accounts of the history of science. Still, this is not only valid in relation to adjusted books. It also means adding emphasis in a thoughtful way to aspects that need more explicit elaboration or highlighting in all book versions. A decision to more frequently include names and pictures, and say something about the interests and driving forces of different scientists who are not heroic or Nobel Prize winners, could provide meaning and context (Hajer and Meestringa 2014; Wikman 2004) and have the potential to make it easier for more students to identify with scientists and science. Living scientists might more easily be able to serve as role models than eighteenth century scientists, especially if care is taken to include both male and female scientists, from different parts of the world.

However, an inclusion of more individuals or a more diverse group of individuals in the textbook narrative is not a single solution (see also Yacoubian, Chap. 13 in this volume). Textbooks and teachers could also include explicit discussions and questions directed towards the culture and traditions that have shaped science. One very rare example of this in both versions of one textbook is:

> Almost all well-known persons in the older history of physics were men. At that time very few women were allowed to engage with science, *and the ones who did were not given the same possibilities or space as men.* Reflect on how it is today. (our translation, our italics mark text that is only available in the general textbook)

In line with what Sheth (2019) argues in relation to "grappling with racism", issues of inequalities ought to be explicitly dealt with and reflected on. It is not enough to include broader images of role models if unequal historical and contemporary legacies are not confronted (Sheth 2019). Thus, to include reflections of this in the teaching of science, as suggested in the example above, might contribute to students' possibilities of taking action against structural hindrances in respect of access to science.

If NOS (including images of scientists – personal characteristics as well as activities in which they are engaged) is viewed as an important part of science education for citizenship then a restriction for specific student groups, such as the heterogeneous group of students in need of supplemental support, can be viewed as an issue of social injustice. The results presented in this chapter urge us to look further into how NOS is taught to different student groups. The results also pose questions as to what a reproduction of stereotypical images of science and scientists in textbooks and science teaching mean for the possibilities of different students to make meaning and engage in science. We believe that such reflections, along with ideas for how to take action against oppressive images, should be key issues for textbook authors and publishers, as well as for science teachers and science education researchers in the quest to accomplish science teaching for social justice.

References

Abd-El-Khalick, F., Myers, J. Y., Summers, R., Brunner, J., Waight, N., Wahbeh, N., et al. (2017). A longitudinal analysis of the extent and manner of representations of nature of science in US high school biology and physics textbooks. *Journal of Research in Science Teaching, 54*(1), 82–120.

Aikenhead, G. S. (2006). *Science education for everyday life: Evidence-based practice*. New York: Teachers College Press.

Allchin, D. (2003). Scientific myth-conceptions. *Science Education, 87*(3), 329–351.

Allchin, D. (2011). Evaluating knowledge of the nature of (whole) science. *Science Education, 95*(3), 518–542.

Brigham, F. J., Scruggs, T. E., & Mastropieri, M. A. (2011). Science education and students with learning disabilities. *Learning Disabilities Research & Practice, 26*(4), 223–232.

Carlone, H. B. (2004). The cultural production of science in reform-based physics: Girls' access, participation, and resistance. *Journal of Research in Science Teaching, 41*(4), 392–414.

Clough, M. P. (2006). Learners' responses to the demands of conceptual change: Considerations for effective nature of science instruction. *Science & Education, 15*(5), 463–494.

Clough, M. P., & Olson, J. K. (2004). The nature of science: Always part of the science story. *The Science Teacher, 71*(9), 28.

Demirdöğen, B., Hanuscin, D. L., Uzuntiryaki-Kondakci, E., & Köseoğlu, F. (2015). Development and nature of preservice chemistry teachers' pedagogical content knowledge for nature of science. *Research in Science Education, 46*(4), 575–612.

DiGiuseppe, M. (2014). Representing Nature of Science in a Science Textbook: Exploring author–editor–publisher interactions. *International Journal of Science Education, 36*(7), 1061–1082.

Erduran, S., & Dagher, Z. R. (2014). *Reconceptualizing the nature of science for science education: Scientific knowledge, practices and other family categories*. Dordrecht: Springer.

Gibbons, P. (2006). *Stärk -språket, stärk lärandet: språk- och kunskapsutvecklande arbetssätt för och med andraspråkselever i klassrummet*. Uppsala: Hallgren & Fallgren.

Hackman, H. W. (2005). Five essential components for social justice education. *Equity & Excellence in Education, 38*(2), 103–109.

Hajer, M., & Meestringa, T. (2014). *Språkinriktad undervisning: En handbok*. Stockholm: Hallgren & Fallgren.

Hedrén, J., & Jidesjö, A. (2010). *Kunskap utan kunskapens användning: En studie av fysikläromedel i grundskolans senare år*. Skolinspektionen.

Henke, A., & Höttecke, D. (2015). Physics teachers' challenges in using history and philosophy of science in teaching. *Science & Education, 24*(4), 349–385.

Hodson, D. (2008). *Towards scientific literacy: A teachers' guide to the history, philosophy and sociology of science*. Rotterdam: Sense Publishers.

Hsieh, H. F., & Shannon, S. E. (2005). Three approaches to qualitative content analysis. *Qualitative Health Research, 15*(9), 1277–1288.
Jönsson, A. (2018). Meeting the needs of low-achieving students in Sweden: An interview study. *Frontiers in Education: Special Educational Needs, 3*(63).
Knain, E. (2001). Ideologies in school science textbooks. *International Journal of Science Education, 23*(3), 319–329.
Kolstø, S. D. (2000). Consensus projects: teaching science for citizenship. *International Journal of Science Education, 22*(6), 645–664.
Leden, L., & Hansson, L. (2017). Nature of science progression in school year 1–9: A case study of teachers' suggestions and rationales. *Research in Science Education, 49*, 591–611. https://doi.org/10.1007/s11165-017-9628-0.
Leden, L., Hansson, L., Redfors, A., & Ideland, M. (2015). Teachers' ways of talking about nature of science and its teaching. *Science & Education, 24*(9–10), 1141–1172.
Lederman, N. G. (2007). Nature of science: Past, present, and future. In S. K. Abell & N. G. Lederman (Eds.), *Handbook of research on science education* (pp. 831–879). Mahwah: Lawrence Erlbaum Associates, Publishers.
Lederman, J. S., & Stefanich, G. P. (2006). Addressing disabilities in the context of inquiry and nature of science instruction. In L. B. Flick & N. G. Lederman (Eds.), *Scientific inquiry and nature of science* (pp. 55–74). Dordrecht: Springer.
Mason, L. H., & Hedin, L. R. (2011). Reading science text: Challenges for students with learning disabilities and considerations for teachers. *Learning Disabilities Research & Practice, 26*(4), 214–222.
McDonald, C. V., & Abd-El-Khalick, F. (Eds.). (2017). *Representations of nature of science in school science textbooks: A global perspective*. New York: Routledge.
Mulvey, B. K., Chiu, J. L., Ghosh, R., & Bell, R. L. (2016). Special education teachers' nature of science instructional experiences. *Journal of Research in Science Teaching, 53*(4), 554–578.
Rommes, E., Overbeek, G., Scholte, R., Engels, R., & De Kemp, R. (2007). 'I'm not interested in computers': Gender-based occupational choices of adolescents. *Information, Community and Society, 10*(3), 299–319.
Sheth, M. J. (2019). Grappling with racism as foundational practice of science teaching. *Science Education, 103*(1), 37–60.
Skolinspektionen [Swedish Schools Inspectorate]. (2010). Fysik utan dragningskraft. En kvalitetsgranskning om lusten att lära fysik i grundskolan. *Kvalitetsgranskning Rapport, 2010*, 8.
Skollagen [The Educational Act]. (2010). Stockholm: Utbildningsdepartementet.
Skolverket. (2014). *Stödinsatser i utbildningen – om ledning och stimulans, extra anpassningar och särskilt stöd*. Stockholm: Skolverket.
Svennbeck, M. (2003). *Omsorg om naturen: om NO-utbildningens selektiva traditioner med fokus på miljöfostran och genus*. Diss. Uppsala: Univ., 2004. Uppsala.
Therrien, W. J., Hughes, C., & Hand, B. (2011). Introduction to special issue on science education and students with learning disabilities. *Learning Disabilities Research & Practice, 26*(4), 186–187.
Van Gorp, B., Rommes, E., & Emons, P. (2014). From the wizard to the doubter: Prototypes of scientists and engineers in fiction and non-fiction media aimed at Dutch children and teenagers. *Public Understanding of Science, 23*(6), 646–659.
Vesterinen, V. M., Aksela, M., & Lavonen, J. (2013). Quantitative analysis of representations of nature of science in Nordic upper secondary school textbooks using framework of analysis based on philosophy of chemistry. *Science & Education, 22*(7), 1839–1855.
Wikman, T. (2004). *På spaning efter den goda läroboken: om pedagogiska texters lärande potential*. Diss. Åbo: Åbo akademi, 2004.
Yacoubian, H. A. (2015) A framework for guiding future citizens to think critically about nature of science and socioscientific issues. *Canadian Journal of Science, Mathematics and Technology Education, 15*(3), 248–260.

Yacoubian, H. A., Al-Khatib, L., & Mardirossian, T. (2017). Analysis of the image of scientists portrayed in the Lebanese national science textbooks. *Science & Education, 26*(5), 513–528.

Zacharia, Z., & Barton, A. C. (2004). Urban middle-school students' attitudes toward a defined science. *Science Education, 88*(2), 197–222.

Chapter 13
Turning Unwanted Stereotypes about Scientists into Nature of Science Learning Experiences that Foster Social Justice

Hagop A. Yacoubian

13.1 Introduction

The students' world is quite rich with stereotypes about science and scientists. Such stereotypes are widespread among science teachers and students, as well as prevalent in science textbooks, educational science programs, science trade books and other educational resources. Stereotypes about science and scientists are problematic because they shape the way students experience science and develop their perceptions about science (Thomson et al. 2019), undermine the students' interest in science (Shapiro and Williams 2012), as well as influence their career choices (Cundiff et al. 2013). In addition to their pedagogical implications, these consequences have socioeconomic implications. For instance, many groups including women, minorities and individuals with disabilities continue to remain underrepresented in STEM-related fields (NSF 2019; Wong 2015). The latter creates a non-equitable distribution of human resources and challenges the self-determination of the members of the marginalized groups.

Stereotypes, generally speaking, are shaped by social and cognitive factors (McGarty et al. 2002). I delimit the discussion to the social factors acknowledging that cognitive factors are also important to attend to. It is in the context of social factors that the present chapter examines the issue of stereotypes about scientists.

Researchers' recommendations on how to challenge those stereotypes mostly focus on a *corrective* approach. Those recommendations are often addressed to pre-service science teacher education programs, textbook authors, curriculum designers and developers of other educational resources. A corrective approach aims, for instance, at highlighting the need for developing not only more authentic, but also

H. A. Yacoubian (✉)
Department of Education, Lebanese American University, Beirut, Lebanon
e-mail: hagop.yacoubian@lau.edu.lb

H. A. Yacoubian, L. Hansson (eds.), *Nature of Science for Social Justice*, Science: Philosophy, History and Education,
https://doi.org/10.1007/978-3-030-47260-3_13

culturally relevant science textbooks and other educational resources. Though one may wish that such stereotypes get reduced and disappear over time, instead of taking a corrective approach and recommending what ought to be presented differently to students, in this chapter a *proactive* approach is proposed that encourages the use of those stereotypes as resources for teaching and learning.

The current chapter makes a case for a *proactive* approach, arguing for turning unwanted stereotypes about scientists into meaningful *nature of science* (NOS) learning experiences that foster social justice. Bell (2016) argues for social justice being a goal and a process. She writes:

> Social justice is both a goal and a process. The *goal* of social justice is full and equitable participation of people from all social identity groups in a society that is mutually shaped to meet their needs. The *process* for attaining the goal of social justice should also be democratic and participatory… (Bell 2016, p. 3)

NOS refers to the epistemology of science, science as a way of knowing, and the values and beliefs inherent to scientific knowledge and its development (Abd-El-Khalick and Lederman 2000; Lederman 1992). Developing informed understandings of NOS is one component of scientific literacy (Norris and Phillips 2003), a major goal of science education (e.g., Kolstø 2001; OECD 2013) that has been associated with building democratic societies (Driver et al. 1996). Scientific literacy for democratic decision-making entails empowering students to develop a critical mindset through which they can question assumptions and political ideologies that underlie science-based social issues (Yacoubian 2018). Democracy, according to Abdi and Carr (2013) is a way of life that should be governed by equality and deliberative discussion. It must also incorporate social justice factors. Carr (2013) highlights the role of education in cultivating a democracy that focuses on political literacy, emancipation, critical engagement, social justice and epistemological liberation.

13.2 Stereotypes About Scientists: A Corrective Approach in Science Education?

Research has shown that school students possess stereotypical images of scientists (Akcay 2011; Archer et al. 2010; Bang et al. 2014; Cakmakci et al. 2011; Chambers 1983; Jane et al. 2007; Sjøberg 2000; Villar and Guppy 2015). For instance, many students consider scientists as white Caucasian (Barman 1999) and males (Akcay 2011; Barman 1999; Jane et al. 2007). They often perceive scientists as weird, bad and mad people (Jane et al. 2007). These stereotypes are related to the personal traits of scientists (scientists being males, white, middle-aged, Westerners, bald, nerd, ugly, disheveled, crazy, unemotional, individualistic, and lonely), what they wear (eyeglasses and white lab coats) and how they do their job (working in a lab and performing dangerous experiments in the natural sciences).

Stereotypical images of scientists are also reported to be present among pre- and in-service science teachers (Elmas et al. 2011; McDuffie 2001; Ünver 2010). For instance, many science teachers possess the stereotypical image of scientists being middle-aged males, intelligent and wearing lab coat (McDuffie 2001).

Similar Stereotypical images are also present in educational science programmes (Long and Steinke 1996), science trade books (Ford 2006) and science textbooks. Many textbooks portray science as a one-person show with minimum emphasis on the social dimension of science (Knain 2001), as a field dominated by males (Vesterinen et al. 2013; Villar and Guppy 2015; Yacoubian et al. 2017) and as masculine and euro-American (Vesterinen et al. 2013). They also depict an imbalance in ethnic diversity of scientists (Brooks 2008).

Many researchers who study the prevalence of stereotypes about scientists present in the students' world (i.e., among students, teachers, educational resources, textbooks, etc.) often provide recommendations on what ought to be done differently. Those recommendations are often intended to challenge the current situation, with the hope of achieving some change regarding how different stakeholders perceive scientists. A number of those recommendations are targeted to curriculum developers, textbook authors, and the media. Turkmen (2015), for example, emphasizes the need of curriculum developers to highlight science careers and depict scientists as everyday people. Good et al. (2010), on the other hand, call for textbook authors to eliminate gender bias.

From another perspective, considering that teachers have a direct influence on their students' learning, many researchers have recommended what teachers can do in science classrooms to challenge the students' stereotypes about scientists. The literature points to at least three directions in terms of what teachers can do.

A common line of recommendation is creating learning experiences for students to interact with scientists. In this way, students will become aware of real scientists (Farland-Smith 2009). This can be achieved through bringing students and scientists close to each other (Leblebicioglu et al. 2011; Turkmen 2015) and creating partnerships between schools and scientists (Avraamidou 2013).

A second common line of recommendation is based on building positive images of scientists (Ünver 2010), and thus changing the messages sent to young girls and boys (Master and Meltzoff 2016).

Third, showing students that those stereotypes are false (Boston and Cimpian 2018). This could be achieved through countering negative views of science and scientists (Nassar-McMillan et al. 2011) and telling stories about scientists using reflective approaches (Sharkawy 2009).

The recommendations described in the previous paragraphs are mostly *corrective* in their approach. What I mean by a *corrective approach* is that they target deleting an unwanted feature, inserting a desired feature, or substituting an unwanted feature with a more desirable one so that students are not exposed to those unwanted stereotypes or are exposed to counter-stereotypes (positive instances).

The underlying assumption in many of these recommendations is that if teachers, textbook authors, curriculum designers or the media can *change* the way they present scientists through *inserting, deleting* or *substituting* certain features, then the perceptions about scientists present among students may also *change*. As a result, stereotypes about scientists would get reduced and hopefully disappear over time.

The corrective approach is certainly important to adopt as it fosters equity and social justice. In fact, it helps in changing the students' world, which could contribute to changing the students' perceptions towards science and scientists. The science education community needs to continue spreading awareness so that unwanted stereotypes get eliminated or replaced by positive instances. Such a change can inevitably have an impact on student learning. Nevertheless, despite its necessity, I do not find the corrective approach to be sufficient.

Although the corrective approach decreases the likelihood of students to be exposed to unwanted stereotypes and increases their chances of being exposed to non-stereotypes, it does not equip the students well with tools needed for challenging the status quo. Those stereotypes continue being present in the society even if as educators, we make a conscious decision to *correct* their world and fill it up with whatever is desired. Emptying their world from stereotypes and replenishing it with non-stereotypes makes their world naïve compared to the real world[1] out there.

By using the corrective approach, students receive an education that only minimally prepares them to be ready to confront the real world out there, which, to say at the least, continues to be racist, sexist and full of injustices. Creating a world devoid of racism, sexism and injustices for students in the science classroom is important but that also brings the undesired consequence of making their world look like a wonderland.

Accordingly, the corrective approach targets social justice as a goal but not as a process as Bell (2016) would argue. The socially just wonderland created in the students' world may influence the real world out there by having, for instance, more members of marginalized groups pursue STEM-related fields. However, to target social justice as a process, there is the need to empower students with the tools needed to engage in a democratic and a participatory process to achieve social justice.

One way to overcome this problem could be through what I propose a *proactive* approach. An educator using a proactive approach acknowledges the fact that stereotypes about scientists are going to be inevitable in the students' world as long as they continue to be present in the world out there. This is simply because the students' world does not exist in vacuum – it is shaped within the context of a broader social, political and cultural world in which those stereotypes are formed and perpetuated.

In addition to fighting against stereotypes present in the students' world, educators can think about using those stereotypes as resources for teaching and learning. The unwanted features of scientists found in the students' world can serve as

[1] While *students' world* includes experiences and interactions that are part of students' everyday environment, the term *real world* is used here to signify the experiences and interactions that people have as part of their everyday lives. The intention is not to separate the two worlds from each other. On the contrary, the *students' world* is regarded as a subset to the *real world*.

platform for teaching them how to challenge those stereotypes. Students can be empowered to develop a critical mindset to challenge those stereotypes. I call this approach a proactive approach and elaborate it further in the subsequent sections.

13.3 From a Corrective to a Proactive Approach

Educators are well aware of the importance of considering a student's background as a starting place for teaching and learning. From a constructivist perspective, stereotypes about scientists present in the students' world could be a plausible starting place to engage them in learning.

Stereotypes in the students' world often reflect those present in the real world. In addition, students are often socialized to adopt similar stereotypes as those present among the public – including their parents, teachers, textbook authors, curriculum designers, and media specialists.

It is important that the students confront their own stereotypes about scientists, question and reflect upon them with the purpose of being metacognitively aware about their origin and source. Students need to learn how to challenge their own stereotypes as well as why they exist and where they come from. Not only should they develop the tools needed in confronting their own stereotypes, but also those that are in their world, with the hope that they can ultimately challenge them in the real world. Accordingly, a proactive approach that the current chapter supports encourages the use of those stereotypes as resources for teaching and learning.

Such a position, or a proactive approach, is based on three justifications. First, it aims at developing a critical mindset among students, with which they can confront stereotypes. Critical thinking (CT) is a fundamental educational ideal and is significant for the preparation of democratic citizens (Siegel 1988). According to Ennis (1996, 2018), CT is a process the goal of which is to produce reasonable and reflective decisions on what to believe or do and which encompasses certain dispositions and abilities. Developing a critical mindset would enable the students to question stereotypes present in their world and the real world as well as question their sources and the reasons for why they exist.

Second, a proactive approach fosters equity and social justice more than a corrective approach. It targets social justice as a process in addition to targeting it as a goal because students learn to be socially responsible and to engage critically in creating a world in which distribution of resources is equitable. Additionally, it facilitates a mindset of challenging stereotypes, scrutinizing social, cultural and political factors that shape such stereotypes, and hopefully bringing some social change instead of perpetuating the status quo. It is naive to assume that science curricula are neutral. It is essential to eradicate the representation of education being neutral and devoid of politics (Carr 2011). Science teachers need to create learning experiences and empower students to bring some of the underlying assumptions and political ideologies into the conscious level by fostering developmentally appropriate learning experiences where future citizens can be guided to critically and explicitly reflect upon them. This would entail empowering them so that they dig deeper into the implicit layers of some of those stereotypes with the purpose of critically and explicitly reflecting upon assumptions and political ideologies underlying them.

Third, a proactive approach keeps the students' world as authentic as possible instead of changing it into a wonderland. Authentic learning situates learning tasks in the context of real-world situations, and allows students to experience the same challenges in the curriculum as they do in the real world (Herrington et al. 2014).

Consequently, a proactive approach would entail starting from the students' stereotypes about scientists to develop learning situations where they confront their own beliefs and perceptions about scientists. Students can be guided to question them and critically analyze why they exist and where they come from, relate them to social norms and values, scrutinize (often implicit) social, cultural and political values that shape some of those stereotypes and take action.

Nevertheless, does the suggested approach mean an additional burden on the science curriculum and added work for the science teacher? In the section that follows, I situate the use of the proactive approach within the context of NOS in school science. I argue that one place in the science curriculum where a proactive approach can be utilized is in the teaching of NOS. I argue that stereotypes can constitute fertile grounds to engage students in *critical* exploration of certain aspects of NOS. This chapter delimits the discussion to one area of NOS, namely how science impacts and is impacted by the social, cultural and political contexts in which it operates.

13.4 NOS in School Science

Science educators differ in their positions in what NOS to address in school science. Lederman (2004), Lederman et al. (2013) and McComas (1998, 2004, 2020) have abstracted, for instance, lists of NOS-related ideas that show some consensus among philosophers, sociologists, historians of science and science educators, which they think are suitable to be addressed in K-12 science classrooms. McComas (2020) highlights several subdomains, one of which is the human elements in science and that of the role of social and cultural influences. Along the same lines, Lederman (2004) and Lederman et al. (2013) underscore the idea that science is embedded in social and cultural contexts.

Critiques of the consensus views highlight the need of providing a more authentic and comprehensive view of how science works. Allchin (2011), for instance, proposes "whole science" (p. 518) and claims that NOS in school science needs to be sensitive to all dimensions of scientific practice. Irzik and Nola (2011), on the other hand, borrow from Wittgenstein's notion of family resemblance to present a picture of NOS that characterizes science under four categories: scientific activities, scientific aims and values, scientific methods and methodological rules, and scientific products.

Erduran and Dagher (2014) and Dagher and Erduran (2016) propose an expanded version of Irzik and Nola's (2011) model, that places emphasis on science as a cognitive-epistemic and social-institutional system. The researchers claim that their framework, particularly in terms of the added categories of "social organizations and interactions", "political power structures" and "financial systems" improve students' understanding of science in relation to society (Dagher and Erduran 2016).

Explicit and reflective methods have been used in teaching NOS to school students (e.g., Akerson et al. 2011; Khishfe and Abd-El-Khalick 2002; Paraskevopoulou and Koliopoulos 2011). As part of engaging students in explicit reflective discussions about NOS, future citizens can be guided to develop their NOS understandings *critically* (Yacoubian 2015).

The use of a proactive approach fits quite well within the context of explicit reflective discussions and can be used as students explore the social, cultural and political influences of science. When an explicit-reflective method embeds a proactive approach in exploring stereotypes about scientists, teachers can have a fertile ground to facilitate the development of a critical mindset among students as well as to prepare advocates of social justice. Students engage in critical exploration of stereotypes, the underlying social, cultural and political values that shape them, as well as how science influences and is influenced by the social, cultural, political aspects in which it operates. This enables them to practice thinking critically about NOS, in an authentic context, develop their NOS understandings as well as their CT-related abilities and dispositions (Yacoubian 2015). In addition, it engages students in critical deliberation without taking the underlying status quo for granted (Yacoubian 2018), thus fostering the development of their political literacy (Carr 2013) and preparing them to take sociopolitical action (Hodson 2003, 2011).

13.5 An Example

As discussed earlier, from a proactive approach, science lessons can be an opportunity to use some of the stereotypes about scientists as a context to guide students to think critically about how science influences and is influenced by the social, cultural and political contexts in which it operates. In this section, I elaborate an example that involves learning experiences where students can critically reflect upon their own stereotypes about scientists, those found in their science textbooks and the social, cultural and political contexts of science.

Elsewhere (Yacoubian 2020), I had elucidated a procedure that can be helpful for teaching NOS critically in a science classroom. The procedure includes the following stages. First, establishing a platform on which critical explorations of NOS can take place. This is important for creating a background context so that discussions about NOS can take place around concrete situations. Second, providing a NOS focus to the lesson with the purpose of directing the students' attention to specific NOS themes within the chosen context. Third, developing a learning activity, which can engage students in critical exploration of the NOS-related ideas in question. Fourth, engaging students in critical exploration of NOS while facilitating explicit reflective discussions.

For example, genetics and molecular biology is a content area relevant to the students as it is covered in high school science curricula. In introducing a unit on genetics and molecular biology, many textbook authors provide an overview of the historical developments of the field. Considering research reports cited earlier that many textbooks portray an imbalance in gender and ethnic diversity of scientists, historical timelines can become fruitful context for science teachers to facilitate a critical discussion about NOS. Studying those timelines and reflecting on them, students can be invited to think about the following issue:

Whether female and non-Western scientists have made important contributions in the field of genetics

At a second stage, a science teacher can provide a NOS focus to the discussions. Let us assume that the focus of the lesson to be placed on the social, cultural and political dimensions of science – *how science impacts and is impacted by the social, cultural and political contexts in which it operates*. The latter is a NOS theme that many science education scholars would consider important to be discussed in high school science classrooms. A NOS question that the students can be invited to explore might include:

How do science and society impact each other?

Students do not engage in exploring the proposed NOS question at the overt level (e.g., What is the job of a scientist? What impact do science and society have upon each other?). Instead, they are encouraged to develop their understandings critically. This brings us to the third stage.

Third, a science teacher can engage the students in a learning activity that can facilitate critical exploration of the NOS-related question. This might include, for instance, critically exploring Rosalind Franklin's contribution to the DNA model and how her work was under-appreciated at the time. For instance, students can be guided to read Brenda Maddox's (2002) book entitled *Rosalind Franklin: The Dark Lady of DNA*. They can be invited to analyze Maddox's ideas and evaluate them, critique the book, relate what they have read to what the situation is about today, etc.... The learning activity (in this case reading Maddox's book) can create the prerequisite platform for students to move into the fourth stage and engage in thorough exploration of the proposed issue and NOS question.

Fourth, students can engage in critical exploration of NOS while engaging in explicit reflective discussions. At this stage, students are encouraged to practice making judgments about NOS. In particular, they practice making judgments about how science and society impact each other. At this stage, they use the learning that they critically constructed in the third stage to explore the proposed issue (Whether female and non-Western scientists have made important contributions in the field of genetics) and the proposed NOS question (How do science and society impact each other?) at some depth. During the process, students develop CT-related abilities and dispositions as well as (critically) construct their NOS understandings.

A proactive approach embedded within explicit reflective discussions gradually empowers students to explore the proposed issue and NOS question at a deeper and more implicit level. Several questions may be raised here to facilitate the process (e.g., Does it surprise you that all the scientists in your textbook are males and westerners, why or why not? Can you research to find more about female and non-western scientists who have made similar contributions to ones we studied and/ or continue making similar contributions today? Why are most scientists in our textbook males and Westerners?)

Gradually, the NOS lesson opens windows through which critical exploration of the broader political, social and cultural systems can take place (e.g. to what extent do you find it fair that our textbook does not include female and non-western scientists?). Here, students can also reflect on notions, concepts and ideas such as gender, race and ethnicity in science; image of science and scientists portrayed in society (and textbooks), and social justice. Finally, the discussion leads into a critical reflection of the process of engaging in such discussions (e.g. Why is it important that we engage in such discussions?).

13.6 Concluding Remarks

This section summarizes the main arguments set forth so far and concludes the chapter with highlighting the value of the proposed proactive approach in fostering social justice and in the preparation of scientifically literate citizenry.

This chapter argued for a *proactive* approach needed in changing unwanted stereotypes about scientists into *NOS* learning experiences. It proposed to empower students with tools that they would need to engage in a process that challenge their own stereotypes, those present in their world and those present in the real world with the purpose of attaining social justice. Compared to a corrective approach that has the goal of attaining social justice, a proactive approach has the potential to target social justice both as a goal and as a process.

Acknowledging that the students' world is shaped by a broader social, political and cultural world in which those stereotypes are formed and perpetuated, the focus was placed on the need to acknowledge that stereotypes about scientists will be inevitable in students' world. Instead of looking for ways to *correct* the students' world – freeing it from stereotypes or counter-balancing it with non-stereotypes, a suggestion was made to focus the spotlight on how those stereotypes could be used as resources for teaching and learning.

When students engage in NOS learning, they can be guided with a proactive approach, within the context of explicit-reflective discussions, to question their and the society's stereotypes about scientists as well as critically analyze underlying social, cultural and political values that shape them. This is important for developing a critical mindset and being prepared to be advocates of social justice, all embedded within authentic learning experiences.

This process entails mentoring students, in developmentally appropriate ways, to engage in critical deliberation and develop a critical mindset to go beyond the overt, to learn to dig beneath the surface to examine what is under that surface, and to evaluate the impact of the overt on the implicit and the implicit on the overt. In addition, the process can also contribute to political literacy, cultivates emancipation and social justice (Carr 2013).

The proposed ideas in this chapter have the potential to contribute towards developing into a full-fledged teaching method that could be used to guide future citizens to engage in critical exploration of science and scientists. The creation of a full-fledged teaching method assumes empirically testing its potential, which was beyond the scope of this chapter. Such a teaching method could be beneficial for contributing to preparing scientifically literate citizenry in the sense described below. In addition, one argument that the current chapter advances is that a proactive approach can have a greater potential for having an impact beyond the classroom and therefore is more conducive for preparing scientifically literate future citizens. This is also open for empirical research.

As discussed previously, developing informed understandings of NOS is one component of scientific literacy (Norris and Phillips 2003). Scientifically literate citizens can contribute to building democratic societies (Driver et al. 1996; Yacoubian 2018). The proactive approach outlined here has the potential to contribute to democratic education as described by Abdi and Carr (2013) and Carr (2013). This is because of its potential to engage students in critical engagement, to foster political literacy, and to aim for emancipation and social justice.

The proposed ideas also have several implications for science teacher education. There is a need to create meaningful learning experiences for science teachers to develop a mindset conducive for appreciating a proactive approach and preparing scientifically literate future citizens in the sense described in this proposal – one that fosters critical engagement and social justice. A talented teacher can change the unwanted features of educational resources into meaningful learning opportunities and encourage students to reflect on their perceptions of science and scientists, challenge stereotypes and question established norms.

References

Abd-El-Khalick, F., & Lederman, N. G. (2000). The influence of history of science courses on students' views of nature of science. *Journal of Research in Science Teaching, 37*, 1057–1095.

Abdi, A. A., & Carr, P. R. (Eds.). (2013). *Educating for democratic consciousness: Counter-hegemonic possibilities*. PeterLang: Bern.

Akcay, B. (2011). Turkish elementary and secondary students' views about science and scientist. *Asia-Pacific Forum on Science Learning and Teaching, 12*(1). Article 5, p. 1.

Akerson, V. L., Buck, G. A., Donnelly, L. A., Narguand-Joshi, V., & Weiland, I. S. (2011). The importance of teaching and learning nature of science in early childhood years. *Journal of Science Education and Technology, 20*, 537–549.

Allchin, D. (2011). Evaluating knowledge of the nature of (whole) science. *Science Education, 95*(3), 518–542.

Archer, L., DeWitt, J., Osborne, J., Dillon, J., Willis, B., & Wong, B. (2010). "Doing" science versus "being" a scientist: Examining 10/11-year-old schoolchildren's constructions of science through the lens of identity. *Science Education, 94*(4), 617–639.

Avraamidou, L. (2013). Superheroes and supervillains: Reconstructing the mad-scientist stereotype in school science. *Research in Science & Technological Education, 31*(1), 90–115.

Bang, E., Wong, S. S., & Jeffery, T. D. (2014). High school students' stereotypic images of scientists in South Korea. *Mevlana International Journal of Education, 4*(1), 96–112.

Barman, C. R. (1999). Students' views about scientists and school science: Engaging k-8 teachers in a national study. *Journal of Science Teacher Education, 10*, 43–54.

Bell, L. A. (2016). Theoretical foundations for social justice education. In M. Adams & L. A. Bell (Eds.), *Teaching for diversity and social justice* (3rd ed.). New York: Routledge.

Boston, J. S., & Cimpian, A. (2018). How do we encourage gifted girls to pursue and succeed in science and engineering? *Gifted Child Today, 41*(4), 196–207.

Brooks, K. (2008). *A content analysis of physical science textbooks with regard to the nature of science and ethnic diversity.* EdD Dissertation: University of Houston.

Cakmakci, G., Tosun, O., Turgut, S., Orenler, S., Sengul, K., & Top, G. (2011). Promoting an inclusive image of scientists among students: Towards research evidence-based practice. *International Journal of Science and Mathematics Education, 9*, 627–655.

Carr, P. R. (2011). *Does your vote count? Critical pedagogy and democracy*. New York: Peter Lang.

Carr, P. R. (2013). Reshaping the democratic truth, and rethinking democracy without elections. In A. A. Abdi & P. R. Carr (Eds.), *Educating for democratic consciousness: Counter-hegemonic possibilities* (pp. 29–49). Bern: PeterLang.

Chambers, D. W. (1983). Stereotypic images of the scientist: The draw-a-scientist test. *Science Education, 67*(2), 255–265.

Cundiff, J. L., Vescio, T. K., Loken, E., & Lo, L. (2013). Do gender-science stereotypes predict science identification and science career aspirations among undergraduate science majors? *Social Psychology of Education, 16*, 541–554. https://doi.org/10.1007/s11218-013-9232-8.

Dagher, Z. R., & Erduran, S. (2016). Reconceptualizing the nature of science for science education: Why does it matter? *Science & Education, 25*(1–2), 147–164.

Driver, R., Leach, J., Millar, R., & Scott, P. (1996). *Young People's images of science*. Philadelphia: Open Uni Press.

Elmas, R., Demirdogen, B., & Geban, O. (2011). Preservice chemistry teachers' images about science teaching in their future classrooms. *Hacettepe University Journal of Education, 40*, 164–175.

Ennis, R. H. (1996). *Critical thinking*. Upper Saddle River: Prentice Hall.

Ennis, R. H. (2018). Critical thinking across the curriculum: A vision. *Topoi, 37*, 165–184.

Erduran, S., & Dagher, Z. R. (2014). *Reconceptualizing the nature of science for science education: Scientific knowledge, practices and other family categories*. Dordrecht: Springer.

Farland-Smith, D. F. (2009). How does culture shape students' perceptions of scientists? Cross-national comparative study of American and Chinese elementary students. *Journal of Elementary Science Education, 21*(4), 23–42.

Ford, D. J. (2006). Representations of science within children's trade books. *Journal of Research in Science Teaching, 43*(2), 214–235.

Good, J. J., Woodzicka, J. A., & Wingfield, L. C. (2010). The effects of gender stereotypic and counter-stereotypic textbook images on science performance. *Journal of Social Psychology, 150*(2), 132–147.

Herrington, J., Reeves, T. C., & Oliver, N. (2014). Authentic learning environments. In J. M. Spector, M. D. Merrill, J. Elen, & M. J. Bishop (Eds.), *Handbook of research on educational communications and technology* (pp. 401–412). New York: Springer.

Hodson, D. (2003). Time for action: Science education for an alternative future. *International Journal of Science Education, 25*(6), 645–670.

Hodson, D. (2011). *Looking to the future: Building a curriculum for social activism*. Rotterdam: Sense Publishers.

Irzik, G., & Nola, R. (2011). A family resemblance approach to the nature of science. *Science & Education, 20*, 591–607.

Jane, B., Fleer, M., & Gipps, J. (2007). Changing children's views of science and scientists through school-based teaching. *Asia-Pacific Forum on Science Learning and Teaching, 8*(1), article 11, 1–21.

Khishfe, R., & Abd-El-Khalick, F. (2002). Influence of explicit and reflective versus implicit inquiry-oriented instruction on sixth graders' views of nature of science. *Journal of Research in Science Teaching, 39*, 551–578.

Knain, E. (2001). Ideologies in science textbooks. *International Journal of Science Education, 23*(3), 319–329.

Kolstø, S. D. (2001). Scientific literacy for citizenship: Tools for dealing with the science dimension of controversial socioscientific issues. *Science Education, 85*, 291–310.

Leblebicioglu, G., Metin, D., Yardimci, E., & Cetin, P. S. (2011). The effect of informal and formal interaction between scientists and children at a science camp on their images of scientists. *Science Education International, 22*(3), 158–174.

Lederman, N. G. (1992). Students' and teachers' conceptions of the nature of science: A review of the research. *Journal of Research in Science Teaching, 29*, 331–359.

Lederman, N. G. (2004). Syntax of nature of science within inquiry and science instruction. In L. B. Flick & N. G. Lederman (Eds.), *Scientific inquiry and nature of science* (pp. 301–317). Dordrecht: Kluwer.

Lederman, N. G., Lederman, J. S., & Antink, A. (2013). Nature of science and scientific inquiry as contexts for the learning of science and achievement of scientific literacy. *International Journal of Education in Mathematics, Science and Technology, 1*(3), 138–147.

Long, M., & Steinke, J. (1996). The thrill of everyday science: Images of science and scientists on children's educational science programmes in the United States. *Public Understanding of Science, 5*(2), 101–119.

Maddox, B. (2002). *Rosalind Franklin: The dark lady of DNA*. New York: HarperCollins.

Master, A., & Meltzoff, A. (2016). Building bridges between psychological science and education: Cultural stereotypes, STEM, and equity. *Prospects, 46*(2), 215–234.

McComas, W. F. (Ed.). (1998). *The nature of science in science education: Rationales and strategies*. Kluwer Academic Publishers.

McComas, W. F. (2004). Keys to teaching the nature of science. *The Science Teacher, 71*(9), 24–27.

McComas, W. F. (2020). *The nature of science in science instruction: Rationales and strategies*. Springer.

McDuffie, T. E. (2001). Scientists – geeks and nerds? *Science and Children, 38*(8), 16–19.

McGarty, C., Yzerbyt, V. Y., & Spears, R. (2002). Social, cultural and cognitive factors in stereotype formation. In C. McGarty, V. Y. Yzerbyt, & R. Spears (Eds.), *Stereotypes as explanations: The formation of meaningful beliefs about social groups* (pp. 1–15). Cambridge: Cambridge University Press.

Nassar-McMillan, S. C., Wyer, M., Oliver-Hoyo, M., & Schneider, J. (2011). New tools for examining students' STEM stereotypes: Implications for women and other represented groups. *New Directions for Institutional Research, 152*, 87–98.

National Science Foundation, & National Center for Science and Engineering Statistics. (2019). *Women, minorities, and persons with disabilities in science and engineering: 2019.* Special report NSF 19–304. Alexandria, VA. Available at https://www.nsf.gov/statistics/wmpd.

Norris, S. P., & Phillips, L. M. (2003). How literacy in its fundamental sense is central to scientific literacy. *Science Education, 87*, 224–240.

Organization for Economic Co-operation and Development (OECD). (2013) *PISA 2015: Draft science framework.*

Paraskevopoulou, E., & Koliopoulos, D. (2011). Teaching the nature of science through the Millikan-Ehrenhaft dispute. *Science & Education, 20*, 943–960.

Shapiro, J., & Williams, A. (2012). The role of stereotype threats in undermining girls' and women's performance and interest in STEM fields. *Sex Roles, 66*(3–4), 175–183. https://doi.org/10.1007/s11199-011-0051-0.

Sharkawy, A. (2009). Moving beyond the lone scientist: Helping 1-st grade students appreciate the social context of scientific work using stories about scientists. *Journal of Elementary Science Education, 21*(1), 67–68.

Siegel, H. (1988). *Educating reason: Rationality, critical thinking, and education*. New York: Routledge.

Sjøberg, S. (2000). *Science and scientists: The SAS study*. Retrieved November 23, 2008, from http://folk.uio.no/sveinsj/SASweb.htm

Thomson, M. M., Zakaria, Z., & Radut-Taciu, R. (2019). Perceptions of scientists and stereotypes through the eyes of young school children. *Education Research International, 2019*. https://doi.org/10.1155/2019/6324704.

Turkmen, H. (2015). Still persistent global problem of students' image. *Asia-Pacific Forum on Science Learning & Teaching, 16*(1), 377–397.

Ünver, A. O. (2010). Perceptions of scientists: A comparative study of fifth graders and fourth year student teachers. *Necatibey Faculty of Education Electronic Journal of Science & Mathematics Education, 4*(1), 11–28.

Vesterinen, V.-M., Aksela, M., & Lavonen, J. (2013). Quantitative analysis of representations of nature of science in Nordic upper secondary school textbooks using framework of analysis based on philosophy of chemistry. *Science & Ed, 22*, 1839–1855.

Villar, P., & Guppy, N. (2015). Gendered science: Representational dynamics in British Columbia science textbooks. *Canadian Journal of Education, 38*(3), 1–24.

Wong, B. (2015). Careers "from" but not "in" science: Why are aspirations to be a scientist challenging for minority ethnic students? *Journal of Research in Science Teaching, 52*, 979–1002. https://doi.org/10.1002/tea.21231.

Yacoubian, H. A. (2015). A framework for guiding future citizens to think critically about nature of science and socioscientific issues. *Canadian Journal of Science, Mathematics and Technology Education, 15*(3), 248–260.

Yacoubian, H. A. (2018). Scientific literacy for democratic decision-making. *International Journal of Science Education, 40*(3), 308–327.

Yacoubian, H. A. (2020). Teaching nature of science through a critical thinking approach. In W. F. McComas (Ed.), *The nature of science in science instruction: Rationales and strategies*. Springer.

Yacoubian, H. A., Al-Khatib, L., & Mardirossian, T. (2017). Analysis of the image of scientists portrayed in the Lebanese national science textbooks. *Science & Education, 26*, 513–528.

Correction to: Nature of Science for Social Justice: Why, What and How?

Lena Hansson and Hagop A. Yacoubian

Correction to:
Chapter 1 in: H. A. Yacoubian, L. Hansson (eds.), *Nature of Science for Social Justice*, Science: Philosophy, History and Education, https://doi.org/10.1007/978-3-030-47260-3_1

The first chapter "Nature of Science for Social Justice: Why, What and How?" of the book has been converted to open access and is available open access under a Creative Commons Attribution 4.0 International License via link.springer.com.

The book has also been updated with this change

The updated version of the chapter can be found at https://doi.org/10.1007/978-3-030-47260-3_1

H. A. Yacoubian, L. Hansson (eds.), *Nature of Science for Social Justice*, Science: Philosophy, History and Education, https://doi.org/10.1007/978-3-030-47260-3_14

In Lieu of a Conclusion

The aim of this volume was to initiate a dialogue on nature of science for social justice (NOS for SJ). As editors, we started this dialogue by arguing for the importance of three questions: Why should school science aimed at SJ address NOS? What NOS-related content, skills and attitudes form the basis when aiming at SJ?, and How can school science address NOS for SJ? We illustrated how previous research contributes to our understanding and argued for the need to continue exploring the three questions along different lines of research. In their respective chapters, the authors have used a variety of approaches and made different recommendations concerning what *NOS for SJ* is and what it entails. They all added to the dialogue by suggesting answers and proposing new questions that in different ways relate to the three overall questions.

At this stage, we don't think it would be a wise decision to converge the dialogue into clear-cut conclusions, recommendations and future directions. This book is the very beginning of a dialogue that we consider important, and not an exhaustive coverage of all relevant perspectives and themes that might fall under the title of *NOS for SJ*. Thus, the construct that we have called *NOS for SJ* is still in its maturing phases. Consequently, a lack of a concluding chapter ensures that the dialogue stays ongoing and divergent. Additionally, it creates space for the community to reflect on different positions, perspectives and frameworks, while deciding how to proceed further.

No concluding chapter also ensures that we stay honest with the central purpose of this project, which was engaging in a genuine dialogue. We believe that our dialogue contributed with a diversity of ideas and approaches that could be helpful to a further understanding of what *NOS for SJ* can mean, as well as raising important issues that need to be taken into consideration in future research. Our hope is that future research in this area would target further refining the construct of *NOS for SJ*, delineating its characteristics, as well as exploring ways of addressing it in science classrooms.

***Hagop A. Yacoubian & Lena Hansson* (editors).**

H. A. Yacoubian, L. Hansson (eds.), *Nature of Science for Social Justice*,
Science: Philosophy, History and Education,
https://doi.org/10.1007/978-3-030-47260-3

The manufacturer's authorised representative in the EU is Springer Nature Customer Service Centre GmbH, Europaplatz 3, 69115 Heidelberg, Germany. If you have any concerns regarding our products, please contact ProductSafety@springernature.com

Printed and bound by CPI Group (UK) Ltd, Croydon, CR0 4YY

30/09/2026

02225505-0010